FRACTAL SPACE-TIME AND MICROPHYSICS

FRACTAL SPACE-TIME AND MICROPHYSICS

Towards a Theory of Scale Relativity

Laurent Nottale

Centre National de la Recherche Scientifique
Observatoire de Paris

Published by

World Scientific Publishing Co. Pte. Ltd.
5 Toh Tuck Link, Singapore 596224
USA office: 27 Warren Street, Suite 401-402, Hackensack, NJ 07601
UK office: 57 Shelton Street, Covent Garden, London WC2H 9HE

Library of Congress Cataloging-in-Publication Data
Nottale, Laurent.
Fractal space-time and microphysics : towards a theory of scale relativity / Laurent Nottale.
p. cm.
Includes bibliographical references.
ISBN-13 978-981-02-0878-3 -- ISBN-10 981-02-0878-2
1. Relativity (Physics). 2. Quantum theory. 3. Scaling laws (Statistical physics). 4. Fractals. I. Title.
QC173.55.N67 1992
530.1'1 -- dc20 92-11167
CIP

British Library Cataloguing-in-Publication Data
A catalogue record for this book is available from the British Library.

PREFACE

After the completion of my review paper on *Fractals and the Quantum Space-Time* which was published in the *International Journal of Modern Physics*, the Editors of *World Scientific* kindly proposed to me to expand the content of this paper into a book. Meanwhile, I further developed the idea of "scale relativity" in the complementary framework of the renormalization group: this led me to write another paper that is now in the press in the same *Journal*.

This book contains and expands the material presented in these two papers, and also presents a first account of several new results. I have tried, in particular, to increase greatly the iconography in comparison with the original papers.

As the reader will see, this book describes a new approach not only to the problems of quantum physics, but also to the multiple problems of "scaling" encountered in modern physics, from the domain of chaotic dynamics to that of cosmology. Its aim is mainly to present in details the ideas and concepts which I suggest are to be set down for the future construction of a genuine theory of fundamental scales and of their emergence in Nature. This is a "snapshot" of a research program still on the way, rather than a completely consistent and fully developed theory. Some of the roads that I suggest to follow may in the end come to a deadlock, but I hope that the leading idea in this book, namely, that Einstein's principle of relativity can be generalized to apply to scale transformations, will remain useful for a real understanding of the physical world.

This book is mainly intended for physicists: I have assumed the reader to have a general background in quantum mechanics and relativity. However, I have attempted to provide in the first chapter an informative introduction, which is comprehensible to readers not particularly familiar with the topics treated. This is also an extended summary of the contents, principles and results that are for the first time extensively presented here. The reader most interested by physical applications may skip, at least partly, Chapter 3 in the first reading, since it is devoted to some mathematical developments of the concept of fractals.

Not being blessed with a native English tongue, I hope that the reader will not suffer too much from my insufficient mastery of the English language: I apologize for possible remaining grammatical flaws.

It is a pleasure to thank all those who have encouraged and supported me since the beginning of this work, among them Suzy Collin, Georges Alecian, Joseph Lemaire, Michel Auvergne, Ana Gomez, and Chantal Balkowski. I greatly benefited from discussions with Jean Heyvaerts, Guy Mathez, Bernard Fort, Jean-Pierre Luminet, Didier Pelat, Yannick Mellier, and Jean Perdang. Students, in particular Yves Lachaud, François Bardou, and Eric Appert, helped to free the explanations of errors and obscurities by their questions and comments. Special thanks go to Jean-Claude Pecker and Jayant Narlikar, who offered me the first opportunities to publish the ideas now developed in this book, to Alain Le Méhauté and Frédéric Héliodore for support and enlightening discussions, and to Thiébaut Moulin for the interesting and stimulating discussions on several topics and for his enthusiasm and open-mindedness. I acknowledge fruitful collaboration with Jean Schneider in the early stages of this work (on the nonstandard analysis approach to fractals), very helpful and stimulating discussions with Silvano Bonazzola, Christiane Vilain, and Joaquin Diaz Alonso. Finally I am greatly indebted to Christophe Kotanyi for his careful reading of part of the manuscript and for his comments and encouragement, to Myriam Marouard, for checking language errors and misprints, and especially to Pierre-Yves Longaretti for his careful and detailed reading of the whole manuscript, his invaluable comments and criticisms, and his many suggestions for improvement.

This book is dedicated to my children Matthieu and Benoît, and to Myriam, Olivier and Nathalie.

L N

Meudon, April 1992

of the new structures suggested in microphysics; then the implication of scale relativity for observational cosmology are analysed. As in microphysics, cosmology is characterized by a fundamental scale dependence of physical laws, as exemplified in the expansion of the Universe. The situation is symmetrical to the microphysical one, and we tentatively reach similar conclusions: there exists a *universal, upper limiting scale* $\mathbb{L}$ in Nature which is *invariant under dilatations*. We identify this scale as yielding the true nature of the *cosmological constant* Λ, $\mathbb{L} = \Lambda^{-1/2}$. In such a perspective, Mach's principle and Dirac's great numbers hypothesis can be reactualized.

Then we jump to a different (but related) subject, by showing that the methods of non-differentiable geometry that were developed in Chapter 5 can also be applied to situations involving dynamical chaos. This is exemplified by a reconsideration of the problem of the distribution of planets in the Solar System. We finally conclude by a prospect about future possible developments of this new field.

Chapter 2

RELATIVITY AND QUANTUM PHYSICS

2.1. On the Present State of Fundamental Physics.

The laws of physics are presently described in the framework of two main theories, namely the theory of relativity (special[1-2] and general,[3] which include classical mechanics) and quantum mechanics[4-6] (developed into quantum field theories). Both edifices are extremely efficient and precise in their predictions; the constraints imposed by special relativity have even been incorporated in a relativistic quantum theory. But these two theories are founded on completely different grounds, even contradictory in appearance, and make use of a completely different mathematical apparatus.

General relativity is a theory based on fundamental physical principles, namely the principles of general covariance and of equivalence. Its mathematical tools come as natural achievements of these principles. On the contrary quantum mechanics is, at present, an axiomatic theory. It is founded on purely mathematical rules which, up to now, are not understood in terms of a more basic mechanism.

This leads to a strong dichotomy in physics: two apparently opposite worlds cohabit, the classical and the quantum. In particular gravitation, so clearly and accurately described by Einstein's theory of general relativity,[3] has escaped up to now any admissible description in terms of the quantum field-particle approach. Conversely, our understanding of the electromagnetic, weak and strong fields has made huge progress in the framework of quantum gauge theories,[7-9] while all classical attempts to unification (e.g., of gravitation and electromagnetism) have ended in failure.

These and other signs indicate, in our opinion, that physics is still in infancy. Several great problems, maybe the most fundamental ones, are still completely open. There is at present no theory able to make predictions about the two "tails" of the physical world, namely *elementarity* and *globality*, i.e., at the smallest and largest time scales and length scales.

At small scales, the "standard model" of elementary particles, based on quantum chromodynamics and electroweakdynamics, is able to include in its framework the *observed* structure of elementary particles and coupling constants (i.e., charges). But it seems, up to now, unable to predict on purely theoretical grounds either the number of elementary particles, or their masses, nor the values of the fundamental couplings. This failure is certainly related to the main failure of electrodynamics (classical and quantum): the divergence of self-energy and charge at infinite energy.[10] Renormalization[11-13] was only a partial solution to the problem. By replacing in calculations the theoretical infinite charges and masses by the observed ones, it allowed physicists to predict with high precision all the other physical quantities of interest. But the problem of masses and charges was left open.

At the other end, that of very large scales, even though the current cosmological theory has known great successes, one must not forget that general relativity, being a purely local theory (its fundamental tool, the metrics element, is differential), tells us nothing about the *global topology* of the Universe.[14] This is, with the problem of sources of gravitation (why does inertia curve space-time?), one of the limiting domains where general relativity is an incomplete theory, as recognized by Einstein himself:[15] an indication of this incompleteness may be its inability to include Mach's principle, except in some particular models, while observations seem to imply that it is effectively achieved by Nature (see Secs. 5.11 and 7.1).

The intermediate classical world is not devoid of open fundamental problems. Recent years have known an impressive burst in the study of dynamical chaos.[16-18] Chaos is defined as a high sensibility on initial conditions which leads to rapid divergence (e.g. exponential) of initially close trajectories, then to a complete loss of predictability on large time scales. Chaos is encountered in equations which look quite deterministic, in a large number of different domains like chemistry, fluid mechanics and turbulence, economics, population dynamics, celestial mechanics, meteorology... The challenge of chaos is that structures are very often observed in domains where chaos has developed, while ordinary methods fail to make prediction because of the presence of chaos itself. The understanding of how "order" (better: "organization") emerges from chaos is the key for the foundation of a future (still not existing) science of classical complexity. This fundamental problem will be addressed in Secs. 3.2, 5.6, 5.7 and 7.2.

We shall attempt to convince the reader that these questions, in the quantum, cosmological and classical complexity domains, may actually be of a similar nature. They all turn around the problem of *scales*, and may be traced back to a still unanswered very fundamental question: *what determines the fundamental scales in Nature*? A theory of scale is needed in physics. We shall propose here that Einstein's principle of relativity applies not only to laws of motion but also to laws of scale, and thus can be used as a basic stone for founding such a theory. But let us first briefly describe the present status of the theories of relativity and of quantum mechanics.

The theory of relativity.

Galilean relativity, Einstein's special and general theories of relativity are successive attempts to make possible the expression of physical laws in more and more general coordinate systems. Let us recall Einstein's statement of the principle of general relativity:[3] "*the laws of physics must be of such a nature that they apply to systems of reference in any kind of motion*". Remark that, under this form, this principle is, strictly speaking, a principle of the relativity of *motion*. This principle is particularly remarkable by the combination of its simplicity and its extraordinary power for deriving the most fundamental constraints which govern the physical world. Take Galileo's statement of the principle: "motion is like nothing". At first sight it may look like a trivial statement.

But the explicit expression of this principle actually imposes strong universal constraints about the possible forms that physical laws can take. Since motion cannot be detected by purely local experiments, only some particular laws of transformation between inertial systems are admissible. This leads to the classical laws of Galilean physics and, adding the postulate of the invariance of some velocity c, to Einstein-Poincaré-Lorentz special relativity. Moreover we shall demonstrate in Sec. 6.4 that this additional postulate is not necessary for deriving the Lorentz transform: i.e., the Lorentz transform may be shown to be the general transformation which achieves the principle of special relativity *in its Galilean form.*

Special relativity leads to the constraint that no velocity can exceed some universal velocity c, which may subsequently be shown to be the velocity of particles of null mass, in particular that of light.[19] Recall that the Minkowskian space-time is characterized by the invariant

$$ds^2 = c^2dt^2 - (dx^2 + dy^2 + dz^2) ,$$

under any change of inertial coordinate system. Then it was one of Mach's main contributions to the evolution of physics to insist on the relativity of *all* motions, not only of inertial ones. From *general covariance* and the *principle of equivalence*, Einstein constructed the theory of general relativity, whose equations are constraints on the possible curvatures of space-time. Einstein's equations

$$R_{\mu\nu} - \frac{1}{2} R\, g_{\mu\nu} - \Lambda\, g_{\mu\nu} = \chi\, T_{\mu\nu} \tag{2.1}$$

are the most general simplest equations which are invariant under any *continuous and differentiable* transformation of coordinate systems.

For a full account of the theory, we send the reader to textbooks as those by Misner, Thorne and Wheeler[22] or Weinberg.[55] Let us only briefly recall here that, in these equations, the $g_{\mu\nu}$'s are tensorial metric potentials which generalize the Newtonian gravitational scalar potential. The general relativistic invariant reads, in terms of Einstein's convention of summation on identical lower and upper indices

$$ds^2 = g_{\mu\nu}\, dx^\mu\, dx^\nu , \quad (\mu, \nu = 0 \text{ to } 3).$$

In general relativity, the curvature of space-time implies that the variation of physical beings (such as vectors or tensors) for infinitesimal coordinate variations depends also on space-time itself. This is expressed by the covariant derivative

$$D_\mu A^\nu = \partial_\mu A^\nu + \Gamma^\nu{}_{\rho\mu} A^\rho$$

which generalizes the partial derivatives. In this expression, the effect of space-time (i.e., of gravitation) is described by the Christoffel symbols

$$\Gamma^\rho{}_{\mu\nu} = \frac{1}{2}\, g^{\rho\lambda}\, (\partial_\nu g_{\lambda\mu} + \partial_\mu g_{\lambda\nu} - \partial_\lambda g_{\mu\nu}),$$

which play the role of the gravitational field. The covariant derivatives do not commute, so that their commutator leads to the appearance of a four-indices tensor, the so-called Riemann tensor $R^\lambda{}_{\mu\nu\rho}$:

$$(D_\mu D_\nu - D_\nu D_\mu) A_\rho = R^\lambda{}_{\rho\nu\mu}\, A_\lambda \,.$$

Contraction of the Riemann tensor yields the Ricci tensor $R_{\mu\nu} = g^{\lambda\rho} R_{\lambda\mu\rho\nu}$:

$$R_{\mu\nu} = \partial_\rho \Gamma^\rho{}_{\mu\nu} - \partial_\nu \Gamma^\rho{}_{\mu\rho} + \Gamma^\rho{}_{\mu\nu} \Gamma^\lambda{}_{\rho\lambda} - \Gamma^\rho{}_{\mu\lambda} \Gamma^\lambda{}_{\nu\rho} \,,$$

while the quantity $R = g^{\mu\nu} R_{\mu\nu}$ is the scalar curvature. Einstein's equations state that the energy-momentum tensor $T_{\mu\nu}$ is equal, up to the constant $\chi = 8\pi G/c^4$, to the geometric Einstein tensor given in the first member of (2.1), in which Λ is the *cosmological constant*. The Einstein tensor and the energy-momentum tensor are conservative in the covariant sense. Einstein's *equivalence principle of gravitation and inertia* is expressed by the fact that one may always find a coordinate system in which the metric is locally Minkowskian, and that in such a system the equation of motion of a free particle is that of inertial motion, $Du_\mu = 0$, where u_μ is the four-velocity of the particle. Written in any coordinate system, this equation becomes the geodesics equation

$$\frac{d^2 x^\mu}{ds^2} + \Gamma^\mu{}_{\nu\rho} \frac{dx^\nu}{ds} \frac{dx^\rho}{ds} = 0 \,.$$

Note that the principle of relativity, in Einstein's formulation, applies to the "*laws of Nature*", which Einstein carefully distinguishes from the *equations of physics*. Laws of Nature are assumed to exist independently from the physicist (this is the first underlying working postulate of any "philosophy of nature") while equations of physics are the mathematical expression of our own (always perfectible) attempts at reaching them. The mathematical translation of the principle of general relativity is Einstein's principle of general covariance:[3] "*the general laws of nature are to be expressed by equations which hold good for all systems of coordinates, that is, are covariant with respect to any substitutions whatever*".

The evolution of the principle of relativity is intrinsically linked to evolution of the concept of space-time. Any interrogation about the physics involved in the transformation of reference frames runs into an interrogation about space-time. Reversely, asking questions about space-time leads one to question relativity. In Galilean relativity, space and time are assumed to be absolute and independent concepts. The special theory[1-2] renounces such views, and introduces the concept of space-time.[20] But the Minkowskian space-time is still absolute, while the analysis by Mach, then by Einstein,[3] clearly shows the general covariance requirement to be inconsistent with the idea of a privileged space or space-time. This leads to the space-time of the general theory which depends on the material and energetic content of the universe.

The way through which space-time properties are related to matter properties is instructive (by "matter" we mean matter and energy). It consists in attributing to space-time those properties of matter which are universal. It is the universality of the Lorentz transform, which applies not only to electromagnetic waves but also to any kind of massive particle or system, that allows the introduction of the Minkowskian space-time. In general relativity, the universal property of matter pointed out by Einstein is the curvature of trajectories. This allows one to understand the crucial role played by the deviation of light rays (and more generally by all effects of gravitation on light) in the construction of the theory[21] and in its final acceptance. The universality of the curved nature of trajectories of particles, whether massive or not, leads to attributing the property of curvature to space-time itself. Then in the curved space-time, free particles follow geodesical lines, in agreement with the equivalence principle.

The power of this approach is appreciable when remembering that, if one completely accepts Einstein's geometrical interpretation, the concepts of forces, of potentials and of field disappear for the benefit of the mere curved space-time: from the principle of equivalence, inertia is found back locally in a freely falling reference system, i.e., one which follows a geodesic of the Riemannian space-time. Space-time, as described by the metric potentials, may be considered as a new mathematical tool, even more profound than that of field; from it the notions of force and of potential may be finally recovered, but as approximations. If we push to its logical ends the argument of the vanishing of the field, we get Einstein's radical interpretation of the nature of gravitation, as being nothing but the manifestation of a universal property of the world, the space-time curvature.[22]

Quantum physics.

At present, quantum mechanics is being based on a completely different approach. Let us briefly recall (and comment) the principles of the nonrelativistic theory (see e.g. Refs. 23-25):

(i) A physical system is defined by a *state function* ϕ. Its coordinate realization, the *complex* wave function $\psi(q,s,t)$, is often used in the non-relativistic theory. It depends on all classical degrees of freedom, q and t, and on additional purely quantum mechanical degrees of freedom s, such as spin. The probability for the system to have values (q, s) at time t is given by $P = |\psi(q,s,t)|^2$, so one may write ψ in the form $\psi = \sqrt{P}\, e^{i\theta}$. Following Feynman, one often rather uses the *probability amplitude* $\psi(a,b)$ between two space-time events a and b. The extraordinary fact about quantum mechanics is that the *full* complex probability amplitude, with modulus $\sqrt{P}$ and phase θ, is necessary to make correct predictions, while only the square P of the modulus is observed. When an event can occur in two alternative ways the probability amplitude is the sum of the probability amplitudes for each way considered separately:

$$\psi = \psi_1 + \psi_2 \quad \Rightarrow \quad P = P_1 + P_2 + \psi_1 \psi_2 + \psi_2 \psi_1$$

The two new rectangular terms additional to the classical terms $P_1 + P_2$ are at the origin of interferences and more generally of quantum coherence. Conversely, when it is known whether one or the other alternative is actually taken, the composition of probabilities takes the classical form $P = P_1 + P_2$.

(ii) If ψ_1 and ψ_2 are possible states of a system, then $\psi = a\psi_1 + b\psi_2$ is also a state of the system.

(iii) Physical observables are represented by linear Hermitian operators, Ω, acting on the state function. For example there corresponds to momentum p_i the *complex* operator $-i\hbar\partial/\partial q_i$. This is another manifestation of the mysterious character of quantum rules that, to a *real* momentum, there corresponds a *complex* operator acting on the complex probability amplitude.

(iv) Results of measurements of physical observables are given by any of the eigenvalues of the corresponding operator, $\Omega\,\psi = \omega_i\,\psi$.

(v) Any state function can be expanded as $\psi = \sum_n a_n \psi_n$ in an orthonormal basis, and $|a_n|^2$ records the probability that the system is in the nth eigenstate.

(vi) The time evolution of the system satisfies Schrödinger's equation

$$H\psi = i\hbar\frac{\partial\psi}{\partial t} \quad ,$$

where the Hamiltonian H is a linear Hermitian operator.

(vii) *Immediately* after a measurement, the system is in the state given by the first measurement. This seventh axiom ("Von Neumann's axiom") is forgotten in many text books though it is necessary to account for experiments: for example after a spin measurement, the spin remains in the state given by the measurement; just after a measurement of position (at $t+\delta t$, $\delta t \rightarrow 0$), a particle is in the position given by the measurement. Its absence may give a false impression of quantum mechanics as a theory where precise predictions can never be done, while this depends on the pure or mixed character of the state of the system. It underlies the phenomenon of reduction of the wave packet.[26]

These axioms have well-known philosophical consequences (or better, they are a self-consistent mathematical transcription of what *experiments* have told us about the microphysical world: the philosophical consequences originate, in the present thinking frame, from observations). Two realizations of the state function, the coordinate and momentum representations, are particularly relevant in this respect. The position and momentum wave-functions may be derived one from the other by reciprocal Fourier transform. From this comes the Heisenberg inequality

$$\sigma_x \, \sigma_p \;\geq \hbar/2 \,,$$

which implies the non-deterministic character of quantum trajectories. Note also that the solution of Schrödinger's equation for a free particle leads to the introduction of the de Broglie length and time: The phase of the wave-function writes $\theta = (p\,x - E\,t)/\hbar$, where p and E are the classical momentum and energy of the particle. The de Broglie *periods*, h/p and h/E, correspond to a phase variation of 2π. Throughout this book, we shall call "de Broglie length and time" the quantities

$$\lambda = \frac{\hbar}{p} \quad ; \quad \tau = \frac{\hbar}{E}$$

such that the quantum phase for a free particle writes $\theta = (x\,/\lambda - t/\tau)$. (Recall that historically, the Schrödinger equation was constructed as the equation whose de Broglie wave, obtained earlier, was a solution). The de Broglie scale may be generalized to more complicated systems: it can be identified with the characteristic transition scale occurring in the quantum phase.

It is clear from the above axioms that most of the quantum mystery may be traced back to the mere question: "*where does the complex plane of quantum mechanics lie*?" We shall in this book propose a solution to this puzzle by showing that a complex plane naturally emerges *in space-time* from the simplest prescription aimed at describing a *non-differentiable* space-time; this allows one to obtain Schrödinger's equation as the form taken by the fundamental equation of dynamics written in such a non-differentiable frame (see Chapter 5).

Quantum mechanics is an axiomatic theory rather than a "theory of principle" as relativity. We do not understand the physical origin of these axioms: we only know that they work, i.e., that the theory developed from them has a high predictive power and is remarkably precise. The mathematical beings of the present quantum theory are defined in an abstract space of state, so that the part played by the standard space-time has apparently been deeply decreased. Quantum physics in its present form describes rather the intrinsic properties of microphysical objects embedded in a space-time which is *a priori* assumed to be Euclidean or Minkowskian. Though it has already been suggested that the structure of the microphysics space-time could be foamy or gruyere cheese-like[27-29], this was assumed to hold only at the level of Planck's length and time. These ideas have now been developed into attempts to build a theory of quantum gravity (see e.g. Ref. 30 and references therein), whose preferential domain of application would be the very early universe.

Geometry and Microphysics

In spite of the advantages one may find in a space-time theory, the several unsuccessful attempts to unify gravitation and electromagnetism from a geometrical approach based on curvature and/or torsion (see e.g. Ref. 31) finally convinced physicists that such an approach has to be given up. The parallel success of quantum gauge theories led to the hope that unification may rather be reached only from the quantized field-particles approach, and that gravitation itself should be quantized in the end. However, among the various causes of the failure of previous geometrical attempts, two may be pointed out in the light of the hereabove remarks:

(i) the observed properties of the quantum world cannot be reproduced by Riemannian geometry;

(ii) the space-time approach cannot be based on particular fields, but on those properties of matter which are universal. It is thus clear that any new insights about the nature of the microphysical space-time may only be gained provided new concepts are introduced.

In this book we shall review the principles and the first results of a new attempt at reconsidering the conclusion that a geometrical approach of the quantum properties of microphysics is impossible. We suggest possible ways towards the construction of a spatio-temporal theory of the microphysical world, basing ourselves on the concept of *fractal space-time* in connection with the suggestion of an extension of the principle of relativity. The crucial new ingredient of our approach with respect to present standard physics (classical and quantum) is that we assume that *space-time is non-differentiable.* Moreover we shall demonstrate (Sec. 3.10) that non-differentiability implies *an explicit dependence of space-time on scale.*

In this quest, our main lead will be Einstein's principle of relativity. But we shall take here relativity as a general method of thinking, rather than as a particular theory. In a relativistic approach to physics, one tries to analyze what, in the expression of physical laws, depends on the particular reference system used, and which properties are independent of it. We shall show that the principle of relativity applies not only to motion, but also to scale transformations, once the resolution of measurements is defined as a state of the coordinate system.

2.2. The Need for a New Extension of Principle of Relativity.

Prior to setting the principle of relativity, there is the definition of coordinate systems and of the possible transformations between these systems. Indeed this principle is a statement about the universality of the laws of physics, whatever the system of coordinates in which they are expressed. So let us try to analyse further what we mean by a system of coordinates. Physics is, above all, a science based on measurements. Its laws apply not to objects by themselves, but rather to the *numerical* results of measurements which have been or may be performed upon these objects. So the definition of coordinate systems should include all the relevant information which is necessary to describe these results and to relate them in terms of physical laws.

It is an experimental fact that *four* numbers are necessary and sufficient to locate an event (i.e., a position and an instant): space-time is of topological dimension 4. This operation of location of an event is found to have the following properties:

(i) It cannot be made in an absolute way. This means that an event can be located only with respect to another event, never to some absolute position or instant. What are measured are always space *intervals* and time *intervals*. This *relativity of events* implies that coordinate systems must be firstly characterized by the setting of an *origin*, *O*.
(ii) Then one needs to define the *axes* of the coordinate system. They may be rectilinear, but more generally curvilinear. This means that space-time is covered by a continuous grid or lattice of lines (i.e., of topological dimension 1). In present physics, this curvilinear coordinate system is also assumed to be differentiable.
(iii) We want to characterize by numerical values the position and instant of a second event with respect to *O*. However length and time intervals are themselves relative quantities: there is indeed no absolute scale in Nature. This second relativity, which we shall call "relativity of scales", is currently translated by the need to use some *units* for measuring length and time intervals. But we shall argue in the following that its consequences for laws of physics may be far more profound.
(iv) A last property of space-time coordinate measurements (and of any measurement) is that they are always made with some finite *resolution*. We claim that resolution should be included in the definition of coordinate systems.[32,33] Being itself a length or time interval, it is subjected to the relativity of scales. This resolution corresponds to the minimal unit which may be used when characterizing the length or time interval by a final number: e.g., if the resolution of a rod is 1 mm, it would have no physical meaning to express a result in Å. This resolution sometimes corresponds to the precision of the measuring apparatus: it may then eventually be improved, this corresponding to an improved precision of the result in classical physics. But it may also correspond to a physical limitation. For example it is probable that the distance from the Sun to the Earth would never be measured with a precision of 1 Å: this would have no physical meaning. And last but not least, resolution of the measurement apparatus plays in quantum physics a completely new role with respect to the classical, since the results of measurements become dependent on it, as a consequence of Heisenberg's relations.

It is well-known that any set of physical data takes its complete sense only when it is accompanied by the measurement errors or uncertainties, and more generally by the resolution characterizing the system under consideration. Complete information about position and time measurement results is obtained when not only space-time coordinates (t, x, y, z), but also resolutions $(\Delta t, \Delta x, \Delta y, \Delta z)$ are given. Though this analysis already plays a central part in the theory of measurement and in the interpretation of quantum mechanics, one may remark however that its consequences for the nature of space-time itself have still not been drawn: we suggest that this comes from the fact that, up to now, resolutions have never appeared explicitly in the definition of coordinate systems, while, as shown hereabove, they are explicitly related to the information which is relevant for our understanding of the meaning of actual measurements.

Once the properties of coordinate systems defined, the next task is to describe the possible transformations that are acceptable between these systems. These transformations change the various quantities which define the state of coordinate systems, i.e., following the above analysis: origin, axes, units and resolution.

First consider changes of origin. The invariance of physical laws under static changes of the origin of coordinates systems is translated in terms of homogeneity of space and uniformity of time (more generally, homogeneity of space-time). This is the basis, under Noether's theorem, for the conservation of momentum and energy. So the very existence of energy-momentum as a fundamental conservative quantity (which is itself identified as the charge for gravitation in general relativity) relies on the first relativity, that of position and instants. Velocity-dependent changes of origin may also be considered in space: this leads to Galileo's relativity of inertial motion. But they may also be subsequently included into static rotations in *space-time*, so this leads us to the second transformation, that of axes.

Axes transformations first include changes of orientation. The invariance of physical laws under rotations in space corresponds to the isotropy of space and yields the conservation of angular momentum. Including rotations in space-time in the transformations considered allows one to describe the relativity of motion as a relativity of orientations in space-time. This yields the Lorentz transform. Finally Einstein's *special theory of relativity* accounts, in terms of the Poincaré invariance group, for the full relativity of positions, instants and axis orientation.

Then including any continuous and differentiable transformation yields Einstein's general relativity: the change to curvilinear coordinate systems introduces not only non-inertial motion but also curvature of space-time that manifests itself as gravitation.

On this road, it is clear that if one still wanted to generalize the class of acceptable transformations, one should give up differentiability, and then, as a last possibility, continuity (recall that non-differentiability *does not* imply discontinuity). Let us call "*extended covariance*" the covariance of the equation of physics under general continuous transforms, including non-differentiable ones. It is also clear that the achievement of such an extended covariance would imply a profound change in the physico-mathematical tool, since the whole mathematical physics is currently founded on integro-differentiation. In this book, we shall try to convince the reader that such an achievement is indeed necessary for our understanding of the foundation of the laws of Nature, in particular in the microphysical domain.

Let us indeed point out what may be considered as remaining defects of the present state reached by physics, which indicate that the principle of relativity of motion itself needs to be extended. Although it became definitively clear after Mach's and Einstein's analysis that the concept of an absolute space-time was to be given up and superseded by a space-time depending on its material and energetic content, the present quantum theory of microphysics still assumes space-time to be Minkowskian, i.e. absolute. This is at variance with the radically new quantum properties of matter and energy, as compared to the classical ones on which the special and general theories of relativity were founded. In other words, the Minkowskian and Riemannian nature of space-time was deduced from the classical properties of objects. In the quantum domain, we know that all objects have quantum properties (all are subjected to Heisenberg's inequalities). We also know that the structure of space-time must depend on its material and energetic content: how, under these conditions, can a space-time whose content is universally quantal be Minkowskian, i.e. flat and absolute?

An additional remark may be made. The goal of a completely general relativity cannot presently be considered as reached, since it is clear that the methods of the present theory of general relativity do not apply to reference frames which would be swept along in the quantum motion, which is continuous but non-differentiable, as discovered by Feynman.[34,35] This non-differentiability of virtual and real quantum paths is one of the key points to our own approach. We shall at length come back on it, showing in particular that it can be described in terms of Brownian motion-like *fractal* properties.[36,33]

Basing ourselves on these considerations, we have suggested that the principle of relativity still needs to be extended.[32,33] Our concept of space-time has evolved from the Galilean independent space and time, to the Minkowskian absolute space-time, then to the Riemannian relative space-time of Einstein's theory. If one wants to include the non-differentiable fractal quantum motion into those described by a theory of relativity, a radically new geometrical structure of space-time must be introduced. Our suggestion is that the quantum space-time is relative and fractal,[33] i.e., *divergent with decreasing scale* (we shall adopt this definition of the word "fractal" here; see Mandelbrot[37,38] for other definitions). We shall indeed demonstrate (see Sec. 3.10) that continuity and non-differentiability implies scale divergence. The same conclusion concerning the fractal stucture of space-time will be reached in the next Section, by basing ourselves on the relativity of all scales in Nature.

In our approach, throughout the present essay, we assume space-time to remain a continuum, even if it is no longer assumed to be differentiable. An ultimate choice for physics would be to give up the hypothesis of continuity itself. Some attempts to introduce discontinuous space-times have been made.[39,27-29] In this respect let us quote one of these attempts, Moulin's "arithmetic relators", which are defined using purely integer numbers.[40] Arithmetic relators are quadratic cellular automata which include internal variables and environment variables. They have proved to be efficient for providing structures, in particular biological ones.[41] Such an ability of making structures emerge from very few conditions is reminiscent of "mappings" often used in the study of dynamical chaos. In particular, arithmetic relators yield a hierarchisation, i.e., the various structures appear at different levels of imbrication.

We shall adopt here a more conservative point of view, by keeping the *space-time continuum* hypothesis and by including the scale dependence in the fundamental principle themselves.

2.3. Relativity of Scales.

We have considered, in the previous Section, the various transformations of the coordinate systems corresponding to changing the origin and axes. The subsequent state of coordinate systems which may be submitted to a transformation are *units*. Some attempts at including such a transformation in physical laws have been made, in particular in the framework of the conformal group.[42,43] Conformal transformations include, in addition to the Poincaré ones, dilatations and special conformal transformations; both of which may be interpreted as related to changes of units. However, while electromagnetic waves are subjected to the conformal symmetry, this is not the case for matter, so that the conformal symmetry cannot be an exact symmetry of nature. Moreover the choice of the unit is in most situations a purely arbitrary one, which does not describe the conditions of measurement, but only their translation into a number.

Let us nevertheless analyse further the physical meaning of units. Their introduction for measuring lengths and times is made necessary by *the relativity of every scales in Nature*. When we say that we measure a length, what we actually do is to measure *the ratio of the lengths of two bodies*. In the same way as the absolute velocity of a body has no physical meaning, but only the relative velocity of one body with respect to another, as demonstrated by Galileo, the length of a body or the periods of a clock has no physical meaning, but only the ratio of the lengths of *two bodies* and the ratio of the periods of *two clocks*.

When we say that a body has a length of 132 cm, we mean that a second body, to which we have arbitrarily attributed a length of 1 cm and which we call the unit, must be *dilated* 132 times in order to obtain the first body's length. Measurements of length and time intervals always amount, in the end, to dilatations. The tendency for physics to define a unique system of units was certainly a good thing, since this was necessary for a rational comparison of measurement results from different laboratories and countries. However, this means that the length of all bodies are referred to a same unique body and the period of all clocks to a same clock, this giving a false impression of absoluteness: such a method masks the actual relation between lengths of all bodies in Nature, which is a two-by-two relation.

The fact which allows us to use a unique unit is the simple law of composition of dilatation, $\rho''=\rho\,\rho'$. This law is certainly extremely well verified in the classical domain: there is no doubt that a body having a length of 21 m also measures 2100 cm. We however claim that we know nothing about the actual law of dilatation in the two domains of quantum microphysics and cosmology, in which explicit measurements of length and time become impossible. In these two domains length and time intervals are deduced from observation of other variables (energy-momentum at small scale, apparent luminosity and diameter at large scale) and from underlying accepted *theories* (quantum mechanics and general relativity) which have been constructed assuming implicitly the hereabove standard law of dilatation. (Compare with the status of velocities before the coming of special relativity, when it also seemed self-evident that their law of composition was $w = u + v$). We shall attack this problem in Chapter 6 (and in Sec. 7.1 for cosmology) and make new proposals based on the principle of scale relativity.

The status of *resolutions* is related to that of units (in particular they are subject to the relativity of scales), but is actually different and of more far reaching physical importance. Changing the resolution of measurement corresponds to an explicit change of the experimental conditions. Measuring a length with a resolution of $1/10^{\text{th}}$ mm implies the use of a magnifying glass; with 10 μm, we need a microscope; with 0.1 μm, an electron microscope; with 1Å, a tunnel microscope. For even smaller resolutions, the measurements of length become indirect, since the atom sizes are reached and exceeded. When we enter the quantum domain, i.e., for resolutions smaller than the de Broglie length and time of a system (as will be specified afterwards), the physical status of resolutions radically changes. While classically it may be interpreted as precision of measurements (measuring with two different resolutions yields the *same* result with different precisions), resolution plays a completely different role in microphysics: the results of measurements explicitly depend on the resolution of the apparatus, as indicated by Heisenberg's relations. This is the reason why we think that the introduction of resolution into the description of coordinate systems (as a *state of scale*) is not trivial, but will instead lead to a genuine theory of scale relativity and the emergence of new physical laws (see Chapter 6).

In present quantum mechanics, the scale dependence is already implicitly present. However it is *explicitly* present neither in the axioms nor in the basic equations. It comes from the interpretation of these equations thanks to a theory of measurements. Specifically, one writes Schrödinger's equation, then solve it. This yields a probability amplitude from which one deduces the probability density; then one may compute the dispersion of the variable considered, and Born's statistical interpretation of quantum mechanics ensures that this will give us the standard error of a statistical ensemble of values resulting from several measurements of this variable. By extension, this dispersion may also be interpreted as the resolution of the measurements: e.g., if one makes position measurements with a resolution Δx, one expects a subsequent dispersion in the values of the momentum given by Heisenberg's relation $\sigma_p \approx \hbar/\Delta x$.

However one may require that a complete physical theory includes in its equations the whole set of physical information yielded by experiment. In other words, it may be demanded that a theory of measurement, instead of being externally added to a given theory, becomes an integral part of it.

Such a requirement of explicit expression of the measurement resolutions in the equations of physics begins to be fulfilled in present physics, even though the interpretation is different from ours. Let us quote two approaches where scale-dependent equations are actually written.

One of these domains is that of the theory of measurement in quantum mechanics, concerning in particular the problem of the so-called reduction of the wave-packet (i.e., sudden collapse of the state vector caused by a measurement). This problem, which underlies that of the quantum-classical dualism (where is the transition from quantum to classical; are classical laws approximations of the quantum ones...?), has recently known a resurgence of interest (see Refs. 44 and 45 and references therein). The basic idea of these works is that reduction of the wave packets originates from an *interaction* of a quantum system *with the environment.* This interaction is described by a *master equation* which is explicitly resolution-dependent, so that the transition from quantum to classical behaviour is found to depend directly on resolution in terms of a *decoherence time scale*[44]

$$\tau_D \approx \left(\frac{\lambda_T}{\Delta x}\right)^2 ,$$

where λ_T is the thermal de Broglie length of the system. We shall come back to this approach in Sec. 5.7.

A second domain where explicitly scale-dependent equations have been introduced is the renormalization group approach.[46-48] First introduced in quantum electrodynamics as the group of transformation between the various ways to renormalize the theoretical divergences, it became under Wilson's influence a general method of description of problems involving multiple scales of length.[48-50] In the renormalization group approach, one writes differential equations describing the infinitesimal variation of physical quantities (fields, couplings) under an infinitesimal variation of scale. The renormalization group will play a leading part in our approach. We shall indeed demonstrate that the renormalization group equations (i) can be interpreted as the simplest lowest order differential equations describing the measure on fractal geometry; (ii) are for scale laws the equivalent of Galileo's group for motion laws. As a consequence we shall propose (in Chapter 6) a generalization of its structure aimed at making it consistent with the principle of scale relativity, which is stated above.

Our first proposal for implementing the idea of scale relativity was to extend the notion of reference system by defining "supersystems" of coordinates which contain not only the usual coordinates but also spatio-temporal resolutions, i.e., $(t,x,y,z;\Delta t,\Delta x,\Delta y,\Delta z)$.[33] The axes of such a reference supersystem would be endowed with a thickness: this corresponds, indeed, to *actual* measurements. Then we proposed an extension of the principle of relativity, according to which the laws of nature should apply *to any coordinate supersystem.* In other words, not only general (motion) covariance is needed, but also *scale covariance.*

Consider now the fundamental behaviour of the quantum world in the light of these ideas. Recall our assumption that when the physicist finds *universal* properties for physical objects, these properties may be attributed to the nature of space-time itself. This analysis applies particularly well to some of the quantum properties, accounting for their universal character: de Broglie's[51] and Heisenberg's[52] relations.

Let us recall indeed that the wave-particle duality is postulated to apply to any physical system, and that the Heisenberg relations are consequences of the basic formalism of quantum mechanics (Sec. 2.1). The existence of a minimal value for the product $\Delta x.\Delta p$ is a universal law of nature. Such a law, in spite of its universality, is considered in the current quantum theory as a property of the quantum objects themselves (it becomes a property of the measurement process because measurement apparatus are in part quantum and precisely because it is universal). But it is remarkable that it may be established without any hint to any particular effective measurement (recall that it arises from the requirement that the momentum and position wave functions are reciprocal Fourier transforms). So we shall assume that the dependence of physics on resolution pre-exists any measurement and that actual measurements do nothing but reveal to us this universal property of nature: then a natural achievement of the principle of scale relativity is to attribute this universal property of scale dependence to space-time itself.

By such a route, we finally arrive at the same conclusion as that at the end of the previous Section, but this conclusion is now reached by basing ourselves on the principle of scale relativity rather than on the extension of the principle of motion relativity to non-differentiable motion. Namely, the quantum space-time is scale-divergent, according to Heisenberg's relations, i.e., by our adopted definition (see Sec. 2.2 and Chapter 3), *fractal.* This idea will be more fully developed in Chapters 4 and 5.

However this first formulation of the principle of scale relativity[33] is not fully satisfactory. It does not incorporate the complete analysis of the relativity of scales (see above), and treats resolutions on equal footing with space-time variables. However, we have shown that resolutions are more accurately described as a *relative state of scale* of the coordinate system, in the same way as velocity describes its state of motion. So, in parallel with Einstein's formulation of the principle of motion relativity, we shall finally set the principle of scale relativity in the form[53]

"*The laws of physics must apply to coordinate systems whatever their state of scale.*"

The full principle of relativity will then require validity of the laws of physics in any coordinate system, whatever its state of motion and of scale. This is completed by a principle of scale covariance (in addition to motion covariance):

"*The equations of physics keep the same form (are covariant) under any transformation of scale (i.e. contractions and dilatations).*"

We shall see in Chapter 6 that in this form the principles of scale relativity and scale covariance imply a profound modification of the structure of space-time at very small scales: we find that there appears a universal, unpassable, limiting scale in Nature, which is invariant under dilatation, and plays for scale laws a role quite similar to that played by the velocity of light for motion laws (i.e., the limiting scale is neither a cut-off, nor a quantization, nor a discontinuity of space-time). Such a modification has observable consequences at presently accessible energies, which may be expressed in terms of 'scale-relativistic' corrections.

2.4. On the Nature of Quantum Space-Time.

All the hereabove arguments indicate that one must give up the absolute Minkowskian space-time postulated in the current quantum theory, and replace it by a space-time which is relative to its material and energetic content and explicitly dependent on scale. Three mathematical methods may be considered to achieve such a program. The first one is geometric: the concept of fractal[37,38] refers to objects or sets which are indeed scale-dependent. It must be generalized to that of fractal space,[37] while the concept of scale invariance must be extended to that of scale covariance. One may also look for an algebraic tool: the renormalization group is thus very well adapted, but must also be generalized in order to satisfy scale covariance. A third method (which will not be considered here) could be to work in the framework of the conformal group, owing to the fact that it already contains dilatations in its transformations. As a consequence, we expect space-time to be described by a metric element based on *generalized, explicitly scale-dependent, metric potentials* $g_{\mu\nu}= g_{\mu\nu}(t,x,y,z;\Delta t,\Delta x,\Delta y,\Delta z)$.[33]

Let us specify the physical meaning of this proposal. The concept of space-time allows one to think of all the positions and instants taken together as a whole. Space-time may be viewed as the set of all events and of the transformations between them. But to the set of all events, $x^{\mu} = -\infty$ to $+\infty$, ($\mu = 0$ to 3), we add the set of all possible resolutions, $ln(\Delta x^{\mu}) = -\infty$ to $+\infty$. Let us call this set "zoom". Hence the geometrical frame in which it will be attempted to work is, strictly speaking, a "zoom-space-time". This means that geometrical structures may be looked for, not only in space-time, but also in the zoom dimension (see Chapter 4). However, as remarked in Sec. 2.3, resolutions and space-time variables do not play an identical role. The "space-time-zoom" is equivalent to phase space rather than an extension of space-time.

The important point to be understood by the reader, since it underlies our whole methods and results, is that we call for a profound change of mentality in the physical approach to the problem of scales. One must give up the "reductionist" view of perfect points whose small scale organization would give rise to the large scale one. One must even go beyond the view of a physics where several particular scales are relevant. We claim that a genuine physics of scale can be constructed only in a frame of thought where all scales in Nature would be simultaneously considered, i.e., when placing ourselves in a continuum of scales. In such a perspective, the standard coordinates themselves lose their physical meaning, and should be replaced by fractal coordinates which are explicitly scale dependent, $X=X(s,\varepsilon)$ (see Chapters 3 and 5).

The geometrical properties and structures of the microphysical space-time remain to be built in details. As stated above, this book reviews an approach of this problem where it is proposed that, on account of its inferred dependence (and divergence) on resolution, one of the main properties of such a geometry would be its fractal character. We recall that we have previously proposed[54] that the concept of fractal should be applied not only to sets or objects embedded in Euclidean space, but to a whole space (more generally space-time) considered in an intrinsic way, i.e., for which curvilinear coordinates, metrics elements, geodesics etc... should be defined. Such an approach may be related to Le Méhauté's,[56] who was able to describe new electromagnetic properties arising in fractal media, by using the mathematical tool of non-integer integro-differentiation.

We indeed think that the concept of fractal space-time allows one to revise the conclusion that the quantum mechanical behaviour cannot be derived from a geometrical theory. Several mathematical properties of fractals go in the right direction, e.g.:

* One of the main characteristics of the fractal geometry is its dependence on resolution. Thus it offers a natural way of actualization of the hereabove suggested extension of the principle of relativity, by the use of a spatio-temporal description.

* It was realized by Feynman[34,35] that a particle path in quantum mechanics may be described as a continuous and non-differentiable curve, while non-differentiability is one of the properties of fractals. More precisely, a particle trajectory (of topological dimension 1) in nonrelativistic quantum mechanics may be characterized by *a fractal dimension* 2 when the resolution becomes smaller than its de Broglie length (see Ref. 36 and Chapter 4). This result, being derived from the Heisenberg relation and equivalent to it (see hereafter), is of a universal character. So, in the same way that general relativity attributes to space-time the universal property of curvature of trajectories, our suggestion is to attribute to quantum space-time the universal property of fractalization owned by quantum mechanical trajectories. The implications of this proposal will be specified throughout this book.

* Infinite numbers arise naturally on fractals, and the occurence of infinite quantities is one of the difficulties of current quantum physics. It is remarkable that the infinities which appeared in quantum electrodynamics precisely concern physical quantities like masses (i.e., self-energy) or charges, which are fundamental invariants built from space-time and quantum phase symmetries. One may wonder whether the need for

renormalization comes from the lack of account of the irreducible space-time infinities which would be a part of the nature of a fractal space-time.

* Extending the general relativistic approach, the particles in a fractal space-time are expected to follow the "geodesical" lines. But the absence of derivative, the folding and the infinite number of obstacles at all scales due to the fractal structure allow one to infer that an *infinity* of geodesics will exist between any two points, so that only statistical predictions will be allowed (see Sec. 5.5).

Additional examples of the adequation of fractals and quantum mechanical properties will be reviewed in this book. We are led throughout this work by the postulate that microphysical space-time is a self-avoiding fractal continuum of topological dimension 4. Fractal 4-coordinates are assumed to be defined on this fractal space-time. (Some ways to deal with their infinite character and with their non-differentiability are proposed in Chapter 3). These fractal coordinates correspond to the ideal case of infinite resolution, $\Delta x^{\mu} = 0$. Then the various supersystems of coordinates which correspond to finite resolution will be obtained by smoothing them with "4-balls" $(\Delta t, \Delta x, \Delta y, \Delta z)$. The classical coordinates (which are independent of resolutions) result from the same smoothing process, but with balls larger than some transitional values λ^{μ}, corresponding to the fractal/nonfractal transition, which we identify with the quantum/classical transition. Note also that the physical being to be used in order to fit with the "zoom-space-time" idea is not only the fractal itself (i.e. the final result of a fractalization process), but mainly the set of all its approximations for all possible values of space-time resolution. Some of its other properties will gradually emerge, while attempts will be made to express the main quantum mechanics results in terms of geometrical fractal structures.

In particular, our application of the principle of scale relativity to the question of the fractal dimension of quantum paths leads us to the conclusion that the constant fractal dimension D=2 obtained from standard quantum physics is only a "large" scale approximation. This Brownian-motion like fractal dimension (see Sec. 5.6) may also be interpreted in terms of a constant anomalous dimension δ=1. But this constant value corresponds to "Galilean" scale laws, while the requirement of scale covariance leads us to the conclusion that the correct renormalization group for space-time must be a Lorentz group (Chapter 6).

Already the new structure of space-time revealed by special motion relativity at the beginning of the century underlies in an inescapable way the Riemannian structure of Einstein's general relativity: Space-time is locally Minkowskian, so that all attempts at constructing a Riemannian theory of gravitation were condemned to fail in the absence of special relativity constraints. In the same way (assuming that the whole approach is correct) if a full theory of scale relativity is to be developed one day in terms of fractal space-time, we think that such a theory will be forced to incorporate in its description the new structure of space-time which is described in Chapter 6: a space-time where the perfect zero point has disappeared from concepts having physical meanings, whose fractal dimension is not constant but scale-dependent and whose local invariance group is Lorentz's, for motion as well as scale transformations.

2 References

1. Einstein, A., 1905, *Annalen der Physik* **17**, 891. English translation in "*The principle of relativity*", (Dover publications), p. 37.

2. Poincaré, H., 1905, *C. R. Acad. Sci. Paris* **140**, 1504.

3. Einstein, A., 1916, *Annalen der Physik* **49**, 769. English translation in "*The principle of relativity*", (Dover publications), p. 111.

4. Schrödinger, E. von, 1925, *Ann. Phys.* **79**, 361.

5. Heisenberg, W., 1925, *Zeitschrift für Physics* **33**, 879.

6. Dirac, P.A.M., 1928, *Proc. Roy. Soc. (London)* **A117**, 610.

7. Salam, A., in *Elementary Particle Physics* (Almqvist & Wiksell, Stockholm, 1968).

8. Weinberg, S., 1967, *Phys. Rev. Lett.* **19**, 1264.

9. Glashow, S., 1961, *Nucl. Phys.* **22**, 579.

10. Dirac, P.A.M., in "*Unification of fundamental forces*", 1988 *Dirac Memorial Lecture*, (Cambridge University Press, 1990), p.125.

11. Feynman, R.P., 1949, *Phys. Rev.* **76**, 769.

12. Schwinger, J., 1948, *Phys. Rev.* **74**, 1439.

13. Tomonaga, S., 1946, *Prog. Theor. Phys.* **1**, 27.

14. Thurston, W.P., & Weeks, J.R., 1984, *Sci. Am.* **251**, 94.

15. Pais, A., '*Subtle is the Lord', the Science and Life of Albert Einstein* (Oxford University Press, 1982).

16. Lorenz, E.N., 1963, *J. Atmos. Sci.* **20**, 130.

17. Ruelle, D., & Takens, F., 1971, *Commun. Math. Phys.* **20**, 167.

18. Eckmann, J.P., & Ruelle, D., 1985, *Rev. Mod. Phys.* **57**, 617.

19. Levy-Leblond, J.M., 1976, *Am. J. Phys.* **44**, 271.

20. Minkowski, H., 1908, *Address delivered at the 80th assembly of german natural scientists and physicians*, Cologne, English translation in "*The principle of relativity*", (Dover publications), p. 75.

21. Einstein, A., 1911, *Annalen der Physik* **35** , 898. English translation in "*The principle of relativity*", (Dover publications), p. 99.

22. Misner, C.W., Thorne, K.S., & Wheeler, J.A., *Gravitation* (Freeman, San Francisco, 1973).

23. Bjorken, J.D., & Drell, S.D., *Relativistic Quantum Mechanics* (Mac Graw Hill, 1964).

24. Schiff, L.I., *Quantum Mechanics* (Mac Graw Hill, 1968).

25. Feynman, R.P., Leighton, R., & Sands, M., *The Feynman Lectures on Physics* (Addison-Wesley, 1964).

26. Balian, R., 1989, *Am. J. Phys.* **57**, 1019.

27. Fuller, R.W., & Wheeler, J.A., 1962, *Phys. Rev.* **128**, 919.

28. Wheeler, J.A., in *Relativity Groups and Topology* (Gordon & Breach, New York, 1964).

29. Hawking, S., 1978, *Nucl. Phys.* **B144**, 349.

30. Isham, C.J., in 11th *International conference on General Relativity and Gravitation*, M. MacCallum ed., (Cambridge University Press, 1986) p.99.

31. Lichnerowicz, A., *Théories Relativistes de la Gravitation et de l'Electromagnétisme* (Masson, Paris, 1955).

32. Nottale, L., 1988, *C. R. Acad. Sci. Paris* **306**,341.

33. Nottale, L., 1989, *Int. J. Mod. Phys.* **A4**, 5047.

34. Feynman, R.P., 1948, *Rev. Mod. Phys.* **20**, 367.

35. Feynman, R.P., & Hibbs, A.R., *Quantum Mechanics and Path Integrals* (MacGraw-Hill, 1965).

36. Abbott, L.F., & Wise, M.B., 1981, *Am. J. Phys.* **49**, 37.

37. Mandelbrot, B., *Les Objets Fractals* (Flammarion, Paris, 1975).

38. Mandelbrot, B., *The Fractal Geometry of Nature* (Freeman, San Francisco, 1982).

39. Brightwell, G., & Gregory, R., 1991, *Phys. Rev. Lett.* **66**, 260.

40. Moulin, Th.,1991, *International Review of Systemic*, in the press

41. Vallet, Cl., Moulin, Th., Le Guyader, H., & Lafrenière, L., 1976, in VIIIth *International Congress of Cybernetics*, Namur (Belgium), p.187.

42. Barut, A.O., & Haugen, R.B., 1972, *Ann. Phys.* **71**, 519.

43. Callan, C.G., Coleman, S., & Jackiw, R., 1970, *Ann. Phys.* **59**, 42.

44. Unruh, W.G., & Zurek, W.H., 1989, *Phys. Rev.* **D40**, 1071.

45. Zurek, W.H., 1991, *Physics Today* vol. **44**, n°10, p.36.

46. Gell-Mann, M., & Low, F.E., 1954, *Phys. Rev.* **95**, 1300.

47. Stueckelberg, E., & Peterman, A., 1953, *Helv. Phys. Acta* **26**, 499.

48. Wilson, K.G., 1975, *Rev. Mod. Phys.* **47**, 774.

49. Wilson, K.G., 1983, *Rev. Mod. Phys.* **55**, 583.

50. de Gennes, P.G., *Scaling concepts in Polymer Physics* (Ithaca, Cornell Univ. Press, 1979).

51. Broglie, L. de, 1923,*C.R. Acad. Sci. Paris* **177**, 517.

52. Heisenberg, W., 1927, *Z. Physik* **43**, 172.

53. Nottale, L., 1992, *Int. J. Mod. Phys.* **A**, in the press.

54. Nottale, L., & Schneider, J., 1984, *J. Math. Phys.* **25**, 1296.

55. Weinberg, S., *Gravitation and Cosmology* (Wiley & Sons, New York, 1972).

56. Le Méhauté, A., *Les Géométries Fractales* (Hermès, Paris, 1990).

Chapter 3

FROM FRACTAL OBJECTS TO FRACTAL SPACES

3.1. Basics of Fractals.

The term "fractal" was coined in 1975 by B. Mandelbrot[1-3] to name objects, curves, functions, or sets "whose form is extremely irregular and/or fragmented at all scales". Such objects have been studied by mathematicians for more than a century.[4-7] They may be characterized by non-integer dimensions and by their non-differentiability. We give in the present section a simple account of the simplest properties of fractal objects, especially those which are particularly relevant to the way we intend to treat them hereafter. More details on several other aspects may be found in Refs. 3 and 8. A compilation of references concerning fractals and their use in various domains is given in Ref. 9.

Let us first recall that two sets have the same topological dimension if and only if one may define a *continuous and one-to-one transformation* between them. Topological dimensions remain integer. But alternative definitions of dimension have been given, such as the Hausdorff-Besicovitch, the box-counting dimension, the covering dimension and the similarity dimension, which are gathered under the term "fractal dimension". Fractals are usually defined as sets of topological dimension D_T and fractal dimension D, such that[3] $D>D_T$.

However, this definition seems to be too restrictive, since some objects or sets that are clearly "fractal" may have either $D=D_T$ or no constant or well-defined dimension. We shall then adopt here a different definition. The sets and spaces that we call fractal throughout this book are characterized by their scale divergence. This means that they are metric spaces for which a finite measure can be defined, but whose standard measure (which corresponds to their topological dimension, i.e. the length of a curve, the area of a surface, etc...) tends to the infinite when resolution tends to zero. This includes usual fractals (power law divergence as ε^{D_T-D}) but also "underfractals" (for example, logarithmic divergence, for which $D=D_T$) and "superfractals" (for example, exponential divergence, for which the formal dimension would be infinite).

A fact to be remarked here is that a fractal may have an integer fractal dimension: e.g. one may define self-avoiding fractal curves of fractal dimension 2 drawn in a 3-dimensional Euclidean space (see hereafter; such curves play an essential role in the present approach to quantum physics). The measure on such curves will be an area, although they do not look at all like a surface. Conversely if the fractal dimension of a continuous fractal equals the dimension of the Euclidean space in which it is embedded, we may be sure from the definition of topological dimension that it will self-intersect (have multiple points). A well-known example of this situation is the Peano curve (see Figs. 5.2 and 5.5), which is a plane-filling curve of fractal dimension 2.

One of the definitions for the new dimensions is related to the Hausdorff measure of a set. It is of practical interest, since it is related to a "physical" method of analysis of the content of a set or object, consisting in covering it with balls of decreasing radii. The length of a segment of "radius" r is $2r$; the surface of a disk of radius r is πr^2; the volume of a sphere is $4\pi r^3/3$... More generally the content of a ball of radius r in a space of dimension d may be defined as:

$$C = (r \, . \, \pi^{1/2})^d \, / \, \Gamma(1 + d/2)$$

This relation is generalizable to dimensions d which are no more integer. Then consider a standard curve: when covering it with 1-dimensional balls (segments), one gets a finite content which is named its length. By the same method, a 2-dimensional set will be given a finite 2-measure called "surface", a 3-dimensional set a finite "volume", etc... Now consider a compact subset F of a metric space M. Assume that F can be covered by $N(\varepsilon)$ balls of radius $\leq\varepsilon$. If one may write

$$0 < lim_{\varepsilon\to 0} \{ N(\varepsilon) (\varepsilon \pi^{1/2})^D / \Gamma(1 + D/2) \} < \infty, \qquad (3.1.1)$$

then the Hausdorff D-measure is generalized to not necessarily integer values. F is said to be of Hausdorff dimension D_H, while the finite number defined by the limit in (3.1.1) is its D_H-measure. A different definition of the measure C leads to the Besicovitch dimension.[3] However the dimensions that are calculated in most physical applications are not Hausdorff dimensions, but rather 'box dimensions':

$$D_B = lim_{\varepsilon\to 0+} sup \frac{log N(\varepsilon)}{log \varepsilon^D} . \qquad (3.1.2)$$

The fractal dimensions considered throughout this book correspond in most cases to this definition. Note that the Hausdorff dimension and the box dimension are in general not equal.[10]

A first important consequence of definition (3.1.2) is that the D_T-measure of a fractal (for which $D > D_T$ by definition) is infinite. This means that the number of points of a fractal "dust" (by definition $D_T = 0$), the length of a fractal curve, the surface of a 2-fractal, the volume of a 3-fractal, etc... are infinite. More may be said about this behaviour, which is closely related to the scaling properties of fractals.

Indeed, a simplified and very effective way to describe the covering of fractals by balls of varying radii is to identify this process with the building of successive approximations of the fractal at various resolutions ε. (Note that throughout this book the word "resolution" stands for the ball radius, which is to be identified with the precision or "error" $\varepsilon = \Delta x$, instead of its inverse, while the current definition, e.g. in spectroscopy and astronomy, is $R = \lambda/\Delta\lambda$). One finds that the d-measure of a subset of a fractal of Hausdorff dimension D is given by

$$V_d = \xi \ \varepsilon^{d-D} , \qquad (3.1.3)$$

where the meaning of the finite quantity ξ will gradually emerge from the following.

The special cases considered hereabove are thus illuminated. When $d = D$, one finds that the D-measure of a set of fractal dimension D is the finite quantity, ξ in (3.1.3). When $d = D_T$, the inequality $D > D_T$ implies a power law divergence of the usual measure (as based on the topological dimension). This is the case most frequently used for physical interpretation and measurement of fractal dimensions: i.e., one naturally tends to measure the number of points of a fractal dust, the length of a fractal curve, the area of a fractal surface, etc..., which depend on resolution as

$$V = \xi \, . \, \varepsilon^{D_T - D} \quad . \tag{3.1.4}$$

We stress here and for future applications the form of this result, which is more than a simple proportionality relation. When ε tends to zero, this D_T-hypervolume (3.1.4) is the product of a finite quantity ξ which is identified with the D-measure of the subset of the fractal considered, and of an infinite quantity, $\varepsilon^{D_T - D}$. This factorization achieves a kind of "renormalization" on fractals, as will be further analysed hereafter.

Conversely a d-measure with $d > D$ vanishes. This happens in particular for the measure corresponding to the topological dimension E of a space in which the fractal has eventually been embedded. For example a fractal curve drawn in a plane such that $1 < D < 2$ has infinite length and infinitesimal area.

An alternative definition for the fractal dimension is the "similarity dimension". It is based on the remark that when one applies a similarity transformation of factor n to a segment, one obtains a new segment $n = n^1$ times longer; to a square, one obtains a new square made of n^2 times the initial square; more generally, to an hypercube in dimension D_T, one obtains a new hypercube made of n^{D_T} initial hypervolumes. Self-similar fractals of similarity dimension D are such that when magnified by a factor of q, one obtains $p = q^D$ versions of the initial fractal. The relation between p, q and D is thus straightforward:

$$D = ln(p) \, / \, ln(q) \quad . \tag{3.1.5}$$

In the simplest cases the Hausdorff and similarity dimensions (and other ones) are identical[3]. This allowed Mandelbrot to introduce the general term "fractal dimension".

Let us illustrate all these notions by effectively constructing one archetype of a class of fractal curves, the von Koch curve.[7] One starts from a segment of length L_0: let us call it F_0. Then one defines F_1 as the curve of Fig. 3.1, made of $p = 4$ segments of length L_0/q, where $q = 3$. F_2 is now defined by scaling F_1 by a factor q^{-1}, and then by replacing each segment of F_1 by its scaled version. This "fractalization" process is then reproduced to infinity, yielding the various approximations F_n and at the end the fractal itself, F. This construction clearly displays one of the most important properties of fractals, i.e., their non-differentiability.

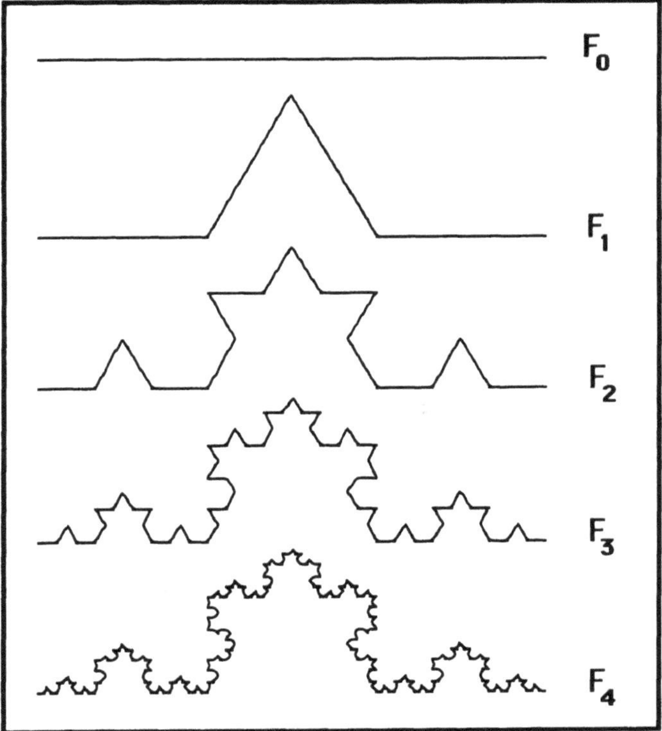

Figure 3.1. Successive approximations (resolutions 3^{-n}, n=0 to 4) of the von Koch fractal curve. Its topological dimension is 1 while its fractal dimension is $log4/log3$.

Note however that the breaking points which join the segments in F_n belong to the final fractal curve whatever the value of n. In this respect the hereabove construction may be considered as a point-by-point construction of the fractal (see Fig. 3.2). The segments introduced play no part in the final fractal object; they only ensure continuity of the various stages of the building process. They could as well be replaced by curves not only continuous, but also differentiable (see also Ref. 11 and hereafter).

The main relations between structure and fractal dimensions may be easily found back on this particular example. The length of the stage F_n of the fractal is

$$L_n = L_0 \ (p/q)^n \tag{3.1.6}$$

with $p/q = 4/3$ for the von Koch curve. F_n may be identified with the curve obtained when smoothing out the fractal with a resolution $\Delta x = q^{-n}$.

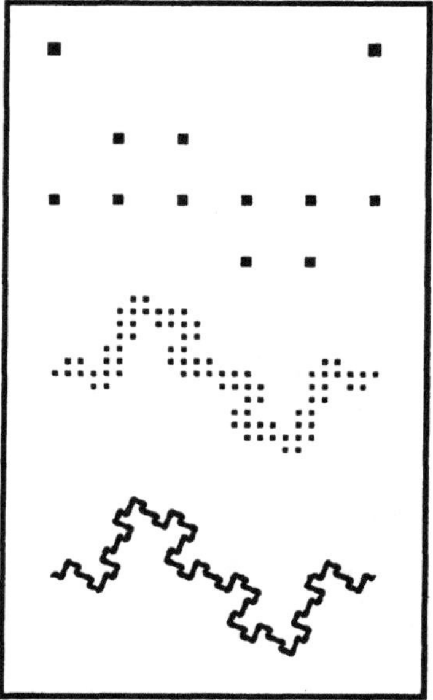

Figure 3.2. Point-by-point construction of a fractal curve ($D=ln9/ln5$). The segments which ensure continuity of the various approximations are omitted.

By writing $p = q^{lnp/lnq}$, equation (3.1.6) may then be written as $L_n = L_o\,(q^{-n})^{1-lnp/lnq}$, and setting $D = \ln p/\ln q$ yields the fundamental relation (3.1.4) since here $D_T = 1$. Moreover, the fact that the fractal construction is carried on to infinity allows the von Koch curve to be completely identical to one-fourth of its magnified version by a factor of 3 (see Fig. 3.1). So its similarity dimension and Hausdorff dimension are indeed identical in this case.

Before further developing the fractal mathematical tool, let us make some additional comments on the origin of fractal phenomena in nature.

3.2. Where Do Fractals Come From?

The extraordinarily huge number of occurrence of fractals in natural (physical and biological) systems is now an unavoidable observational fact. This was established by Mandelbrot[1-3] (fractal behaviour has been suggested for coast lines, the distribution of galaxies, turbulence, the structure of the lungs) and in subsequent studies (asteroids, moon craters, sun spectrum, the brain, the blood and digestive systems, hadron jets, dielectric pulling, growth phenomena).[8,12,13] The list is now so large that it becomes very difficult to be exhaustive.

The now definitive confirmation of the universality of fractals requires, more urgently than in the past, an explanation of their physical *origin.*[14] From which underlying laws of Nature do they arise ? We shall in this section consider some possible roads towards an answer to this question. But we shall see that the final answer may well be that fractal geometry is simply *more general* than the Euclidean and Riemannian geometries. Fractals would then be the structural manifestation of the fundamental non-differentiability of Nature. The recognition that fractal phenomena finally dominate non-fractal ones may well lead to the recognition that, at the end, the laws of Nature are non-differentiable.

Multiple (contradictory) optimization.

One of the possible ways to understand fractals would be to look at the fractal behaviour as the result of an *optimisation process*. However, at this level of the discussion, one must distinguish between two of the usual characterizations of fractals, which do not need to always coexist: namely self-similarity and divergence in terms of resolution. Indeed, while most fractal models or sets which have been studied (or found) up to now are both self-similar and of constant fractal dimension, we shall see that self-similarity is the simplest (lowest order) case of a more general behaviour: to be more specific, self-similarity is a scale-invariant property, which can be generalized into the concept of *scale covariance* (see Sec. 3.8).

Consider first fractal dimensions. Most fractals are characterized by the fact that their fractal dimension is larger than their topological dimension D_T, so that their D_T-measure (length of a line, area of a surface...) is divergent, while at the same time they remain bounded. Such a combination of divergence and boundedness may come from a process of *optimization under constraint*, or more generally of optimization of several quantities, sometimes apparently contradictory. Assume for example that some process leads to maximizing surface while minimizing volume, then a solution which optimizes both constraints is a fractal of dimension larger than 2 but smaller than 3 (infinite surface and infinitesimal volume).

What may now be the origin of self-similarity ? Self-similarity means that the object or phenomenon under consideration is found to remain (locally) identical to itself after application of a dilatation or contraction. While this is true of any dilatation or contraction for a standard object as a straight line or a plane, for most current fractals one finds *exactly* the initial configuration only for some given generic dilatation and its powers. In other words, when looked at in terms of the resolution variable (in logarithm form), a self-similar fractal is a *periodic* system in the "zoom" dimension. If indeed some fractalization process comes from an optimization law, the interpretation of such a periodic self-similarity is that, after the first generic dilatation, *the system is taken back to the previous state of the problem*, up to the scale factor of the generator.

Let us illustrate these two aspects of fractals, namely multiple optimization and coming back to previous state, by a simple model example. Assume that a system needs to increase its energy input from the environment, which is achieved through a membrane. The energy input is proportional to the surface of the membrane; so the systems needs a maximization of a D=2 dimensional quantity. The simplest solution would be to exchange energy through the outer limiting surface of the body; then the optimization leads to an increase of its size. However the balance of such a process is negative: the increase of the size by ρ increases the surface by ρ^2, but also the volume by ρ^3, so that the energy per unit volume *decreases* as $1/\rho$. Moreover the thermal *losses* of energy also occur through the outer surface, so that it is not efficient to increase it too much (this, with the gravity constraint, leads to a well-known limit on the size of living systems on earth).

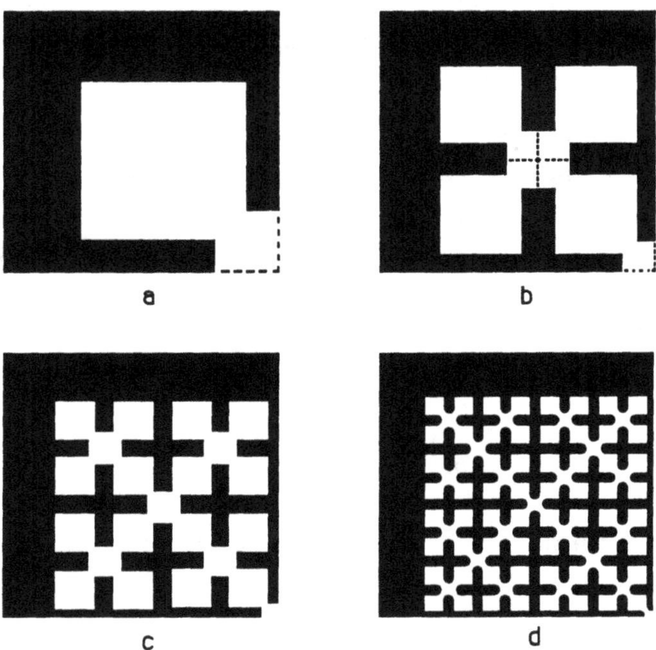

Figure 3.3. Model of self-similar increase of a surface of exchange (see text).

So the problem is now set in a different way: is it possible to increase the surface of exchange without increasing the volume (equivalently the outer surface) ? The only solution is clearly to increase the surface *inside* the body by "invagination" (Fig. 3.3a).

However this increase will soon also find a limit, when the internal increase is limited by the outer surface as may be seen in Fig. 3.3a. The solution is the same: new invaginations, but in the inverse direction (Fig. 3.3b). The comparison of Figs. 3.3a and 3.3b shows that we have now reached the basic condition for self-similarity: the system is now brought back in its preceding state, up to a scale factor. Self-similarity is now ensured by the principle of causality. The same causes leading to the same effects, fractalization is expected to go on along self-similar structures. The process stops when other constraints apply to the system (in our example, thickness of the membrane and diameter of the various openings). The lung, with its ≈20 levels of fractalization (in base 2) from the windpipe to the cells (see also Mandelbrot[3]) is an example of a natural system that such a model could describe (very roughly).

Renormalization group.

Another line of attack of fractal properties relies on the renormalization group formalism. Wilson's many scale of length approach was also designed to deal with explicit scale dependence.[15] The similarity between fractals and the renormalization group has already been pointed out by Callan (see Mandelbrot[16]). Consider indeed a relevant field; its scale dependence is described by a renormalization group equation

$$\frac{d\varphi}{d\,ln\frac{\lambda}{r}} = \beta(\varphi) = \delta\,\varphi \quad ,$$

where δ is its *anomalous dimension*. The (lowest order) solution to this equation is a power law

$$\varphi = \varphi_0\,(\lambda/r)^{\delta} \quad ,$$

which has exactly the form of the divergence of the D_T-measure of a fractal of dimension D:

$$\mathcal{L} = \mathcal{L}_0\,(\lambda/r)^{D-D_T} \quad .$$

This allows us to identify the renormalization group anomalous dimension with the difference $D-D_T$:

$$\delta = D-D_T \quad .$$

Conversely this means that the definition of an infinitesimal generator for fractals and of subsequent differential equations of which they would be solutions may be obtained in a renormalization group-like approach.

However one must point out an important difference between fractals and the renormalization group (we shall make this point again in following sections). The renormalization group approach consists in describing a system at *small* scale (e.g., an ensemble of elementary quantum spins), then dilating the system by some scale factor, and looking at the way the various quantities (fields, couplings) have changed (have been renormalized). Up to the renormalization group transformation, one is brought back to the previous problem and may iterate the process. One may then reach scales very large with respect to the initial ones (e.g. from $\approx 10^{20}$ elementary magnets to the global macroscopic magnetization of a body) in a logarithmic ($\approx$20 steps) rather than a linear way (10^{20}). But when going from one scale to the next, one usually replaces the information on the system by an average, so that it is impossible to come back to the smaller scale. In other words, there is no inverse transformation in the renormalization group, i.e., it is only a semi-group.[15] On the other hand fractals are usually built from the large scale to the smaller. One defines a generator, then builds smaller and smaller structures by successive application of the generator after rescaling. So fractals may actually come as an *inverse transformation* for the renormalization group. We shall come back later on these remarks, and use them explicitely in our first development of the special (linear) theory of scale relativity (Chapter 6).

The renormalization group is also related to self-similarity. In a recent paper by Yukalov[17], it was demonstrated that a self-similar β function results from an optimization condition. In other words, the fastest approach to an attractor corresponds to the motion along the self-similar trajectory. This may help solve one of the most important questions of present theories of critical phenomena and high energy particle physics:

namely, the determination of the β functions which are most of the time known only as expansions having a limited domain of validity.

Dynamical chaos and fractals.

One of the most lively domains in which fractals play a leading role is that of dynamical chaos.[18-21] Chaos is defined as a high sensibility to initial conditions and is often described by an exponential divergence of initially close trajectories:

$$x = x_0 + \delta x_0 \; e^{t/\tau}$$

where τ^{-1} is the so-called Lyapunov exponent. Note the explicit intervention of resolution in this relation, $x=x(t,\delta x)$. Trajectories may tend in phase space towards strange attractors which are often characterized by fractal properties. But fractals may intervene directly in space or space-time and actually characterize, e.g., the distribution of chaotic trajectories. Celestial mechanics offers several examples of this connection of chaos and fractals: fractals are not explained by chaos, nor chaos by fractals, but a same underlying phenomenon (namely resonances between orbital periods) gives rise to both the fractal structure and the chaotic trajectories, which finally leads to unpredictability. Let us give two examples of celectial mechanical chaos.

Consider first a three-body (1+2) problem, i.e., the relative evolution of two interacting small bodies in the potential of a large one. It has been shown by Henon[22,23] in numerical simulations that the distribution of the relative impact parameters which separate the various kind of relative motion of one body in the other's frame (in particular escape towards one direction or its reverse) was a fractal dust (of the Cantor set and "devil staircase" type).

Chaos may also arise in the 3 (2+1) body problem, i.e., two large bodies and a test particle. This corresponds for example to the combination of the Sun, Jupiter and a smaller planet or asteroid. This is an interesting physical problem, since one may explicitly and analytically describe the source of chaos and its fractal properties.[24] Chaos arises from resonances between orbits, which occur when the periods of Jupiter, T, and of the test object, τ, are fractions composed of small integers.[25] Assume both orbits to be circular and consider a body of period double that of Jupiter (2:1 resonance). Each two revolutions, the body passes at a distance $(2^{2/3}-1)R$ from Jupiter, where R is the Jupiter-Sun distance (following Kepler's third law $t^2/r^3=cst$) and is submitted to a strong "kick" which perturbs its trajectory. It may in the end even be ejected from its trajectory and cross the orbits of other planets: this explains some of the so-called Kirkwood gaps in the asteroid belt which lies between Mars and Jupiter.[25]

Consider now all possible circular orbits of radius between 0 and 1 in units of R. Dynamical chaos is related to the *distribution of rational fractions* $\tau/T=(n/d)$ among reals in the segment [0,1], i.e., in terms of radius, of $(n/d)^{3/2}$. An analytical expression has been obtained for the *fluctuation* of the *average* force exerted by Jupiter (circular orbits approximation):[24]

$$F = \frac{1.025}{d}\ \left[1-\left(1-\frac{k}{d}\right)^{2/3}\right]^{-2}\ exp\left(\frac{k}{5}+\frac{k-2}{d}\right)\ cos\,(2\pi kt/T_0)$$

where $\tau<T$, $k=d-n$ and $T_0=n\,T=d\,\tau$.

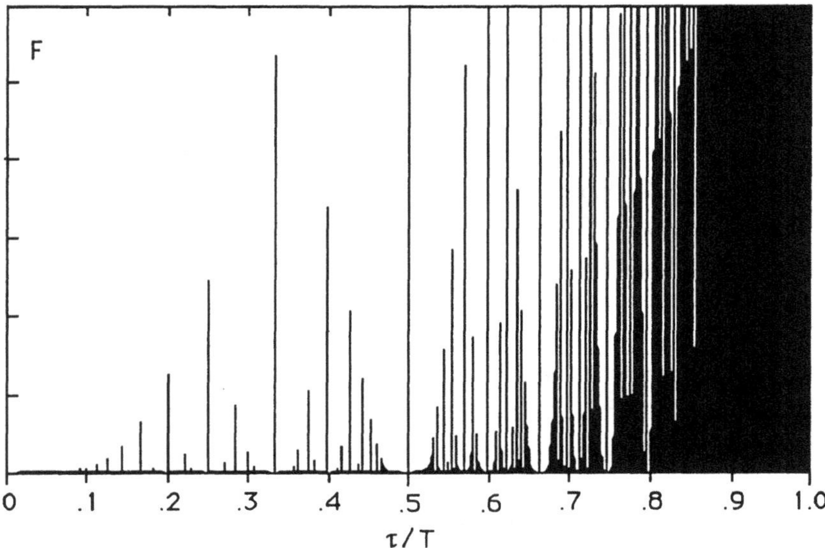

Figure 3.4. Relative amplitude of the fluctuation of the average force exerted on a test body by Jupiter (distance R), in terms of $\tau/T=(r/R)^{3/2}$ (circular approximation).

The amplitude of this force is plotted in Fig. 3.4. This is a fractal function which explicitely depends on the resolution at which it is plotted. This opens the possibility to write a new equation of dynamics, $F(r;\varepsilon)=m\gamma$, with F now being a fractal rather than a standard function. Such an equation becomes highly nonlinear, since its definitive expression depends on its solutions: the value of ε to be injected in F must be in accordance with the temporal and spatial characteristic scales of evolution of the trajectory.

Such an equation is explicitly fractal, one of its members being a fractal function. However fractals may also occur as (still unknown) solutions of completely classical equations. This is still a mostly unexplored domain, which may however yield interesting results. We shall see for example in Sec. 3.8 that a derivative of fractal functions may be introduced which , once renormalized, may in some case be quasi-identical to the initial fractal. Does the equation $dy/dx=y$, which was known to yield the exponential function as solution, also have fractal solutions ?

We shall come back in Sec. 7.2 on the question of chaos, in particular in the Solar System: new methods will be suggested to deal with unpredictability at large time scales.

Fractals and non-differentiable geometry.

We now come to the leading point of view of the present book. When applied to the structure of space-time, the concept of fractal is justified mainly because it may be defined as a non-differentiable generalization of Riemannian geometry. Let us specify the meaning of this proposal.

The idea of a fractal space (and space-time) stands as a basic cornerstone of the present approach. Up to now (and to our knowledge), even if the words "fractal geometry" have been extensively used, the mathematical domain of fractals has not yet reached a status comparable to Euclidean geometry or Riemannian geometry. Fractals in the present literature are either a characterization of some properties of some phenomena in nature, or objects or sets most of the time seen as embedded in an underlying Euclidean space.

Our approach here is different. The discovery by Gauss of a curved geometry which definitely proved Euclide's fifth axiom to be independent of his other axioms, was made effective when he became able to characterize a surface with totally *intrinsic* methods. Thus not only already known surfaces like spheres could be defined and seen from the inside, but this moreover allowed him to prove the logical existence of surfaces which cannot be embedded in Euclidean space, like the hyperbolic one, and which deserved the name of a new geometry. Once such a "Pandora box" was opened, it was once again the merit of Gauss to partially close it by restricting himself to spaces which remain *locally Euclidean*, an hypothesis which was kept by Riemann in his generalization to any topological dimension. One knows the fundamental part played by Gauss's hypothesis in the use of Riemannian geometry as a mathematical tool for the achievement of the general theory of relativity. Gauss's hypothesis is the basis for the *mathematical transcription of the equivalence principle.* Indeed assuming local flatness of space-time means that it is locally Minkowskian, while the laws of physics in a Minkowskian space are known to be the special relativistic laws of inertia.

Now the Pandora box is once again opened with fractals. Making the fractal hypothesis may be seen as giving up Gauss's hypothesis: instead of getting flatness and rectifiability back at a small scale, on the contrary the never ending occurrence of new details as the scale gets smaller and smaller implies that the curvature of a fractal space increases up to infinity as resolution Δx approaches zero: this will be detailed in subsequent sections (3.6 and 3.10). Fractal surfaces may be built which are flat at the larger scale (once the fractal / non-fractal transition is crossed), and whose curvature at the infinitesimal level is everywhere infinite, but with a fractal distribution of negative and positive signs.

So, as already stated in Chapter 2, two different and complementary roads lead us to fractals. The extension of the principle of motion relativity to the quantum motion requires a non-differentiable description (see Chapter 4). The principle of scale relativity requires an explicitly scale-dependent geometrical tool. Fractals satisfy both constraints. More precisely, what we shall call a *fractal space-time* throughout the present book is, by definition, a geometry which is *both non-differentiable and explicitly scale-dependent*. We shall demonstate in Sec. 3.10 that *non-differentiability everywhere actually implies scale divergence.*

In such a framework the question: "why is space-time fractal?" gets a status similar to the question which may be asked in general relativity: "why is space-time curved?". In this last case one may take two possible positions.

One possibility is to place oneself in a "causal" approach to general relativity, in which its structure is stated as: "matter and energy curve space-time (following Einstein's equations); then in the curved space-time, free particles follow the geodesics". In this approach, the particular property which is owned by matter and energy (i.e., by the inertial content of objects itself) and which manifests itself at large scale as curvature of space-time is not described by the theory. This is the problem of the sources of gravitation, and, as recognized by Einstein himself, the theory is incomplete in this respect (another domain of incompleteness of general relativity is the global topology of the universe: this comes from the local character of its main tool, the metric invariant). A very interesting approach to this problem has been Sakharov's, who interpreted gravitation as elasticity of space that arises from particle physics.[26]

The fractal approach suggests another route for tackling such questions, as will be seen in Secs. 5.3 and 5.11. Mass, energy and momentum may be identified as fractal structures themselves, while smoothing out to large scales of fractalization leads to curvature. In this framework fractals may explain curvature, but we have only moved the problem back: the question "why fractals ?" should find its solution in a still larger theory (the only generalization beyond non-differentiability would be *discontinuity*, while thoughout this book, a space-time *continuum* remains assumed).

Another view of general relativity is to remark, with Einstein, that Einstein's equations are both the simplest and the most general which are *invariant under general continuous and differentiable coordinate transformations*, i.e. which satisfy general covariance. Then general covariance alone (actually with the principle of equivalence) introduces universal structures in physical equations: these structures may be shown to be nothing but curvature of space-time, and finally what we call gravitation is the various manifestations of this curvature.[26]

In the theory of general relativity, curvature is a consequence of the principle of relativity and the principle of equivalence. In the fractal approach, we are very far from such an achievement. We may however define it as the main goal of this theoretical attempt: find a system of equations which would be *invariant under general continuous (not necessarily differentiable) coordinate transformations*, in motion and in scale. We conjecture that such a programme of implementation of the principles of *extended motion covariance* and *scale covariance* will yield fractal structures of space-time as an unavoidable consequence, being aware that such general fractal structures imposed by physics at the fundamental level are expected to be, perhaps, very different from the usual, currently known fractals (see Chapter 6 in this respect). The quantum behaviour of microphysics would hopefully be a manifestation of this fractal structure. But before launching such a huge programme, one must ensure that it can be justified by currently known experimental data and by the present structure of the theories which have proved to be efficient; this is one of the goals of the present book.

3.3. Fractal Curves in a Plane.

Let us now come to our first attempts to define fractals in an intrinsic way and to deal with infinities and with their non-differentiability. We first consider the case of fractal curves drawn in a plane. The von Koch construction may be generalized in the complex plane by first giving ourselves a base (or "generator"[21]) F_1 made of p segments of length $1/q$. The coordinates of the p points P_j of F_1 are given, either in Cartesian or in polar coordinates (see Figs. 3.5 and 3.6):

$$Z_j = x_j + i\, y_j = q^{-1} . \rho_j . e^{i\,\theta_j} \quad , \quad j = 1 \text{ to } p.$$

Let us number the segments from 0 to p-1. Then another equivalent representation would be to give ourselves either the polar angle of the segment j, say ω_j, or the relative angle between segments j–1 and j, say α_j. We further simplify the model by choosing a coordinate system such that F_o is identified with the segment [0,1]. The length of the individual segments is now $1/q$, and the fractal dimension will be given by

$$D = \ln p / \ln q .$$

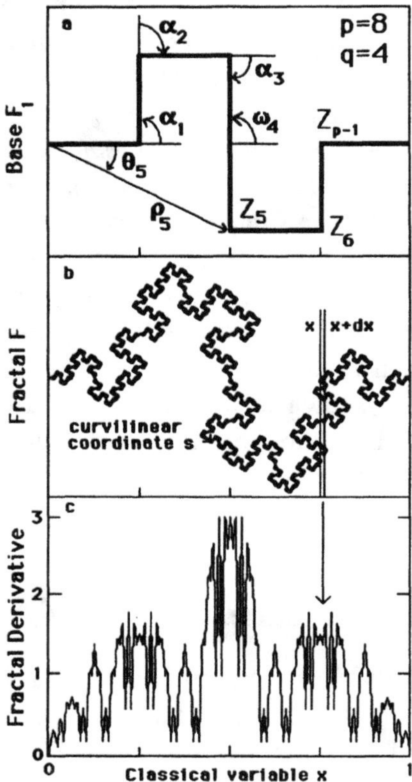

Figure 3.5. Construction of a fractal curve from its generator (or base) F_1. Figure **a** defines the structural constants used in the text. A curvilinear coordinate s is defined on the fractal curve. In (**c**) its fractal derivative is plotted at approximation ξ_4 (see Sec. 3.7).

The following additional relations hold between our "structural constants":

$$\alpha_j = \omega_j - \omega_{j-1} \quad ; \quad \sum_{j=0}^{p-1} \alpha_j = 0$$

$$\omega_j = \omega_o + \sum_{k=0}^{j} \alpha_k \tag{3.3.1}$$

$$q = \sum_{j=0}^{p-1} e^{i\omega_j} \quad ; \quad Z_{j+1} - Z_j = q^{-1} \; e^{i\omega_j}$$

These relations include the case when $\omega_o \neq 0$ (for j=0, the first equation writes $\alpha_o = \omega_o - \omega_{p-1}$). The ω_j 's and α_j 's are complete and independent sets of parameters, so that ω_o in the second relation must be expressable in terms of the α_j 's. This is indeed achieved by solving the equation $\Sigma sin\,\omega_j = 0$.

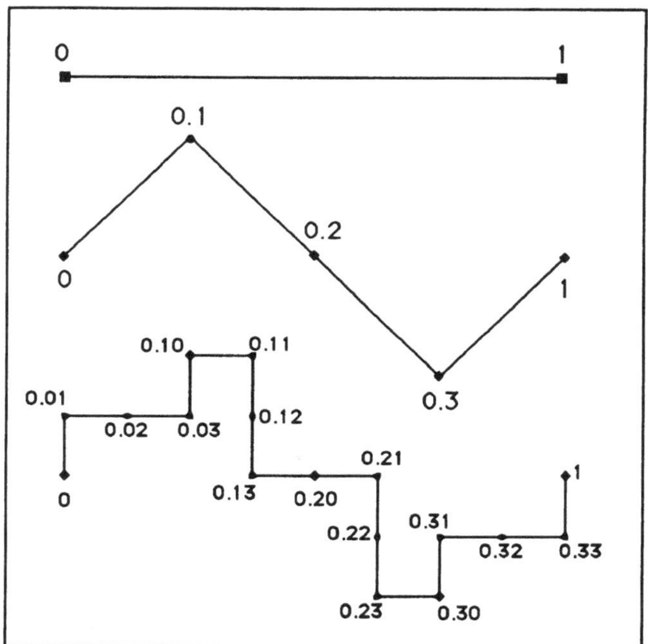

Figure 3.6. Parametrization of a fractal curve in the counting base p (in the case shown here, p=4, $q=2\sqrt{2}$, so that D=4/3; the generator of this fractal has a nonzero slope at origin, so that the slope on the fractal is never defined in this case : see also Fig. 3.9).

A parameter s may now be defined on the fractal and written in terms of its expansion in the counting base p (see Fig. 3.6):

$$s = 0.s_1s_2...s_k... = \Sigma_k s_k p^{-k} \quad , \tag{3.3.2}$$

with each s_k taking integer values from 0 to $p-1$. This parameter is a normalized curvilinear coordinate on the fractal curve. The hierarchy of its figures reproduces the hierarchical structure of the fractal. This allows us to write the fractal equation in the form[11]

$$Z(s) = Z_{S_1} + q^{-1}\ e^{i\omega_{s_1}} \Big[\, Z_{S_2} + q^{-1}\ e^{i\omega_{s_2}} \left[\, Z_{S_3} + \ldots \right]\Big] \ .$$

We now set

$$\varphi_{S_k} = \omega_{S_1} + \omega_{S_2} + \ldots + \omega_{S_{k-1}} + \theta_{S_k} \quad ,$$

and the parametric equation of the fractal becomes

$$Z(s) = \sum_{1}^{\infty} \rho_{S_k}\ e^{i\varphi_{s_k}}\ q^{-k} \quad . \tag{3.3.3}$$

This equation may still be generalized to the case where some additional transformation is applied to the generator (for example some fractals are constructed by alterning the orientation of the generator). This may be described by an operator $S_j = e^{\sigma_j}$, so that one obtains a generalized equation:

$$Z(s) = \sum_{1}^{\infty} \rho_{S_k}\ e^{\Sigma_j \sigma_{s_j} + i\varphi_{s_k}}\ q^{-k} \quad .$$

The above equations yield an "external" description of the fractal curve, in which, for each value of the curvilinear coordinate s, the two coordinates $x(s)$ and $y(s)$ in the plane may be calculated ($Z(s) = x(s) + i\, y(s)$). In terms of s, $x(s)$ and $y(s)$ are *fractal functions,* for which successive approximations $x_n(s)$ and $y_n(s)$ may be built. Though only one value of x and y corresponds to each value of s (while the reverse is false), their fractal character is revealed by the divergence of their slope when $n \to \infty$. Their fractal dimension is the same as that of the original fractal curve. This is illustrated in Fig. 3.7 for the fractal curve of Fig. 3.5.

The structure of Eq. (3.3.3) is remarkable, since it evidences the part played by p on the fractal and q in the plane: ($s = \sum s_k p^{-k}$) $\leftrightarrow$ [$Z(s) = \sum C_k(s) q^{-k}$]. An "intrinsic" construction of the fractal curve may also be made.[11] Placing ourselves on F_n, we only need to know the change of direction from each elementary segment of length q^{-n} to the following one. On the fractal generator F_1, these angles have been named α_i. The problem is now to find $\alpha(s)$.

The points of F_n which are common with F (those relating the segments) are characterized by rational parameters s written with n figures in the counting base p, $s=0.s_1 s_2...s_n$. Let us denote by s_h the last non-null digit of s, i.e.

$$s = s_1/p + s_2/p^2 + ... + s_h/p^h \ .$$

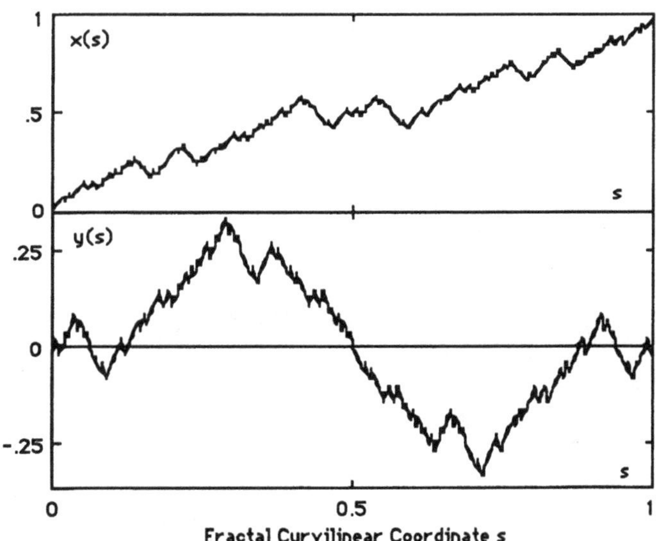

Figure 3.7. The coordinates x and y of the fractal curve of Figs. 3.5 and C1 in terms of the normalized curvilinear coordinate intrinsic to the fractal, s. The function $x(s)$ and $y(s)$ are fractal functions which themselves vary with scale.

It is easy to verify, provided that $\alpha_0 = \omega_0 - \omega_{p-1} = 0$ (which is a necessary condition for self-avoidance), that the relative angle between segment number $(s.p^n - 1)$ and segment number $(s.p^n)$ on F_n is given by[11]

$$\boxed{\alpha(s) = \alpha_{s_h}} \ . \qquad (3.3.4)$$

This formula completely defines the fractal in a very simple way, uniquely from the (p-1) structural angles α_i, and independently of any particular coordinate system in the plane (x,y). Drawing of fractal curves based on (3.3.4) is at least 5 times faster than from (3.3.3).

Let us end this section by considering another generalization of the fractal construction. The von Koch-like fractals are discontinuous in scale, because of the discreteness of the method that consists in applying generators. Continuity of the construction may be recovered by introducing intermediate steps between F_n and F_{n+1}. We define a sublevel of fractalization, k, such that $0 \leq k < 1$, and we generalize (3.3.4) in the following way: placing ourselves on F_n,

$$\text{if } s_n = 0 \quad \text{then} \quad \alpha(s) = \alpha_{s_h} \quad ,$$
$$\text{if } s_n \neq 0 \quad \text{then} \quad \alpha(s) = k\, \alpha_{s_n} \quad .$$

The result is illustrated in Figs. 3.8 and C2, in which a fractal curve is plotted in terms of the space variable x and of the scale variable $ln\delta x$.

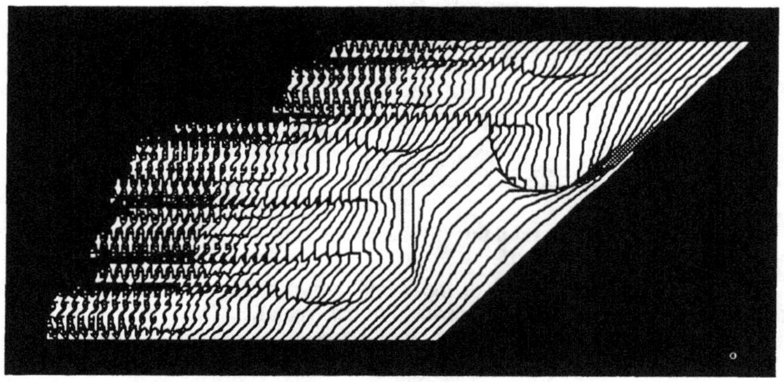

Figure 3.8. The fractal curve of Fig.3.5 in terms of resolution (see also Fig. C2).

We may now use this method to illustrate the periodic self-similarity of fractals: we show in Fig. 3.9 the result of a zoom on two different fractals, one with zero slope at the origin of its generator, the other with a nonzero slope. In the first case, zooming amounts to a translation, while in the second case there is a never-ending *local* rotation at the origin, while the *global* shape of the fractal is conserved.

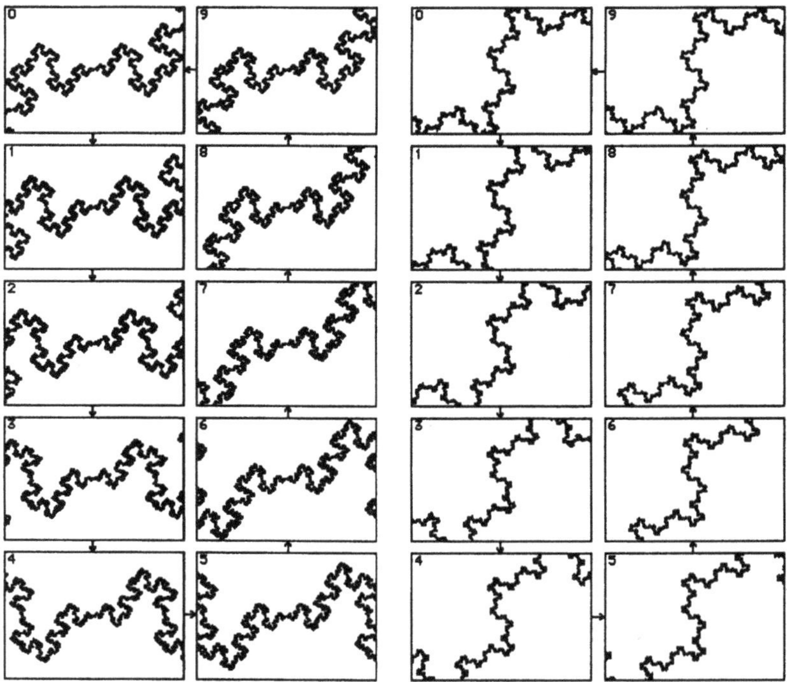

Figure 3.9. Zoom on two fractal curves. Figures 0 to 9 constitute a movie; the subsequent Fig. 10 is exactly identical to Fig. 0.

3.4. Non-Standard Analysis and Fractals.

We have proposed[11] to deal with the infinities appearing on fractals, and then to work effectively on the actual fractal F instead as on its approximations F_n by using Non-Standard Analysis (NSA).

It has been shown by Robinson[27] that proper extensions $^*\mathbb{R}$ of the field of real numbers $\mathbb{R}$ could be built, which contain infinitely small and infinitely large numbers. The theory, first evolved by using free ultrafilters and equivalence classes of sequences of reals[28], was later formalized by Nelson[29] as an axiomatic extension of the Zermelo set theory. We do not intend to give here a detailed account of this field which is now developed as a genuine new branch of mathematics; we shall just recall the results which we think to be most relevant for application to fractals.

Let us briefly recall the ultrapower construction of Robinson. Though less direct than the axiomatic approach (which actually has compacted into additional axioms all the essential new properties of Robinson's construction), it allows one to get a more intuitive contact with the origin of the new structure. Indeed the new infinite and infinitesimal numbers are built as equivalence classes of sequences of real numbers, in a way quite similar to the construction of $\mathbb{R}$ from rationals. So, in the end, some of the ideal character of the new numbers is found to be already present in real numbers.

Let $\mathbb{N}$ be the set of natural numbers. A free ultrafilter $\mathcal{U}$ on $\mathbb{N}$ is defined as follows.
$\mathcal{U}$ is a non empty set of subsets of $\mathbb{N}$ $[\mathcal{P}(\mathbb{N}) \supset \mathcal{U} \supset \varnothing]$, such that:

(1) $\varnothing \notin \mathcal{U}$
(2) $A \in \mathcal{U}$ and $B \in \mathcal{U} \Rightarrow A \cap B \in \mathcal{U}$.
(3) $A \in \mathcal{U}$ and $B \in \mathcal{P}(\mathbb{N})$ and $B \supset A \Rightarrow B \in \mathcal{U}$.
(4) $B \in \mathcal{P}(\mathbb{N}) \Rightarrow$ either $B \in \mathcal{U}$ or $\{j \in \mathbb{N}: j \notin B\} \in \mathcal{U}$, but not both.
(5) $B \in \mathcal{P}(\mathbb{N})$ and B finite $\Rightarrow B \notin \mathcal{U}$.

Then the set $^*\mathbb{R}$ is defined as the set of the equivalence classes of all sequences of real numbers modulo the equivalence relation:

$$a \equiv b \ , \ \text{provided} \ \{j : a_j = b_j\} \in \mathcal{U} \ ,$$

a and b being the two sequences $\{a_j\}$ and $\{b_j\}$.

Similarly, a given relation is said to hold between elements of $^*\mathbb{R}$ if it holds termwise for a set of indices which belongs to the ultrafilter. For example:

$$a < b \Leftrightarrow \{j : a_j < b_j\} \in \mathcal{U} \ .$$

$\mathbb{R}$ is isomorphic to a subset of $^*\mathbb{R}$, since one can identify any real $r \in \mathbb{R}$ with the class of sequences $Cl(r,r,...)$. It is the axiom of *maximality* (4) which ensures $^*\mathbb{R}$ to be an ordered field. In particular, thanks to this axiom, a sequence which takes its values in a finite set of numbers is equivalent to one of these numbers, depending on the particular ultrafilter $\mathcal{U}$. This allows one to solve the problem of zero divisors: indeed the fact that $(0,1,0,1,...).(1,0,1,0,...) = (0,0,0,...)$ does not imply that there are zero divisors, since axiom (4) ensures that one of the sequences is equal to 0 and the other to 1.

That $^*\mathbb{R}$ contains new elements with respect to $\mathbb{R}$ becomes evident when one considers the sequence $\{\omega_j = j\} = \{1,2,3,.. \ n, \ ...\}$. The equivalence class of this sequence, ω, is larger than any real. Indeed, for any $r \in \mathbb{R}$, $\{j : \omega_j > r\} \in \mathcal{U}$, so that whatever $r \in \mathbb{R}$, $\omega > r$. It is straightforward that the inverse of ω is infinitesimal.

Hence the set $^*\mathbb{R}$ of hyper-real numbers is a totally ordered and non-Archimedean field, of which the set $\mathbb{R}$ of standard numbers is a subset. $^*\mathbb{R}$ contains infinite elements, i.e. numbers A such that $\forall n \in \mathbb{N}$, $|A| > n$ (where $\mathbb{N}$ refers to the set of integers). It also contains infinitesimal elements, i.e. numbers ε such that $\forall n \neq 0 \in \mathbb{N}$, $|\varepsilon| < 1/n$. A finite element C is defined: $\exists n \in \mathbb{N}$, $|C| < n$.. Now all hyper-real numbers may be added, substracted, multiplied, divided; subsets like hyper-integers $^*\mathbb{N}$ (of which $\mathbb{N}$ and the set of infinite hyper-integers $^*\mathbb{N}_\infty$ are subsets), hyper-rationals $^*\mathbb{Q}$, positive or negative numbers, odd or even hyper-integers, etc...may be defined, and more generally most standard methods and definitions may be applied in the same way as for the standard set $\mathbb{R}$. But the different sets or properties are classified as being either internal or external.[28,29]

An important result is that any finite number a can be split up in a single way as the sum of a standard real number $r \in \mathbb{R}$ and an infinitesimal number $\varepsilon \in \mathfrak{I}$: $a = r + \varepsilon$. In other words the set of finite hyper-reals contains the ordinary reals plus new numbers (a) clustered infinitesimally closely around each ordinary real r. The set of these additional numbers $\{a\}$ is called the monad of r. More generally, one may demonstrate that any hyper-real number A may be decomposed in a single way as $A = N + r + \varepsilon$, where $N \in {}^*\mathbb{N}$, $r \in \mathbb{R} \cap [0,1[$ and $\varepsilon \in \mathfrak{I}$.

The real r is said to be the "standard part" of the finite hyper-real a, this function being denoted by $r=st(a)$. This new operation, "*take the standard part of*", plays a crucial role in the theory, since it allows one to solve the contradictions which prevented previous attempts, such as Leibniz's, to be developed. Indeed, apart from the usual strict equality "=", one introduces an equivalence relation, "≈", meaning "infinitely close to", defined by $a \approx b \Leftrightarrow st(a-b)=0$. Hence two numbers of the same monad are infinitely close to one another, but not strictly equal.

A practical consequence is that a very large domain of mathematics may be reformulated in terms of NSA, in particular, concerning physics, the integro-differential calculus.[28] The method consists in replacing the Cauchy-Weierstrass limit formulation by effective sums, products and ratios involving infinitesimals and infinite numbers, and then taking the standard part of the result. Hence the derivative of a function will be defined as the ratio

$$df/dx = st\left\{ [f(x+\varepsilon) - f(x)] \,/\, \varepsilon \right\} \;,$$

with $\varepsilon \in \mathfrak{I}$, provided this expression is finite and independent of ε. The integral of a function is defined from an infinitesimal partition of the interval $[a,b]$ in an infinite number ω of bins, as a sum

$$\int_a^b f(x)\, dx = st\left(\sum_{i=1}^{\omega} f(x_i)\, \delta x_i \right) \;,$$

provided it is finite and independent of the partition. The number of bins, infinite from the standard point of view, is assumed to be a given integer belonging to $^*\mathbb{N}_\infty$. It is said to be *-finite. Such a method allows, more generally, a new treatment of the problem of infinite sums. A summation from 0 to ∞ may be replaced by a summation over an *-finite number of terms ranging from 0 to $\omega \in {}^*\mathbb{N}_\infty$. The sum will be said to converge if for different ω's, its standard part remains equal to the same finite number.

Thanks to its ability to deal properly with infinite and infinitesimals, Non-Standard Analysis is particularly well adapted to the description of fractals. To this purpose we have proposed[11] to continue the fractalization process (F_0, F_1, ..., F_n, ...F) up to an *-finite number of stages ω. This yields a curve F_ω, from which the fractal F is now defined as

$$F = st\,(F_\omega)\,\cdot\,. \tag{3.4.1}$$

This means that we define an *-curvilinear coordinate from its expansion in the base p:

$$^*s = 0.s_1 ... s_k ... s_\omega = \sum_{k=1}^{\omega} s_k\, p^{-k}$$

and that the equation of F_ω is given by Eq. (3.3.3) now summed from 1 to ω, $Z(s)$ being its standard part. One of the main interests of the introduction of the curve F_ω is that it contains and sums up all the properties of each of the approximations F_n, and also of their limit F. Moreover a meaning may now be given to the length of the fractal curve. The length of F_ω is a number of $^*\mathbb{R}_\infty$:

$$L_\omega = L_o\,(p/q)^\omega = L_o\; q^{\omega(D-1)},$$

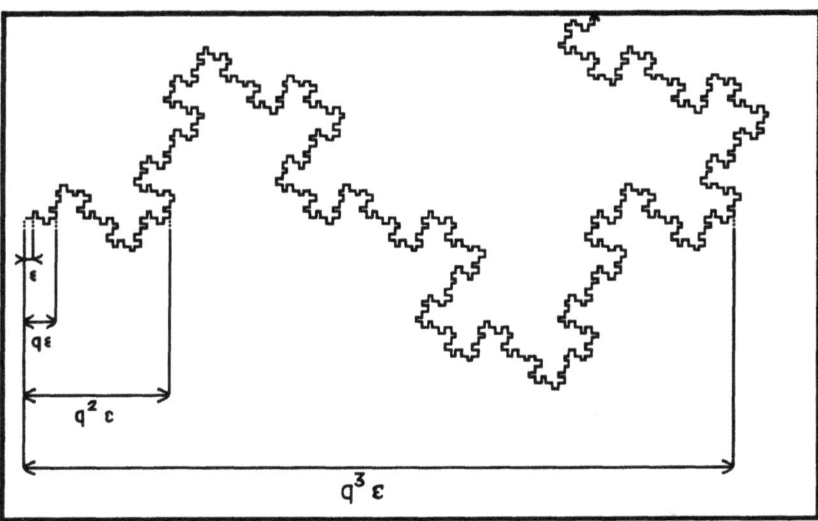

Figure 3.10. Infinite magnification of a non-standard fractal curve F_ω. The elementary segment has a length $\varepsilon = q^{-\omega}$. The whole figure is an internal structure of the zero point.

and the (non-renormalized) curvilinear coordinate on F_ω is $\xi = {}^*s.L_\omega$. Between 0 and s, it is made of $s.p^\omega$ segments of length $q^{-\omega}$. The "surface" of F_ω may now also be defined as

$$S_\omega = L_0 \, (p/q^2)^\omega = q^{\omega(D-2)} \quad ,$$

which is an infinitesimal number when $D<2$. When taking the standard part of all these quantities, one finds again that the length is undefined (infinite) and the surface is null for $D<2$ and finite for $D=2$ (i.e. a curve in $\mathbb{R}^2$ becomes plane-filling).

The non-differentiability of the fractal is now visualizable in a new way, by the fact that any standard point of the fractal may be considered as structured: when viewed with an infinite magnifier, it is found to contain all the values of the slopes owned by the complete fractal[11] (see Fig. 3.10). However we have made the remark that a kind of differentiability can be defined for fractals, which we have called ε-differentiability.[11] It consists in imposing that any part of the fractal magnified by q^ω be differentiable: indeed there is an infinite number of curves F_ω, the standard part of which is the same fractal F; an ε-differentiable one may be *a priori* chosen. This is in fact equivalent, from a practical viewpoint, to imposing that each approximation F_n be differentiable, which is always a possible choice.

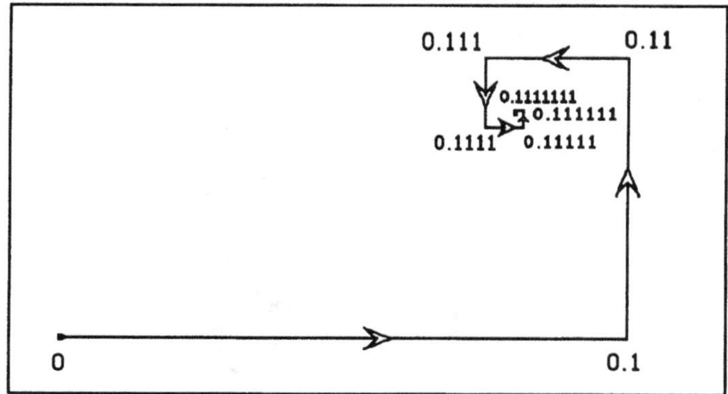

Figure 3.11. Schematic representation of the way to the point of intrinsic coordinate $0.11111... = 1/(p-1)$ on a fractal curve: it is reached by an infinite spiral, so that the slope of the fractal curve cannot be defined for this point.

This concept of ε-differentiability has been recently clarified and generalized by Herrmann.[30] The consequence for applications to physics is not negligible: this means that the non-differentiable fractal may be built as the limit of a family of differentiable curves, for which the usual integro-differential formalism may then be recovered (see Sec. 3.8).

Let us close this section by an additional illustration of the nature of the non-differentiability of fractals. Consider the point of curvilinear coordinate $s = 1/(p-1) = 0.1111..$ on a fractal curve such that $\omega_0=0$ and $\omega_1=\pi/2$. As shown in Fig. 3.11, this point is reached as the limit of an infinite spiral; this is also the case of most points of a fractal, with most of the time far more complicated patterns.

3.5. Fractal Curves in Space.

The above results concerning the equations of fractal curves in the plane are easily generalizable to curves drawn in higher dimensional space (see Figs. 3.12 and C3). In $\mathbb{R}^3$ the rotation complex operators $e^{i\omega_k}$ are simply replaced by 3-dimensional rotation matrices R_k. The generator F_o is defined by the coordinates of its p points, $U_k = (X_k, Y_k, Z_k)$ and the point of parameter s will be defined by the vector[11,31]

$$U(s) = \sum_k R_{s_k} R_{s_{k-1}} .... R_{s_1} U_{s_{k+1}} q^{-k} \quad . \tag{3.5.1}$$

In the same way, if one gives oneself the relative rotation matrices A_k on the generator, the intrinsic equation of the fractal is given by the last non-null figure (of rank h) of the s -expansion :

$$A(s) = A_{s_h} \quad . \tag{3.5.2}$$

A particularly interesting subclass of such fractal curves, concerning the physical aims of the present book, is the class of curves of fractal dimension 2, since, as will be seen in the following sections, this is the universal fractal dimension of particle paths in quantum mechanics. They are built from $p=q^2$ segments of length $1/q$ (in the case of perfect self-similarity). The case of orthogonal generators in $\mathbb{R}^3$ is particularly simple. For example, with $q=3$ (and thus $p=9$), one can construct, among others, the following generators:

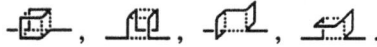

Thanks to the relation $2(3^2-1) = 4^2$, one can combine them to obtain a large class of ($q=4$, $p=16$) generators, for example:

and several other combinations. We give in Fig. 3.12 some examples of such fractal curves drawn to higher order approximations.

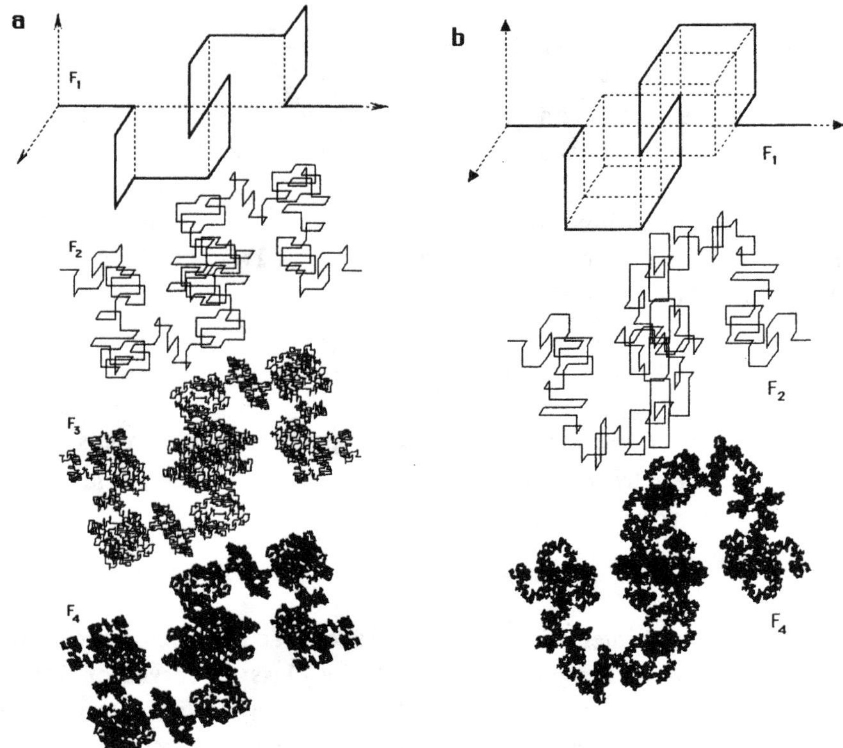

Figure 3.12 a and b

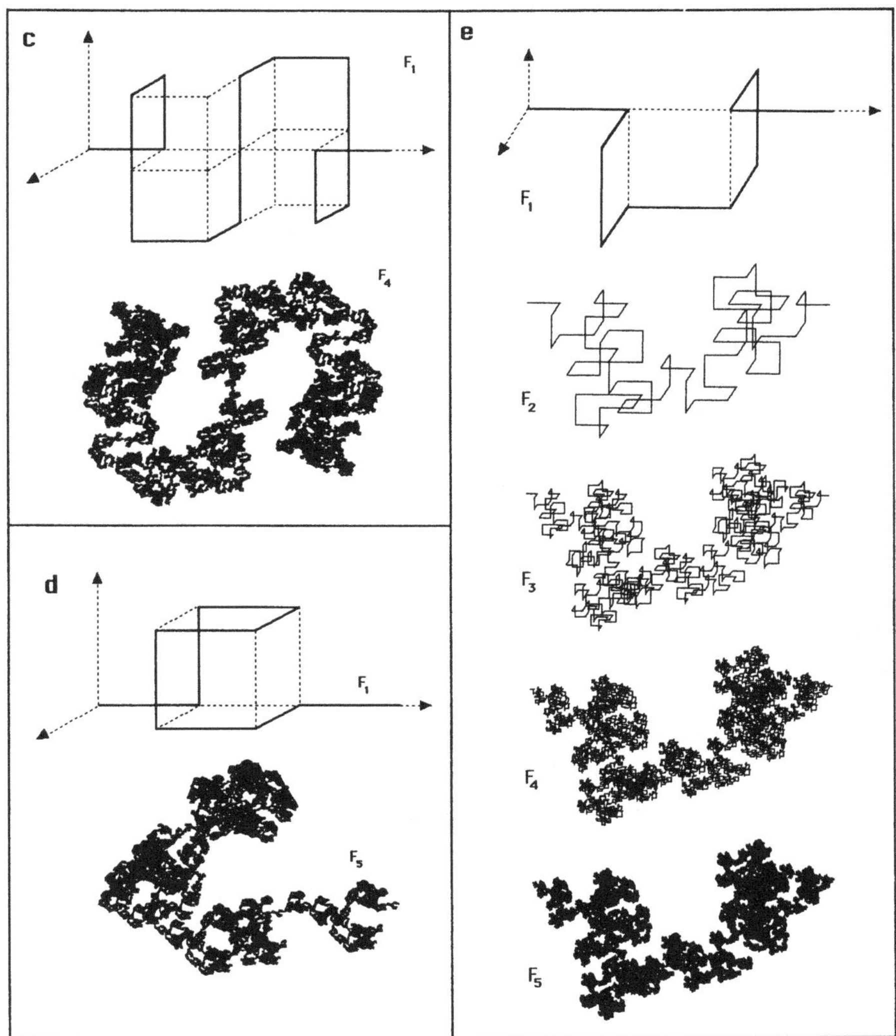

Figure 3.12 (cont.). Drawing of successive approximations of fractal curves of fractal dimension 2 in space (R^3). The curves (**a**), (**b**) and (**c**) are based on q=4 and p=16 (i.e. their generators are made of 16 segments of length 1/4), while (**d**) and (**e**) are based on q=3 and p=9. Note that the generator of (**a**) is built from a symmetrisation of (**e**), with its last segment excluded (see text).

3.6. Fractal Surfaces.

Although the study of fractal curves is already instructive for the understanding of the properties of fractal spaces, since the geodesical lines of fractal spaces or space-times are particular fractal curves, problems more specific for spaces (as compared to functions or applications) begin to be encountered when studying surfaces (i.e., fractals of topological dimension 2). We illustrate this point hereafter by a preliminary study of a particular class of fractal surfaces made of orthogonal sides.

Let us attempt to obtain a first member of this class by a two-dimensional generalization of the well-known fractal curve shown in Fig. 3.5. F_0 being a square of side 1, one gets the generator F_1 of Fig. 3.13, made of 48 new squares of length 1/4. If one now wants to build F_2 by the usual "fractalization" method, i.e. replacing each one of the 48 squares by a scaled version of F_1, a new and specific difficulty appears: matching of the structures at their boundary. While F_1 may be matched to itself when there is no rotation from one face to the adjacent one, this is no longer the case when the relative rotation is $\pm\pi/2$. Hence a complete description of such a fractal implies not only the giving of one generator, but also of the various matching conditions from one side to the other. More generally a generator of that type may be built by giving ourselves, in a Cartesian system of coordinates (x,y,z), the altitudes Z_{ij} (i=0 to $q-1$, j=0 to $q-1$) of the horizontal faces in F_1. Then the total number of faces is

$$p \;=\; q^2 + \Sigma_{ij} \,|\, Z_{i+1,j} - Z_{ij} \,|\, + \Sigma_{ij} \,|\, Z_{i,j+1} - Z_{ij} \,| \;\; .$$

If this number is conserved whatever the connections, the fractal dimension is $D=ln(p)/ln(q)$. Note that there would be an additional term if F_1 contained faces turning backwards, in which case the number of horizontal faces would be larger than q^2.

As shown in the previous example, even when one wants to ensure the highest level of self-similarity, the generator cannot be unique, except in the special case where the external faces are all of zero altitude, i.e. $Z_{0j} = Z_{q-1,j} = Z_{i0} = Z_{i,q-1} = 0$ (see Figs. 3.14 and C4).

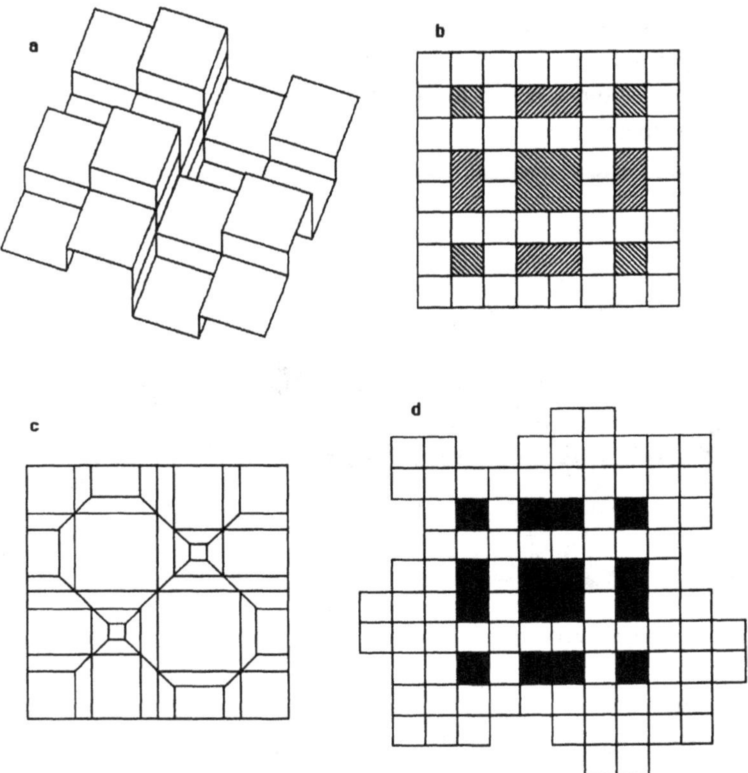

Figure 3.13. Various representations of the generator of an orthogonal fractal surface, which generalizes the fractal curve of Figs. 3.5 and 3.7. (**a**) The generator in 3-space $\mathbb{R}^3$ (see also Fig. **d**). Its is made of 48 squares of side 1/4. (**b**) An unfolded plane version of the same fractal surface. The surface of Fig.**a** may be obtained by making holes in the hachured squares of Fig.**b**, then by folding it into the shape of the generator of Fig. 3.5a, in both x and y directions. The points related by the hachures of the degenerated faces come in contact. (**c**) Lattice representation of the same fractal surface generator. The metrics on this lattice is such that each segment has the same length 1/4. (**d**) An improved version of the same generator in the representation (**b**), (but now with q=6: note that such a fractal would still have multiple points). It allows connection of orthogonal faces to the next order of the fractal construction, F_2. The right side of (**d**) may be matched to its upper side if there is no rotation between the faces. It may be matched to the same side rotated of π, i.e., to the left side of (**d**) if the two faces are orthogonal.

A solution which allows one to keep at maximum the same structure whatever the position and resolution consists in having only the outer distribution of faces changed (thus defining a "connecting box"), while the internal structure is kept invariant (see Fig. 3.13d: this is a particularly interesting case since the same border is used whatever the relative angle of the faces, but a rotation of 180° must be imparted to one face in order to connect it to the other one with a relative angle of $\pi/2$).

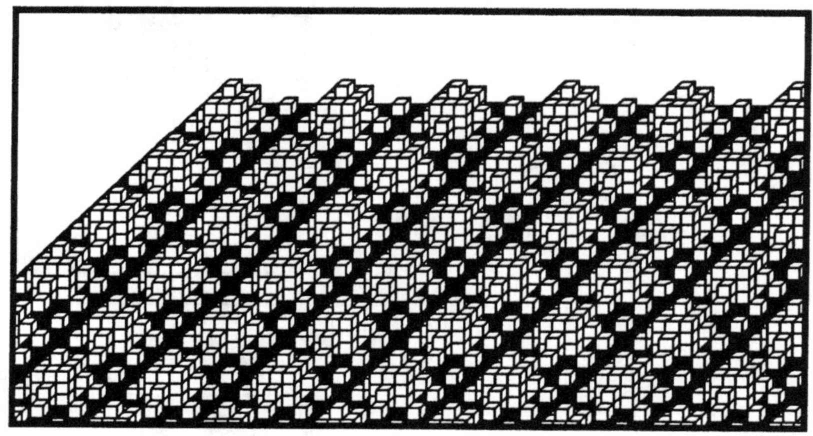

Figure 3.14. Periodic fractal surface (approximation F_2).

The effective construction of such a fractal is also instructive. It may indeed be obtained from the folding of a lacunar fractal drawn in the plane. For example the structure of Fig. 3.13a is obtained from a 8x8 square in which the regular pattern of holes of Fig. 3.13b has been made. This result is generalizable to any fractal surface of this type. This planar diagram may be seen as the representation of the generator in a particular curvilinear coordinate system, (x, y), in terms of which the metric may be degenerated. While the metric inside the uncut faces is Euclidian, $ds^2 = dx^2 + dy^2$, the metric inside the holes may be $ds = 0, dx, dy, dx + dy, dx - dy$, depending on the way the faces are connected after folding. Finally, pursuing to infinity the fractalization process will yield, in terms of such coordinates, a metric on the fractal surface alternatively Euclidian and degenerated, this in a fractal way.

An alternative and maybe more powerful representation is the drawing of a lattice in the plane (Fig. 3.13c). The metrical properties of this structure are set by the requirement that all segments are of equal length, but the definition of a 2-dimensional coordinate system is more difficult. The various points may be connected to 3, 4, 5 or 6 adjacent points. This remark leads us to a first and quick look at the question of curvature on fractal surfaces. For such orthogonal fractals (always assuming continuity and self-avoidance), there are clearly 4 possibilities concerning the connections of faces around a given point, as shown in Fig. 3.15. The standard one is 4 faces and corresponds to flatness. A point belongs to only 3 faces at the vertex of a cube, and this corresponds to infinite positive curvature: lets us call it $\{+1\}$. Finally there are two cases of infinite negative curvature, with 5 and 6 faces, which we call $\{-1\}$ and $\{-2\}$, i.e. $\{C = 4 - S\}$, where S is the number of segments starting from the point considered. (It may be remarked that $\Sigma C_i = 0$ on the vertical lines in the absence of backward turning faces). These results, straightforward when considering the generator F_1, apply as well to any point of the fractal F, since F may be considered as made of all the vertex points of its various approximations F_n. Hence the curvature of a fractal surface is *a fractal alternance of infinite positive and negative curvatures.*

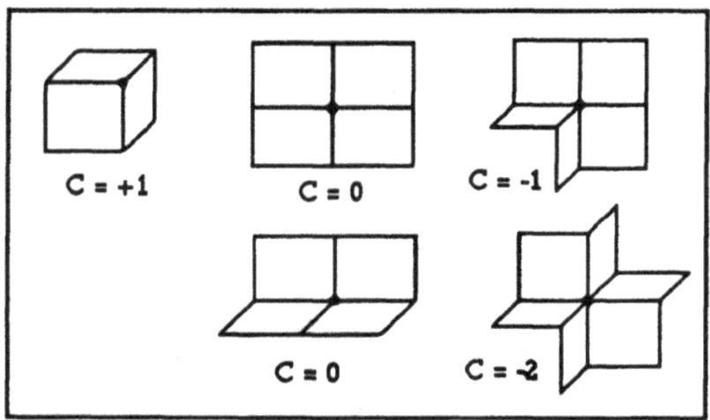

Figure 3.15. The 3 possible cases of infinite curvature on orthogonal fractal surfaces, C = −2, −1, and +1 (+ two flat cases C = 0).

Additional illustrations of fractal surfaces, some having differentiable generators, are given in Figs. 3.16, C5 and C6. For a more general mathematical description of fractal surfaces based on two variable fractal functions, see Massopust.[32] We shall also consider in Sec. 3.10 the general case of fractal spaces, of which fractal surfaces are the particular 2-dimensional achievement.

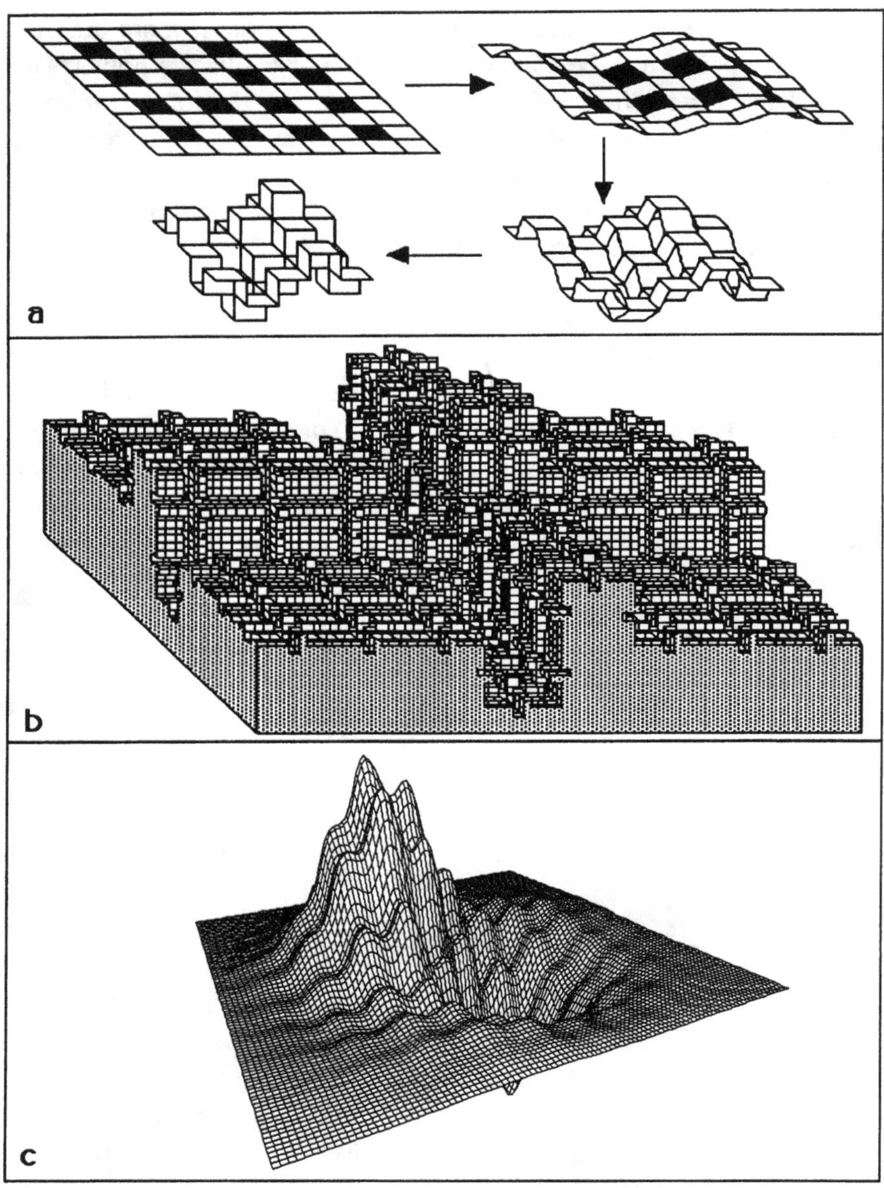

Figure 3.16. Fractal surfaces. Figure **a**: steps in the folding of a lacunar generator yielding the generator of a continuous fractal surface. Figure **b**: construction to second order of an orthogonal fractal surface. Figure **c**: second order approximation of a surface built up from a differentiable generator. (See also Figs. C4-6).

3.7. Fractal Derivative.

Consider a fractal F of Hausdorff-Besicovitch dimension D and topological dimension D_T embedded in an E-dimensional Euclidean (or Riemannian) space $\mathbf{E}$, where a system of curvilinear coordinates $(x_1, x_2,..., x_E)$ has been defined. We assume for simplicity that E-1 is the integer part of D. Note however that the embedding in space $\mathbf{E}$ is *not necessary* for the following construction (this comes from the possibility, shown hereabove, of an intrinsic description). It is only a simplifying hypothesis allowing easier understanding of the new concepts we will define here.

The fractal F can be considered as the standard part $F = st(F_\omega)$ of a non-standard set F_ω built up with infinitesimal hypervolumes $(q^{-\omega})^{D_T}$, where q is a finite real number and ω an infinite integer.[11] In other words the fractal is binned with a regular infinitesimal non-standard lattice, from which its metrical properties will be studied. A finite Hausdorff measure $\zeta = \zeta(x_i)$ of dimensionality $[L^D]$ may be defined for any finite subset of the fractal included in any domain of $\mathbf{E}$. Thanks to the NSA approach we are able to define for this subset a "number of points", a "length", a "surface", etc..., more generally an "N-hypervolume" given by

$$V_N = \zeta \ (q^\omega)^{D-N} \quad . \tag{3.7.1}$$

This hypervolume is a NS infinite quantity for $N<E$, while it is infinitesimal for $N=E$. In particular its D_T infinite hypervolume is

$$V = \zeta \ (q^\omega)^{D-D_T} \quad . \tag{3.7.2}$$

We stress once more that the NSA formulation is equivalent to the standard one in terms of varying resolution. The hereabove equation, written for an infinitesimal resolution $\Delta x = q^{-\omega}$, is equivalent to saying that, for *any* finite resolution Δx, the N-hypervolume is measured to be $V(\Delta x)=\zeta \, \Delta x^{N-D}$.

Let us now define the "fractal derivative." This new concept is itself related to the problem of the relation between fractal coordinates and classical coordinates. Our proposal is that the classical coordinates (and the classical physical behaviour) are obtained when smoothing the fractal space-time with balls of radii larger than some critical (possibly periodic) scale. As a consequence, while curvilinear coordinates on the fractal should represent the proper coordinates (i.e. those which actualize homogeneity of spacetime), the classical variables with which measurements are actually performed should be identified with (some of) the x_i's, more precisely with the x_i's along which the fractal spacetime extends from $-\infty$ to $+\infty$. The other x_i's have no direct classical spatio-temporal meaning; along them the fractal space-time extension is of the order of the critical periods, which will be identified in Chapter 5 with the de Broglie length and time.

One is now led to wondering about the nature of a subset of the fractal included in an *infinitesimal* subset of **E**, and in particular to subsets included in infinitesimal intervals of the variables which are to be identified with the classical ones. For this purpose, one only needs to describe the subset of the fractal included in the domain $(x_i, x_i + dx_i)$. As already known, the intersection of the fractal by the hypersurface $x_i = cst$ is *another fractal* of topological dimension $D_T' = D_T - 1$ and of fractal dimension "almost surely equal"[3] to D–1. On this new fractal $F'(x_i)$, which we call the *derived fractal*, a measure $\xi(x_i)$ may be defined and its infinite D_T'-hypervolume is given by

$$V' = \xi(x_i) \quad (q^{\omega})^{D-D_T} \quad . \tag{3.7.3}$$

We may now choose to take $dx_i = q^{-\omega}$ without any loss of generality of the final results in *standard* space, since in any case this NS infinitesimal binning is finer than any standard binning $(x\text{-}x_0) \to 0$. With this choice the meaning of V' and ξ should be specified. We have defined the fractal F as the standard part of a given set F_ω, $F = st(F_\omega)$, which is made of $\zeta\, q^{\omega D}$ elementary infinitesimal hypervolumes $(q^{-\omega})^{D_T}$. Each of these elementary hypervolumes may be replaced by a single point: we get a set of NS points P_ω, and the properties of NSA ensure that the standard fractal is still given by $F = st(P_\omega)$. Now (3.7.3) should be understood as meaning that in the interval $(x_i, x_i + q^{-\omega})$, one finds $\xi(x_i)\, (q^{\omega})^{D-1}$ points of the set P_ω.

The D_T-hypervolume of the subset of F included inside (x_i, x_i+dx_i) is thus nothing but $(V'q^{-\omega})$. Then F may be built back by summing all these subsets from x_i^1 to x_i^2 , i.e., by performing the following infinite sum over $n = q^{\omega}(x_i^2- x_i^1)$ terms :

$$V = \sum V' q^{-\omega} = \sum \xi(x_i)\, (q^{\omega})^{D-D_T}\, q^{-\omega} = (q^{\omega})^{D-D_T} \sum \xi(x_i)\, q^{-\omega} \quad .$$

We now recognize $\sum \xi(x_i)\, q^{-\omega}$ as the non-standard equivalent of the classical integral $\int \xi\, dx_i$, and comparison with the hereabove expression $V = \zeta(q^{\omega})^{D-D_T}$ yields

$$\zeta = \int \xi(x)\, dx \quad .$$

This result may be easily extended to the whole infinite hypervolume and we may write $V=\int V' dx$. Then the fractal F is deduced from the fractal F' by an integration, so that one may inversely define *F' as the derivative of F with respect to x_i*, which I call hereafter the "fractal derivative" $đ/dx$:

$$V' = đV/dx = (\, đ\zeta/dx)\, (q^{\omega})^{(D-1)-(D_T-1)} = (\, đ\zeta/dx)\, (q^{\omega})^{D-D_T} \quad . \tag{3.7.4}$$

The exponent is once again $D-D_T$, since both the topological and fractal dimensions have decreased by 1.

In other words, the fractal derivative is nothing but the ratio over dx of the sum of all hypervolume elements included in $[x, x+dx]$ and dx:

$$đV/dx = (\textstyle\sum_i dV_i)\, /\, dx \quad . \tag{3.7.5}$$

This may be made clearer by describing a detailed example of construction of the fractal derivative for the case of a fractal curve drawn in the plane $\mathbb{R}^2$ (see Sec. 3.3). We have identified the curvilinear coordinates on the fractal with the new "proper" variables. Classical variables are obtained by smoothing of the fractal curve with balls of radius $\Delta x \geq \lambda_0$. In the case of a fractal curve in a plane, because of the constraint $\Sigma_0^{p-1}\, q^{-1}\, exp(i\omega_k) = 1$ (Eq. 3.3.1), the classical variable should be identified with the variable x. Then the relation from the fractal variable s to the classical variable x is (see Figs. 3.5, 3.7 and 3.17)

$$x(s)= \sum_k \rho_{s_k} \cos(\varphi_{s_k})\, q^{-k} \quad . \tag{3.7.6}$$

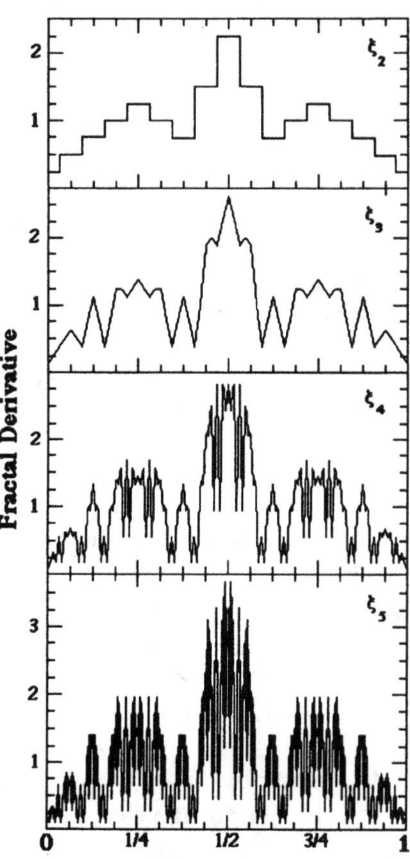

Figure 3.17. Successive approximations of the fractal derivative of a fractal curve (see its generator in Fig. 3.5).

In this form, the relation from fractal to classical coordinates *no longer depends on any embedding of the fractal space into a classical space.* Moreover, one may remark that the hereabove equation is sufficient to completely define the initial planar fractal curve (but could apply to an infinity of non-planar curves) and that the fractal derivative may also be computed from it alone. It is constrained by the equation

$$x \leq \sum_k \rho s_k(x) \cos [\varphi s_k(x)] \, q^{-k} < x + q^{-\omega} .$$

The solutions are $s^d(x)=0.\ s^d{}_1\, s^d{}_2 \ldots s^d{}_n \ldots$ and form a set $\mathbf{D}(x)=\{\ s^d(x)\ \}$. This set contains an infinite number of elements, i.e., the infinite number of points of the NS fractal F_ω included in the interval $[x,\ x+q^{-\omega}]$. What we have called the "fractal derivative" is then defined as the NS infinite number:

$$\eth s/dx = \xi_\omega(x)\ q^{\omega(D-1)} = Card[\ \mathbf{D}_\omega(x)].$$

Another equivalent definition is possible. We do not consider an *a priori* binning of the curvilinear coordinate s by equal segments of length $q^{-\omega}$, but simply state that between x and $x+q^{-\omega}$ we find an infinity of infinitesimal sections of the curvilinear coordinate, ds_i, and that the fractal derivative is given by (3.7.5):

$$\eth s/dx = \sum_i ds_i \ /dx.$$

For any approximation F_n of the fractal curve corresponding to measurements with a resolution $\Delta x = q^{-n}$, the number of solutions becomes finite and proportional to $q^{n(D-1)}$. We are now able to make the announced clarification. The initial fractal cannot be built again from just a knowledge of its fractal derivative, since information is lost in the projection. The mathematical object which contains this complete information is the set of proper infinitesimal vectors $\{\ e^{i\omega(s^d)} / \cos\omega(s^d)\ \}$ which are defined at the points of the "derived fractal" having an abscissa x and an ordinate:

$$y^d(x) = \sum_k \rho s^d{}_k(x)\ \sin [\ \varphi s^d{}_k(x)]\ q^{-k} \ .$$

Strictly speaking the slope on the fractal $tg\omega(s)$ is undefined, but it may be defined for any of its finite approximations. But even for such an approximation, the slope for a given value of x remains undefined, since it becomes a multivalued variable. This naturally leads to a probabilistic description, using e.g. the rate of the values of the slope for a given value of x, $P_n[\omega(s^d(x))]$, which remains defined on the fractal F, or its total number $N_n[\omega(s^d(x))]$, from which the fractal derivative may be recovered as $đs/dx = \sum N[\omega(s^d(x))] / \cos\omega(s^d(x))$. To exemplify these new concepts, various approximations of the fractal derivative of the fractal curve of Fig. 3.5 are given in Fig. 3.17 (see also Fig. C7). Other examples of fractal derivatives can be found in Fig. 5.11.

The derivatives dx/ds, dy/ds and *a fortiori* the slope dy/dx are still *not* defined on the fractal limit. The new fractal derivatives $đs/dx$ and $đs/dy$ do not change this property of fractals, i.e., their basic non-differentiability. Let us also remark that the normalized fractal derivative $\xi(x)$, taken as a function of x, may also be a fractal, but not of the same type as the fractal curve F. $\xi(x)$ is a fractal *function*, as are $x(s)$ and $y(s)$ (see Fig. 3.7). We shall develop this concept in Sec. 3.8. For any value of x there is only one value of ξ, so that its "fractal derivative" would be =1 everywhere. Its length is indeed infinite, but this divergence of length is due to the fact that the slope is infinite everywhere, rather than to the folding shown by fractal curves of the F type. In the case where the "fractal derivative" is a fractal function of x, it should not be confused with the "derived fractal", which has been defined (for $D_T=1$) as a fractal dust corresponding to one particular value of x.

Let us conclude this section with a remark concerning physical applications. The finite part of the fractal derivative $\xi(x)$ is not necessarily a fractal function; it may also be a standard curve, as exemplified by the case of the Peano curve: $\xi(X) = x$ for $0 \le X \le 1/2$ and $1-x$ for $1/2 \le X < 1$, or even $\xi =$ *constant* in the case of the inclined Peano curve (see Fig. 5.2). This means that the underlying periodicity, which remains apparent in the original fractal or in the fractal functions $x(s)$ and $y(s)$, may completely disappear in the projection.

3.8. Fractal Functions.

We shall define a *fractal function* as a family of functions depending on a parameter ε, (that describes resolution), modulo some equivalence relation $\mathcal{E}$:

$$y = f(x;\varepsilon) \quad ,$$

$$f \equiv g \Leftrightarrow \quad \forall\, \varepsilon,\ \forall\, x,\ \ |f(x;\varepsilon)-g(x;\varepsilon)| < \varepsilon \quad . \qquad (3.8.1)$$

The function f is assumed to be *differentiable* for any $\varepsilon \neq 0$. Only the limiting fractal function $f(x;0)$ that would correspond to the usual definition is non-differentiable. Note that, in the above definition, ε is a resolution of the y coordinate: $\varepsilon = \delta y$. The transitivity of the equivalence relation is ensured by the fact that the inequality must be true whatever the value of ε. If for a given value of ε one has three functions such that $|f-g|<\varepsilon$, $|g-h|<\varepsilon$ but $|f-h|>\varepsilon$, one expects that, for $\varepsilon' << \varepsilon$, either $|f-g|>\varepsilon'$ or $|g-h|>\varepsilon'$. This definition of the equivalence relation may be made more general. Assume that some distance $\mathcal{D}[x,y;x_1,y_1]$ has been defined in the plane, we may set

$$f \equiv g \Leftrightarrow \quad \forall\, \varepsilon,\ \forall\, x, \qquad \exists\, x_1, \qquad \mathcal{D}[x,f(x;\varepsilon);\, x_1,g(x_1;\varepsilon)] < \varepsilon \quad .$$

There is a still more economical definition which allows one to refer ε to the x axis (and so could be more adapted to some physical problems):

$$f \equiv g \ \Leftrightarrow\ \forall\, \varepsilon, \forall\, x, \exists\, x_1, |x-x_1| < \varepsilon, \quad f(x;\varepsilon) = g(x_1;\varepsilon) \text{ or } f(x_1;\varepsilon) = g(x;\varepsilon) \quad .$$

At this stage, the hereabove definitions may apply to standard functions as well. The fractal behaviour of the family of functions should now be characterized. Let us add two additional relations:

$$\forall\, \varepsilon,\ \forall\, \eta<\varepsilon,\ \ \exists\, \phi_{\varepsilon\eta},\ f(x;\varepsilon) = \int \phi_{\varepsilon\eta}(x,y)\, f(y;\eta)\, dy\ (mod\, \mathcal{E}) \quad , \qquad (3.8.2)$$

$$\frac{\partial f}{\partial x}(x;\varepsilon) = \varphi(x)\ \left[1 + \xi(x;\varepsilon) \left(\frac{\lambda}{\varepsilon}\right)^{\delta} \right] \quad . \qquad (3.8.3)$$

with $<\xi(x;\varepsilon)>=1$. The first relation, in which $\phi_{\varepsilon\eta}(x,y)$ is a smoothing function, expresses the fact that the function at resolution ε is a smoothed version of the function at a smaller resolution η. This is a local version (in terms of resolution), of the method that consists in smoothing the limiting fractal itself $f(x,0)$: but such a method assumes that we have already been able to define this limiting fractal, which may not be the case. From the point of view of physics, which is our final aim here, the problem is still more radical: the strict limit $\varepsilon=0$ may have no physical meaning at all (see Chapter 6).

The second relation expresses the fractal character of the function, for the case where the fractal dimension is defined ($D>D_T$). This can be generalized to any type of scale divergence, by replacing $(\lambda/\varepsilon)^{\delta}$ by $\theta(\varepsilon)$ with $\theta\to\infty$ when $\varepsilon\to0$. We assume f to be a *finite* fractal function whose fractal behaviour is seen in the divergence of its derivative as ε approaches zero: we verify that $f(x,0)$ is non-differentiable. The form of (3.8.3) expresses the facts that the derivative is defined for any $\varepsilon\neq0$, but undefined for $\varepsilon=0$, and that there is a fractal-non fractal transition around $\varepsilon=\lambda$. The quantity $\xi(x,\varepsilon)$ is a "renormalized" derivative for the function f and is itself a new finite fractal function. For $\varepsilon>>\lambda$, $\xi\,(\frac{\lambda}{\varepsilon})^{\delta}$ vanishes and $\partial f/\partial x$ reduces to a standard function $\varphi(x)$.

If one neglects the residual fluctuation $\xi-1$, $\partial f/\partial x$ is a solution of a renormalization group-like equation:

$$\frac{\partial}{\partial ln(\varepsilon/\lambda)}\frac{\partial f}{\partial x} = \delta\,(\varphi - \frac{\partial f}{\partial x})\ ,$$

where δ is an *anomalous dimension*. How is the anomalous dimension δ related to the fractal dimension D ? The answer depends on whether ε is a resolution of the variable x or of the variable y. Then we first need to know the relation between δx and δy on a fractal function.

A very interesting tool in this respect in the concept of fractional derivative, which has been particularly developed by Le Méhauté in conjonction with fractals.[8,33,34] The differential element of a fractal of dimension D may be written as $dx^{1/D}$, i.e. the searched relation between δx and δy is

$$\delta y = \delta x^{1/D}\ .$$

Indeed, as soon as $\delta y >> \delta x$, the differential element on the fractal becomes $\delta \mathcal{L} = (\delta x^2 + \delta y^2)^{1/2} \approx \delta y$. The total length diverges as $\delta y^{1-D} \approx \delta \mathcal{L} / \delta x \approx \delta y / \delta x$, and we finally obtain the above relation.

We can now come back to our initial problem. When $\varepsilon = \delta y$, the anomalous dimension is related to the fractal dimension by

$$\delta = D - 1 \;\Rightarrow\; D = 1 + \delta$$

while, when $\varepsilon = \delta x$, we obtain

$$\delta = 1 - \frac{1}{D} \quad \Rightarrow\; D = \frac{1}{1-\delta} \quad .$$

This result will play an important role in the physical applications to follow.

Let us quote a theorem about ξ, a complete demonstration of which will be given elsewhere: *the fractal dimension of the renormalized derivative of any fractal function is (formally) infinite.* This means that

$$\frac{d\xi}{dx}(x;\varepsilon) = \zeta(x;\varepsilon)\left(\frac{\lambda}{\varepsilon}\right) \quad ,$$

with ζ finite and $\varepsilon = \delta x$, so that $\delta = 1 \;\Rightarrow D = \infty$ from the above formula. This result will hold for any of the following derivatives. Indeed, whatever the value of ε, ξ is non-vanishing, since $<\xi> = 1$, and is alternatively positive and negative, so that its slope is finally proportional to ε^{-1}.

• Another property of ξ can be obtained from a reintegration of (3.8.3). From the hypothesis that f is finite, we conclude that $(\lambda/\varepsilon)^{\delta} \int \xi(x;\varepsilon)dx$ is itself finite, so that the integral of ξ is vanishing as ε^{δ}, contrary to the integral of $|\xi|$ which is finite and non-null.

Two kinds of fractal functions that have already been considered in previous sections are particularly relevant for physical applications. Recall first that, once the curvilinear coordinate s is defined and normalized on a fractal curve, the classical coordinates x^i become fractal functions $x^i(s,\varepsilon)$ with the same fractal dimension as the original fractal, as illustated in Fig. 3.7. Another interesting case of fractal function is the fractal derivative, which we introduced in the previous section.

Let us study in more detail the fractal derivative of the well-known fractal curve of generator ⌐⌐⌐. Its equation may be written in a recursive (or mapping-like) form (see also Refs. 35 and 36 for a development of the concept of "iterative function systems"). Let the coordinate x be written in the counting base q=4 as

$$x = 0.\,x_1 x_2 \ldots x_n \ldots, \qquad x_i \in \{0,1,2,3\} \ .$$

From one approximation (n) to the following (n+1), the fractal derivative changes by the inclusion of points coming from an application of the generator to horizontal (h) and vertical (v) segments. The total normalized number of segments is the sum

$$\xi_n = v_n + h_n \ .$$

Depending on the value of the digit x_n , we obtain in the case x_n =0,1 or 2:

$$\begin{bmatrix} h_{n+1} \\ v_{n+1} \end{bmatrix} = \begin{bmatrix} 1/2 & \delta_0^{x_n} \\ x_n/2 & 1-x_n/2 \end{bmatrix} \begin{bmatrix} h_n \\ v_n \end{bmatrix} \ .$$

The case x_n =3 is more complicated, since it implies a contribution from backward segments, originating from vertical segments of the subsequent coordinate $x = 0.x_1x_2...x_n + q^{-n}$. Let us call formally these contributions $v_n(4)$ and $h_n(4)$, then

$$h_{n+1}(3) = v_n(4) + h_n/2 \ , \quad v_{n+1}(3) = v_n(4)/2 + h_n/2 \ .$$

These terms express the non-locality of the fractal derivative. In the case $x_i \neq 3$, $\forall i$, we obtain, in terms of the above matrix M,

$$\begin{bmatrix} h_n \\ v_n \end{bmatrix} = \mathrm{M}_{x_n}\, \mathrm{M}_{x_{n-1}} \,\cdots\cdots\, \mathrm{M}_{x_1} \begin{bmatrix} 1 \\ 0 \end{bmatrix} \ .$$

This matrix product takes a simpler form for some values of the coordinate. For example $(\mathrm{M}_0\,\mathrm{M}_2)^k = (5/4)^k \begin{bmatrix} 1 & 0 \\ 4/5 & 0 \end{bmatrix}$. This means that the fractal derivative is divergent as $(\sqrt{5}/2)^n$ for the point of abscissa 0.20202...

This demonstrates that, though finite nearly everywhere, the fractal derivative contains infinite spikes (see Fig. 3.5).

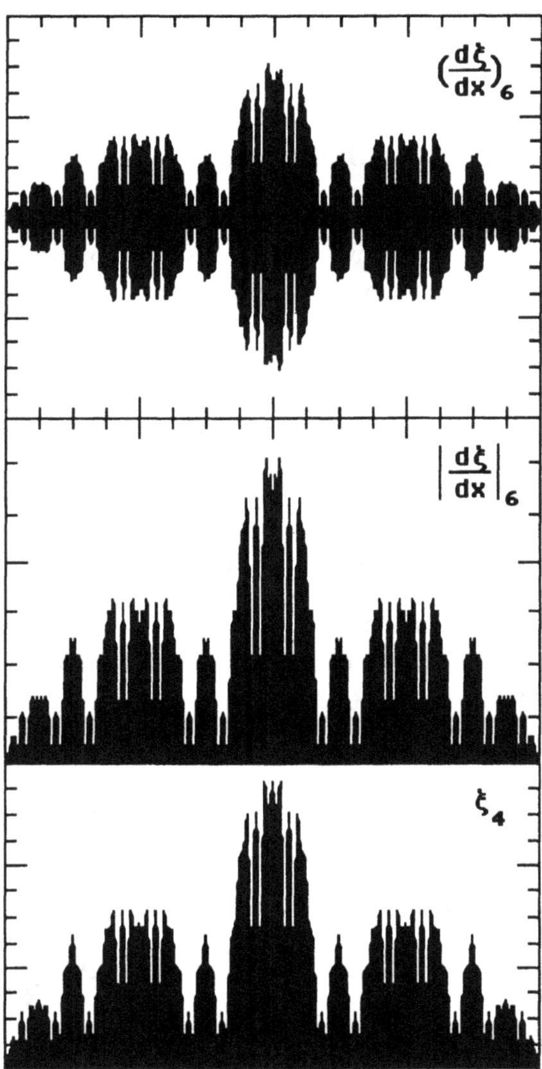

Figure 3.18. Comparison between the 'fractal derivative' of the fractal curve of Fig. 3.5 (see also Fig. 3.17) and its own derivative (in the usual generalized meaning, see text).

Consider finally the derivative of this fractal function in the sense introduced in Eq. (3.8.3): we have plotted in Fig. 3.18, $|\partial\xi(x,\varepsilon)/\partial x|$ for $\varepsilon=q^{-6}$. The similarity with ξ itself is remarkable. This opens the hope that usual differential equations might have still unknown fractal solutions.

3.9. Variable Fractal Dimension.

The case of constant fractal dimension and perfect self-similarity that we mainly considered up to now corresponds to a simplifying assumption which proves to be insufficient for physical applications (see Chapters 6 and 7). We shall briefly exemplify here two different situations: fractal dimension varying with the curvilinear coordinate and fractal dimension varying with scale.

Let us introduce a simple generalization of the von Koch curve that may have any fractal dimension between 1 and 2, based on a generator that depends on one parameter, . The parameter q is given by

$$q = 2\,(1 + \cos\beta) \ ,$$

so that the fractal dimension is $D(\beta) = 2/(1+ln_2(1+\cos\beta)$, and, conversely, $\beta(D)=Arccos[2^{(2/D)-1}-1]$. We can now construct a fractal with variable dimension by generalizing Eq. (3.3.4)

$$\alpha(s) = \alpha_{s_h}[\beta(s)] \ .$$

An example of curve with variable fractal dimension is given in Fig. 3.19: it begins with D=1 and becomes plane-filling in the end (D=2).

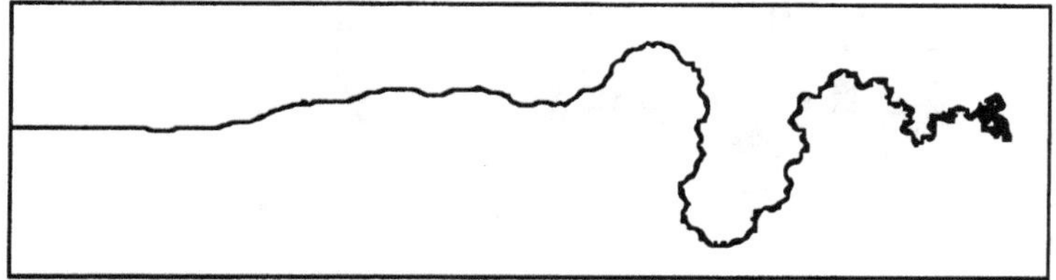

Figure 3.19. Fractal curve of dimension varying with the curvilinear coordinate.

The same type of variable generator can be used to construct models with dimension varying with scale. In the current definition,[3] only the limit where $\Delta x \to 0$, if it exists, of such a variable dimension would be named fractal dimension. However we are mainly interested here in physical applications, for which the meaningful object is the fractal seen at all its scales simultaneously. So the idea that the various quantities on a fractal are described by fractal functions $f(x, \Delta x)$ can be extended to the fractal dimension and anomalous dimensions themselves, $\delta = \delta(x, \Delta x)$. This concept will become particularly meaningful in Chapter 6. To achieve such a fractal, we build up F_n by including in F_{n-1} a generator depending on the level of fractalization n, after having chosen a function $D=D(\varepsilon)$. An example of such a construction is given in Fig. 3.20.

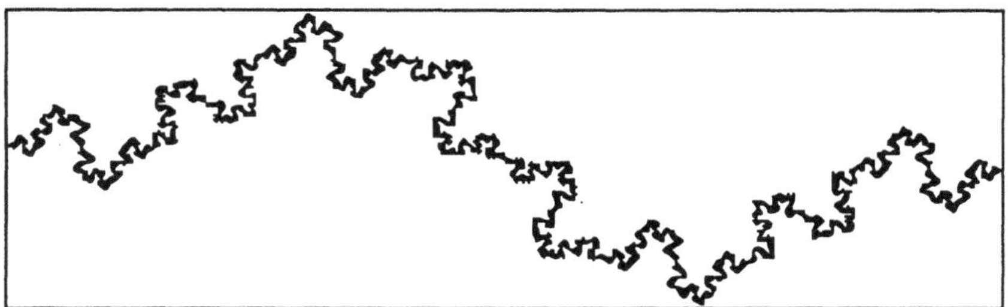

Figure 3.20. Curve of "fractal dimension" varying with resolution.

3.10. Towards the Definition of a Fractal Space-Time.

Up to now, the basic non-differentiability of fractals, a property which may be their main interest as a new mathematical tool for physics, seemed also to have prevented physicists from working with fractal spaces in much the same way they work with Euclidean or Riemannian spaces (i.e. by using concepts like curvilinear coordinates, metrics, geodesical lines, etc...). Some ways to deal with the infinite quantities appearing on fractals and to define curvilinear coordinates on them by using Non-Standard Analysis have been indicated and reviewed in the previous sections.

Before more thoroughly studying the concept of fractal space, let us make more precise the relation between non-differentiability and fractality. We shall first demonstrate that continuity and non-differentiability imply scale-dependence. Consider a continuous function $f(x)$ between two points A_o $(x_o, f(x_o))$ and A_Ω $(x_\Omega, f(x_\Omega))$. Since f is non-differentiable, there exists a point A_1 of coordinates $(x_1, f(x_1))$ with $x_o < x_1 < x_\Omega$, such that A_1 is not on the segment A_oA_Ω. Then the total length $\mathcal{L}_1 = \mathcal{L}(A_oA_1) + \mathcal{L}(A_1A_\Omega) > \mathcal{L}_o = \mathcal{L}(A_oA_\Omega)$. We can now iterate the argument and find two coordinates x_{01} and x_{11} with $x_o < x_{01} < x_1$ and $x_1 < x_{11} < x_\Omega$, such that $\mathcal{L}_2 = \mathcal{L}(A_oA_{01}) + \mathcal{L}(A_{01}A_1) + \mathcal{L}(A_1A_{11}) + \mathcal{L}(A_{11}A_\Omega) > \mathcal{L}_1 > \mathcal{L}_o$. By iteration we finally construct successive approximations $f_o, f_1, \ldots f_n$ of $f(x)$ whose lengths $\mathcal{L}_o, \mathcal{L}_1, \ldots \mathcal{L}_n$ increase monotonically with resolution.

However, at this stage $\mathcal{L}_n$ may still be bounded when $n \to \infty$. Let us outline a demonstration, in the framework of non-standard analysis, of a stronger theorem: continuity and non-differentiability everywhere actually implies scale divergence of the length. Let us first recall a theorem demonstrated in Stroyan and Luxemburg (Ref. 28, p. 87): if $f'(x)$ is defined on a neighbourhood of c but discontinuous at c, then $f'(x)$ has infinite variation, that is, there exists a *-finite sequence $x_1, \ldots, x_\omega$ all infinitely close to c satisfying $\Sigma_{i=1}^{\omega} |f(x_{i+1}) - f(x_i)|$ infinite. The demonstration makes use of the definition of discontinuity: "there is an ε finite so that on any neighbourhood of c, $f'(x)$ is at least ε from $f'(c)$", and applies it to the sequence $x_1, \ldots, x_\omega$. Now the discontinuity of the derivative at any point will read: "$\forall\, x_o$, there exists ε finite so that $\forall \eta$, $|x - x_o| < \eta \Rightarrow |f'(x) - f'(x_o)| > \varepsilon$", so that the Stroyan and Luxemburg demonstration can be extented to this case. The infinity of $\Sigma_{i=1}^{\omega} |f(x_{i+1}) - f(x_i)|$ finally implies the infinity of the length of the curve $f(x)$ on a finite interval.

Since we intend to describe a fractal space (more generally space-time), the difference between a fractal object and a fractal space should be specified. The first point is that only some particular types of fractals might be adapted to the description of space-time. In particular, we shall impose it to be a self-avoiding continuum of topological dimension 3+1 (signature +,–,–,–). Moreover, contrary to fractal objects, a fractal space-time should not be limited or bounded. Though one may imagine a fractal space for which the "fractalization" (i.e. the building process implying successive levels of resolution) extends as well towards small resolutions as towards the large ones, one of the constraints imposed by physics to any attempt of modelling the micro-space with fractals is the existence of an upper "resolution" λ^{μ} beyond which the fractal structure ends. We shall see in Chapter 4 that this scale may be identified with the quantum-classical transition, which is given by the de Broglie scale.

On the other hand, one of the most original features of the use of fractals suggested here is the absence of any minimal cut-off for the fractal behaviour, in agreement with the universality of the Heisenberg relations. This is at variance with all the applications of fractals to natural systems that have been proposed up to now[3,9], which are characterized by lower and upper cut-offs in resolution. It is expected that this introduction in physics of the absence of scale limit for the fractal structure will account for some of the quantum properties which are clearly irreducible to classical ones.

The upper transition is given by the de Broglie length and time of the physical system under consideration (see Chapters 4 and 5). The fundamental and universal existence of this transition leads us to propose to build a fractal space as a periodic or pseudo-periodic reproduction of what are presently known as fractal objects.[31] These periods will then be reinterpreted in our approach as the first "structural constants" of the microphysics space-time. They depend on the energy and momentum of the system, as would be expected of a relativistic theory.

A general description of the structures of fractal spaces relevant to physical application is still an open problem. Here we will only try to indicate a possible way towards such a description. As described in above sections for fractal curves and surfaces, the fractalization process consists in giving ourselves a first set of points (F_1), defining a new reference system tied to each point (segments for curves, squares for the particular class of surfaces studied in Sec. 3.6), and then reproducing the same construction after scaling in each of these new frames. The examples given hereabove correspond to the choice of Cartesian coordinate systems. Indeed it has the advantage of making obvious the absence of well-defined slope and curvature.

But, as remarked in Secs. 3.3 and 3.8, fractals may also be obtained as the limit of curves, surfaces, volumes, etc..., which are differentiable. Secondly, while we have, up to now, given examples mainly of discrete processes of fractalization, they may clearly be replaced by a continuous process. This may be shown by the following argument: Assume we have been able to define a fractal F. Then cover it with balls of varying radii ε, we obtain a continuous set of approximations of the fractal, $F(\varepsilon)$, which tend to the fractal itself when ε tends to 0 and which cease varying (i.e. looses its fractal character) when ε becomes larger than a transitional value λ. At the end the discrete construction should be replaced by a construction using infinitesimal differential generators.

In fact the anomalous dimension $\delta = D - D_T$, and the constant q which gives the ratio of the segment lengths in the generator F_1 over F_0, may be considered as the first elements of a more general set of structural constants defining a fractal structure, not necessarily self-similar (see Ref. 11). An increase of the information on the structure will be obtained with successively enlarged subsets of these constants. Hence knowing only δ, a general information on the increase of the length is known (we consider the case $D_T=1$): $L \propto (\Delta x)^{\delta}$. Knowing q implies information on the transition region: setting $\Delta x = \lambda\, q^{-\nu}$ (where now ν is a continuous parameter measuring resolution which generalizes the discrete parameter n of the successive approximations F_n), we may write the increase of length as $L = L_0\,(1 + q^{\nu})$, which is a solution of the differential equation[31]

$$dL/d\nu - A\,L + B = 0 \ ,$$

where $B/A = L_0$ and $A = ln(q)$. As already remarked previously, this is nothing but a renormalization group equation.

A possible way towards a more general definition of fractal spaces consists in applying such a differential approach to families of Riemannian spaces. How can we actually build such a new tool? Our construction of fractal functions as families of functions $f(x,\varepsilon)$ can be generalized to spaces, in defining a fractal space-time as a family of Riemannian space-times. Let us be more specific and justify this choice. Recall first that one of the best ways to work on fractals and to deal in a consistent mathematical manner with their non-differentiability and their infinities is to use Non-Standard Analysis.

In Ref. 11 (as recalled in Sec. 3.4), we have generalized the construction of fractals by first defining a non-standard fractal F_ω obtained from the application of a generator ω times (thus getting a scale factor $q^{-\omega}$), ω being an infinite integer, and then taking its standard part: $F = st(F_\omega)$. While strictly F is non-differentiable, one may replace F_ω by some non-standard set G_ω whose standard part is always F, but for which a kind of derivative, called ε-differentiability above, may be defined: to simplify the argument, it is equivalent to saying that the same standard F may be obtained by applying an infinite reduction factor q^ω to some differentiable set (itself infinite).

The interest of this approach is that, once seen under such an infinite magnifying glass, a fractal space recovers the usual properties of differential geometry; we may apply to such a geometry Gauss's hypothesis and assume that it becomes locally Euclidean, i.e. flat on a scale $\ll q^{-\omega}$. In other words, one may define a fractal continuous space-time *as becoming Riemannian once magnified by a factor* q^ω.

There is a complementary and equivalent approach to the same problem. Consider the successive approximations F_n of a fractal F. They may be obtained in several ways. The first one is by successive application of a generator G. In case of perfect self-similarity, this generator is unique and independent of scale: $F_n = G^{(n)}(F_0)$. But the generator may also change with scale: $F_n = G_n[G_{n-1}[...[G_1(F_0)]]]$ (Sec. 3.9). Indeed, as recalled previously, self-similarity is not an obligatory property of fractals, but only a simplifying assumption. It corresponds to scale invariance: but scale invariance itself is a particular case of scale *covariance*. Reversing the argument leads to the conclusion that scale covariance will be implemented by a kind of self-similarity of the form of the equations themselves, but not necessarily of the objects to which they will be applied. (Compare to the covariance of the laws of motion: the *form* of equations, Einstein equations, geodesics equations..., is the same in any coordinate system as it is in inertial systems[37]). By taking the continuum limit in the zoom-space (the space of scales), we are led to defining an infinitesimal and local generator for the fractal and to renormalization group-like equations of the form $\partial F/\partial \ln\varepsilon = \beta(F(\varepsilon))$.

The possibility of replacing the non-differentiable fractal F by an equivalence class of families of differentiable functions can be now applied to geometry. For any given resolution ε, we are now back in the framework of differentiable geometry, so that the approximation of the fractal space may be taken to be Riemannian. We then define a fractal space as the equivalence class of a family of Riemann spaces characterized by a family of Riemann tensors $R_{ijkl}(\varepsilon)$. The fractal limit itself is undefined, being non-differentiable and then non-Riemannian. Its curvature will be everywhere infinite, as shown in Sec. 3.6 for specific cases of fractal surfaces. The curvature of the various Riemann space-times, $\mathcal{R}(\varepsilon)$, is expected to fluctuate in a chaotic way with decreasing scale, unceasingly jumping from increasingly large negative to positive values: there lies the expected new physics (since a smooth variation of curvature would manifest itself as gravitation-like laws). For each resolution ε, a coordinate system is given (of which ε is the "state of resolution" or "state of scale"), $x^i(\varepsilon)$, in which metric potentials $g_{ij}(\varepsilon)$ are defined, and from which the Riemann tensor coefficients may be computed by the usual expressions of Riemannian geometry (see e.g. Weinberg[37]). The hereabove relations obtained for fractal functions give some hints about the way the various spaces at different ε are related: If such a theory is to be implemented in the future, it is to include in its foundation a generalized (possibly second order) renormalization group equation for space-time, formally:

$$\phi \left[\frac{\partial^2}{\partial(\ln\varepsilon)^2} , \frac{\partial}{\partial(\ln\varepsilon)} \right] g_{ij}(\varepsilon) = \beta\,[g_{ij}(\varepsilon)] \quad .$$

In such a space-time, the geodesics equation would also be explicitly scale dependent:

$$\frac{d^2x^i(\varepsilon)}{ds^2} + \Gamma^i_{jk}(\varepsilon)\,\frac{dx^j(\varepsilon)}{ds}\,\frac{dx^k(\varepsilon)}{ds} = 0 \ ,$$

yielding an infinite family of geodesics. But a detailed analysis of the properties of such a geometry exceeds the scope of this book and remains to be developed.

3 References

1. Mandelbrot, B., *Les Objets Fractals* (Flammarion, Paris, 1975).
2. Mandelbrot, B., *Fractals* (Freeman, San Francisco, 1977).
3. Mandelbrot, B., *The Fractal Geometry of Nature* (Freeman, San Francisco, 1982).
4. Weierstrass, K., 1895, *Mathematische Werke* (Berlin: Mayer & Muller)
5. Cantor, G., 1883, *Mathematische Annalen* **21**, 545
6. Peano, G., 1880, *Mathematische Annalen* **36**, 157.
7. Koch, H. von, 1904, *Archiv für Mathematik, Astronomi och Physik* **1**, 681.
8. Le Méhauté, A., *Les Géométries Fractales* (Hermès, Paris,1990)
9. Hurd, A.J., 1988, *Am. J. Phys.* **56**, 969.
10. Massopust, P.R., private communication.
11. Nottale, L., & Schneider, J., 1984, *J. Math. Phys.* **25**, 1296.
12. Heck, A., & Perdang, J.M. (Eds.), *Applying Fractals in Astronomy* (Springer-Verlag, 1991).
13. Feder, J., & Aharony, A., (Eds.), *Fractals in Physics* (North-Holland, Amsterdam, 1990).
14. Kadanoff, L.P., 1986, *Physics Today* **39**(2), 6.
15. Wilson, K.G., 1979, *Sci. Am.* **241** (August), 140.
16. Mandelbrot, B., *The Fractal Geometry of Nature* (Freeman, San Francisco, 1982), p.331.
17. Yukalov, V.I., 1991, *J. Math. Phys.* **32**, 1235.
18. Eckmann, J.P., & Ruelle, D., 1985, *Rev. Mod. Phys.* **57**, 617.
19. Perdang, J.M., in *Applying Fractals in Astronomy,* Heck, A., & Perdang, J.M., Eds., (Springer-Verlag, 1991), p.1.
20. Bergé, P., Pomeau, Y., & Vidal, Ch., *L'Ordre dans le Chaos* (Hermann, 1984).
21. Wolfram, S., 1983, *Rev. Mod. Phys.* **55**, 601.
22. Hénon, M., 1988, *Physica* **D33**, 132.
23. Petit, J.M., & Hénon, M., 1986, *Icarus* **66**, 536.
24. Nottale, L., Bardou, F., & Appert, E., in preparation.
25. Wisdom, J., 1987, *Icarus* **72**, 241.
26. Misner, C.W., Thorne, K.S., & Wheeler, J.A., *Gravitation* (Freeman, San Francisco, 1973).
27. Robinson, A., 1961, *Proc. Roy. Acad. Sci. Amsterdam* **A 64**, 432.

28. Stroyan, K.D., & Luxemburg, W.A.J., 1976, *Introduction to the theory of infinitesimals* (Academic Press, New York).

29. Nelson, E., 1977, *Bull. Amer. Math. Soc.* **83**,1165.

30. Herrmann, A., 1989, *J. Math. Phys.* **30**, 805.

31. Nottale, L., 1989, *Int. J. Mod. Phys.* **A4**, 5047.

32. Massopust, P.R., 1990, *J. Math. Anal. Applic.*, **151**, 275.

33. Le Méhauté, A., 1990, *New J. Chem.* **14**, 207

34. Héliodore, F., Cottevieille, D., & Le Méhauté, A., 1991, *Rev. Scien. Tech. Défense*, in the press.

35. Barnsley, M.F., *Fractals Everywhere* (Orlando Fl: Academic Press, 1988).

36. Barnsley, M.F., & Demko, S., 1985, *Proc. Roy. Soc. London* **A399**, 243.

37. Weinberg, S., 1972, *Gravitation and Cosmology* (John Wiley and Sons, New York).

Chapter 4

FRACTAL DIMENSION OF A QUANTUM PATH

4.1. Introduction.

Feynman's path integral.

The discovery that the typical "trajectories" of a quantum particle are fractal should be certainly attributed to Feynman.[1,2] While the standard ("Copenhagen") interpretation of quantum mechanics, as advocated by Bohr, has completely abandoned the concept of trajectory, Feynman's path integral approach was in some sense coming back to realism. In Feynman's approach, though undeterminism of trajectories is kept as an essential property of quantum mechanics, this does not prevent one from considering all the possible trajectories that the particle may have followed: this allows one to study their structures and properties, and to find that, in the end, the possible trajectories are dominated by those which are fractal.

Dirac's[3] and Feynman's[1,2] discovery is that the wave function (more generally, the probability amplitude between two points) is given by an integral over all possible paths of $exp(iS_{cl})$, where S_{cl} is the classical action for each path. Hence the *probability* of each path is a constant (each path is equiprobable), while its probability amplitude is not (it has only a phase term for each given path). The paradox is that, while all paths are equiprobable when considered at zero resolution (ideal infinite precision), the more realistic consideration of paths *at a given resolution* shows disappearance of paths which are far from the classical trajectory: indeed in the "tube" corresponding to the resolution considered, the fast variation of S_{cl} leads to destructive interferences and to a very low resulting probability. On the other hand, there will be some domain around the classical trajectory where interferences are constructive (see Sec. 5.6). In such a domain, the great majority of trajectories are fractal (the ratio of the number of non-differentiable over differentiable trajectories is infinite).

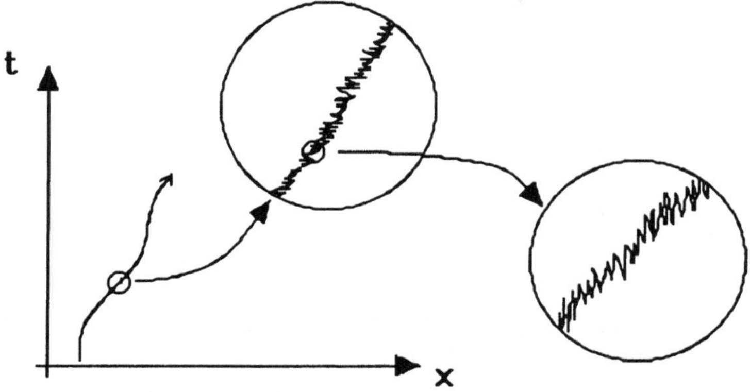

Figure 4.1. Sketch of the typical path of a quantum-mechanical particle in space-time.

Let us quote Feynman and Hibbs,[2] who include an explicative figure from which Figure 4.1 is adapted:

"The important paths for a quantum-mechanical particle are not those which have a definite slope (or velocity) everywhere, but are instead quite irregular on a very fine scale."

"Typical paths of a quantum-mechanical particle are highly irregular on a fine scale, as shown in the sketch. Thus, although a mean velocity can be defined, no mean square velocity exists at any point. In other words, the paths are non-differentiable."

Feynman and Hibbs demonstrate that (with $\varepsilon = \delta t$ here)

$$\left\langle \left(\frac{x_{k+1} - x_k}{\varepsilon} \right)^2 \right\rangle = -\frac{\hbar}{im\varepsilon} \langle 1 \rangle \ . \qquad (4.1.1)$$

"This equation says that the transition element of the square of the velocity is of the order $1/\varepsilon$, and thus becomes infinite as ε approaches zero".

It is clear from the formula, figure and comments that Feynman discovered the fractal structure of quantum paths, even though he did not use the word, which was coined by Mandelbrot in 1975 only. Figure 7-1 of Feynman and Hibbs[2] (p. 177) is exactly the kind of figure one now uses for explaining fractals and self-similarity. The same is true of the characterizations used in the text (highly irregular on a fine scale; non-differentiable). The explicit scale dependence of the velocity is also described: "If some average velocity is defined for a short time interval Δt, as for example $[x(t+\Delta t)-x(t)]/\Delta t$, the "mean" square value of this is $-\hbar/im\Delta t$. That is, the "mean" square value of a velocity averaged over a short time interval is finite, but its value becomes larger as the interval becomes shorter". Let us add a few comments to these results.

Note first that Eq. (4.1.1) apparently does not allow the existence of a classical velocity for a particle: it seems to be a purely quantum relation. When ε tends to infinity, $<v^2>$ tends to zero. However null velocity means rest frame, and a Lorentz transformation may lead to any velocity ($\leq c$). (The existence of a classical velocity is apparent in Feynman's Fig. 7-1: see Fig. 4.1 above). We shall see in the next Section that a direct fractal approach, as pioneered by Abbott and Wise[4], leads to both the quantum fractal behaviour and the classical non-fractal behaviour, the transition occurring at the de Broglie scale. More will be said on classical-quantum transition in Sec. 5.7. A realistic (and relativistic) evolution of the velocity with scale is described in Sec. 5.3.

Our second comment concerns the fractal dimension. We shall see in the next section that it is found to be $D=2$. What is the relation of this result with Feynman's ? The equation (4.1.1) above implies $|dx/dt| \approx <v^2>^{1/2} \propto (\delta t)^{-1/2}$. A fractal dimension 2 is usually characterized by the relation $|dx/dt| \propto (\delta x)^{-1}$. This apparent discrepancy is resolved from our result of Sec. 3.8: Recall that we have demonstrated that, starting from some fractal function $x(t)$ of dimension D, the derivative of this function was diverging as $(\delta x)^{1-D}$, but also as $(\delta t)^{(1/D)-1}$, depending on the choice of the variable which defines the resolution. This comes from the relation $\delta x \propto (\delta t)^{1/D}$ (see in this respect Le Méhauté and his development of the fractional derivative approach to fractals)[5]. Then both results have finally the same meaning, that the fractal (box counting) dimension of quantum trajectories is $D=2$.

Similarity dimension of the hydrogen atom.

Let us close this introductive section by showing in a simplified description that in some situations, a similarity dimension can be calculated for a quantum system.

Consider an hydrogen (or hydrogenoid) atom. The average distance of an electron from the proton is given, in terms of the two quantum numbers n and ℓ, by[6]

$$<r> = \frac{3}{2}\, n^2 - \frac{1}{2}\, \ell\,(\ell+1) \ .$$

Assume that we want to measure the position of the electron in the atom. This will perturb its state in such a way that it will jump to another state. Assume that the electron jumps successively on all the successive states. The total length covered will then be of the order of

$$\mathcal{L} \approx \sum_n \sum_{\ell=0}^{n-1} \{\frac{3}{2}\, n^2 - \frac{1}{2}\, \ell\,(\ell+1))\} \ .$$

Since $\sum_\ell \ell^2 \approx n^3$, we get $\mathcal{L} \approx \sum_n n^3 \approx n^4$. Finally, with $<r> \approx n^2$ we obtain

$$\mathcal{L} \approx r^2 \ .$$

This rough calculation may be interpreted as a similarity dimension 2 for the whole structure of orbitals in the hydrogen atom. We shall come back in a subsequent section (4.4) on the hydrogen structure from the view-point of temporal phenomena.

Before coming to a general demonstration that quantum systems are characterized in a *universal* way by a fractal dimension 2, let us add some additional arguments in favor of an internal self-similarity of atomic orbitals. Consider the radial wave functions of hydrogenoid atoms in the special case $\ell = n-1$. These states correspond to quasi-spherical "trajectories", the probability of presence depending on the radial variable (in Bohr radius unit) as[6,7]

$$P_n = \frac{1}{(2n)!}\ \frac{8}{n^3}\ \left(\frac{2r}{n}\right)^{2n-2}\ e^{-2r/n} \quad .$$

Here the probabilities are normalized by $\int Pr^2 dr = 1$. The average distance is $<r> = n\ (n + 1/2)$ and the dispersion around it $\sigma_n = \frac{1}{2} n \sqrt{2n + 1}$, so that $\sigma / <r>$ decreases as $\approx n^{-1/2}$ when n increases. Let us normalize the distances by setting $x = r/<r>$; then define

$$\chi\ [x(r)] = x^2\ e^{2(1-x)} = e^{2(1-x+lnx)} \quad .$$

To a very good approximation one finds that P_n is a power law:

$$P_n \approx \frac{1}{\sigma_n}\ [\chi\,(x)]^n \quad .$$

Now setting $k(x) = 2(1 - x + lnx)$, we get the symmetrical form

$$\sigma_n\, P_n \approx e^{k.n} \quad .$$

One can now place oneself in a given state (n fixed) and consider the variation with r of the probability in terms of the variable k: it is given by $(e^n)^k$; or one may place oneself at a given value of k and consider the variation of the probability between different states, which is given by $(e^k)^n$. Hence there is an approximate similarity between the variation inside a state and the variation between states, indicating a global self-similarity of the hydrogen atom.

4.2. Fractal Dimension of a Nonrelativistic Quantum Path.

Abbott and Wise[4] have computed the fractal dimension of the *observed* path of a particle from quantum mechanical predictions. They find a specific result, $D=2$, as a direct consequence of the Heisenberg uncertainty relation. This was confirmed in an analytical calculation for a particle in an harmonic operator potential by Campesino-Romeo *et al.*[8] (see also Refs. 9 and 10).

Let us first recall their argument in its most simplified form. The position of a nonrelativistic particle is measured N times with a resolution Δx. During the time Δt between two measurements, the particle travels an *average* distance $<\Delta l>$, so that the mean length of the particle path is $<l> = N <\Delta l>$. The average velocity along the trajectory is $<|v|> = <|p|>/m$. For a particle at rest in the mean, i.e. $<p> = 0$, one has $<|p|> \approx \Delta p$. Then the uncertainty principle, $\Delta x.\Delta p \approx \hbar$ allows one to write $<\Delta l> = <|v|>.\Delta t \approx \hbar\Delta t / m\Delta x$, so that for a given total time $T = N.\Delta t$, we get $<l> \approx m^{-1}\hbar\, T.(\Delta x)^{-1}$. The length of a fractal curve of dimension D being expected to diverge with decreasing resolution Δx as $l = l_0(\Delta x)^{1-D}$, one finds D=2, while a classical path is obviously of (fractal and topological) dimension 1. Abott and Wise then present a more complete calculation yielding essentially the same result, but valid only provided that $<\Delta l>$ is large compared with Δx. They finally describe the quantum to classical transition, which rapidly occurs around $\lambda = \hbar/2p$, i.e. about the de Broglie's length (recall that, throughout this book, we call 'de Broglie length' $\lambda_{dB} = \hbar/p$, and 'de Broglie wavelength' the full period h/p).

Let us present another derivation of the Abbott-Wise result in order to point out the precise way it originates from Heisenberg's relation. Assume we want to "know completely", at a given resolution Δx, the trajectory of a particle. To this purpose, the *maximal* number of measurements should be made since there is no way to interpolate the trajectory between each measurement. This means that the average path length between two measurements must be $<\Delta l> \approx 2\Delta x$, if we identify Δx with the radius of a "resolution ball"with which the trajectory is covered. Though this case was excluded from their calculation, since they assume $<\Delta l> >> \Delta x$, it will be shown that even in that case the Abbott-Wise result still holds. Indeed a simple inspection of the problem shows that the computation of the fractal dimension comes *not* from the relation between Δp and Δx, but instead from the relation between $<|p|>$ and Δx, since p and v are proportional in the nonrelativistic case. Fractal dimension 2 is obtained only because $<|p|> \approx \sigma_p$ in the quantum case, so that the Heisenberg relation implies $<|p|>.\Delta x \approx \hbar/2$, while in the classical case (i.e., $<|p|> >> \sigma_p$) one gets $<|p|> \approx$ *constant*, independent of σ_p, so that the trajectory length becomes independent of Δx. Hence the complete translation in terms of fractals of the Heisenberg relation reads: the fractal dimension of a particle's trajectory jumps from $D=1$ to $D=2$ when the resolution becomes smaller than its de Broglie length.

Let us specify this behaviour by looking at the one-dimensional case of a Gaussian wave-function $\varphi(p)$. One may write in the momentum representation[6,7]

$$<|p|> = \int |p| \; |\varphi(p)|^2 \, dp \quad ,$$

$$<p> = \int p \; |\varphi(p)|^2 \, dp \quad ,$$

$$\sigma_p{}^2 = \int (p-<p>)^2 \; |\varphi(p)|^2 \, dp \quad ,$$

$$\sigma_x{}^2 = \hbar \int |d\varphi(p)/dp|^2 \, dp \quad .$$

Let us set $Y=<p>/\sigma_p = \sigma_p{}^{-1} \int p \, |\varphi(p)|^2 \, dp$. When $\varphi(p)$ is a Gaussian function, one finds

$$<|p|>/\sigma_p = \sqrt{2/\pi} \; exp(-Y^2/2) + Y \{ -1+\sqrt{2/\pi} \int_{-\infty}^{Y} exp\,(-z^2/2)\, dz \} \quad .$$

Then we get $<l>$ from the Heisenberg relation: $<l> \approx (\hbar/2)T(<|p|>/\sigma_p)\, \sigma_x{}^{-1}$. This is illustrated in Fig. 4.2. It is clearly seen from this equation that :

*For $Y<1$, $<|p|> \approx \sigma_p$. For example in the Gaussian case and in the limit of the particle at rest (i.e. $<p>=0$), we get $<|p|> \approx \sqrt{2/\pi}.\sigma_p \approx 0.8\,\sigma_p$. This corresponds to a fractal dimension 2.

*For $Y>1$, $<|p|> \approx <p>$, independent of Δx. The topological and fractal dimensions now coincide, both being equal to 1; in other words the behaviour is no longer fractal. The transition occurs for $<p>/\sigma_p \approx 1$, i.e., for $\Delta x = (\hbar/2).<p>^{-1}$, which is half the de Broglie length.

The constraint $<\Delta l> >> \Delta x$ in the Abbott-Wise calculation was set in order to compute $<\Delta l>$ as $\int |x| \; |\psi(x)|^2 \, d^3x$. This is an integral of the same kind as that used to compute $<|p|>$, so the condition comes in fact from the same transitional behaviour as that described by the hereabove equation. Due to the rapidity of the transition, the limiting case $<\Delta l> \approx 2\Delta x$ already corresponds to the Abbott-Wise condition $<\Delta l> >> \Delta x$ (indeed for Y=2 the ratio $<|p|>/\sigma_p$ takes the value 2.016), so that the Abbott-Wise result still holds in this case, as announced above.

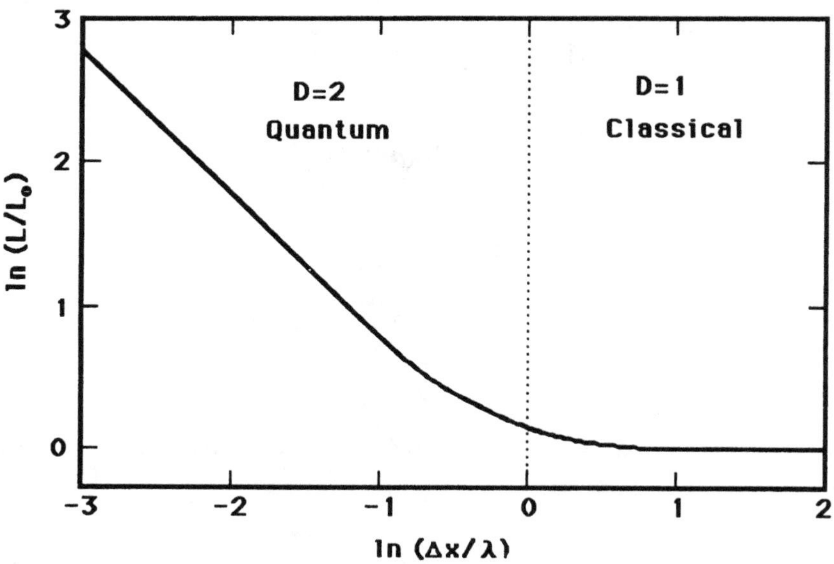

Figure 4.2. The fractal/non-fractal transition around the de Broglie length $\lambda=\hbar/p$, which is identified with the quantum/classical transition for a particle. As a consequence of Heisenberg's relation, the length of a quantum path diverges as $L \propto \delta x^{-1}$ in the quantum regime $\delta x<\lambda$.

This result suggests the following remarks:

(i) Clearly one *cannot* reduce even the measured properties of a quantum mechanical path to a simple fractal curve. Successive position measurements with decreasing Δx on a fractal curve allow one to approximate it more and more precisely. This is not the case in quantum mechanics, where each new measurement is obtained by an interaction of the particle with a measurement apparatus, and the smaller the precision, the larger is the perturbation. We do not measure in quantum mechanics a pre-existing fractal curve, since the trajectory is in part defined by interaction with the apparatus.

(ii) Moreover, even for a given precision, the relation $<l> \approx m^{-1} \hbar T(\Delta x)^{-1}$ is obtained as a mean over a set of successive experiments of N measurements *under the same initial conditions*. Each experiment will yield a *different* trajectory. This will remain true even if one keeps only those experiments in which the initial positions and the final positions respectively coincide (thus simulating the problem of geodesics between two given points). Attributing a fractal dimension to a quantum "trajectory" does not mean coming back to determinism: fractal dimension 2 characterizes any one of the infinite number of possible quantum paths. As in Feynman's approach, we may consider, study and analyse the possible paths, then attribute to them universal properties, if any, without pretending that a well determined path has been followed by the particle.

(iii) The difference between a well defined fractal curve and a quantum mechanical path is seen to be even stronger when one realizes that the $(\Delta x)^{-1}$ divergence is obtained for a fractal curve only when joining together balls of radius Δx (i.e. one has necessarily $<\Delta l> \approx 2.\Delta x$), while it is obtained whatever the $(<\Delta l>, \Delta x)$ relation is in quantum mechanics: i.e., when $<\Delta l> >> \Delta x$, the various positions where the particle is detected are connected by linear, non-fractal segments of diverging lengths. We may attribute this difference to the measurement process. One should, in quantum mechanics, make the difference between one unique measurement made on a prealably prepared system and two or successive measurements made on a given system.

In the first type of experiment, one prepares the system, then makes one measurement. Then one prepares again the system under exactly the same conditions (as far as knowable) as previously, then makes again the same measurement. After n similar experiments, one may study the statistics of the measurement results.

In the second type of experiments, one makes successive measurements on the same system without "resetting" the initial conditions between each measurement. In that case it is clear that the perturbation brought to the system by the measurement $n-1$ (and all previous ones) should be accounted for in our understanding of the results of the measurement *n*.

Our postulate in the present approach is that a single measurement indeed brings us informations on the unperturbed system. The fact that after the measurement the system is definitely perturbed and the precise way it is perturbed are tentatively attributed to those internal structures of the system which are tested by a single measurement (see Fig. 5.3). Those properties are shared by the system and the measurement apparatus, which also has quantum properties. We shall come back in more detail on this question in Chapter 5, where it will be shown how the fractal model allows one to understand the results of the above two types of experiments.

(iv) The $(\Delta x)^{-1}$ divergence would be given by a "trajectory" in 1-dimension as well: this shows that each of the three spatial coordinates of the particle trajectory is characterized by a fractal dimension 2. This agrees with the results of Chapter 3. Start with a fractal curve of dimension D drawn in a N dimensional space, then each of its coordinates $X_i(s)$ are fractal functions of the same dimension in terms of a curvilinear coordinate s on the fractal.

Our conjecture here is that this observed behaviour of *classical length* measurements on quantum particles is the external manifestation of underlying fractal variables, and that all these differences between fractals and quantum mechanics will be understood when comparing a quantum mechanical path, not to a particular and well-defined fractal curve, but *to the geodesics of a fractal space-time* (see Chapter 5). In this framework, the classical-quantum dualism cannot be avoided, since it is interpreted here as the manifestation of a fundamental transition of physics from a fractal to a non-fractal nature of space-time: we shall also come back to this point in Sec. 5.7.

4.3. Fractal Dimension of A Quantum Relativistic Path.

As explained above, the fractal dimension of a quantum mechanical path derives from the $<|v|>(\Delta x)$ relation, which is itself the result of two relations in the nonrelativistic case:

(i) the $<|p|>(\sigma_p)$ relation: it is easy to show that there exists σ_z such that $<|p|>^2 = \sigma_p{}^2 - \sigma_z^2$ when $<p> = 0$, so that $<|p|> \leq \sigma_p$ in the quantum (non-classical) case;
(ii) the $\sigma_p(\sigma_x)$ relation, always constrained by the Heisenberg inequality $\sigma_p.\sigma_x \geq \hbar/2$. Now, one should ask oneself what happens in the special relativistic situation. Relativistic effects can manifest themselves in two ways. The first and straightforward one is when the velocity of the particle is assumed to be close to the light velocity. The second one is a consequence of the Heisenberg relation. Indeed if we try to measure the position of a particle with a resolution $\Delta x < \hbar/mc$, the dispersion of momentum measurements will become larger than $\approx mc$, thus implying relativistic velocities for the particle.

Let us illustrate this point by a detailed example of such a measurement. Assume that someone wants to measure the position of an electron at rest by using photons of wavelength λ (i.e. we look at them with a microscope). Let θ be the diffusion angle, p the photon momentum before diffusion, and p', after diffusion when its wavelength becomes $\lambda + \Delta\lambda$, and P the electron momentum after the interaction. Letting $\Delta\lambda$ be as large as we want, let us define the measurement precision as $\Delta x = \sqrt{\lambda(\lambda+\Delta\lambda)}/sin(\theta/2)$ (see e.g. Ref. 7). The Compton effect formula yields for the photon

$$\Delta\lambda = 2\,(h/mc)\,sin^2(\theta/2) \quad .$$

The electron momentum is given by the relation

$$P^2 = p^2 + p'^2 - 2\,p\,p'\,cos\theta \quad .$$

Then P may be expressed exactly in terms of Δx and of the (non-reduced) Compton wavelength $\lambda_c = h/mc$:

$$P = (2\,h\,/\,\Delta x)\,[\;1 + (\lambda_c\,/\,\Delta x)^2\;]^{1/2} \quad . \qquad (4.3.1)$$

Then, while one finds again $P \propto \Delta x^{-1}$ (corresponding to fractal dimension 2) for $\Delta x > \lambda_c$, a new result is obtained when $\Delta x < \lambda_c$: $P \propto \Delta x^{-2}$, with a new fast transition between the quantum mechanical and special relativistic quantum mechanical cases. This result agrees with the remark by Landau and Lifchitz[11] that in the rest frame of an electron the minimum error on the measurement of its coordinates is $\Delta x \approx h/mc$, beyond which the notion of effective measurement of a precise position loses its physical meaning.

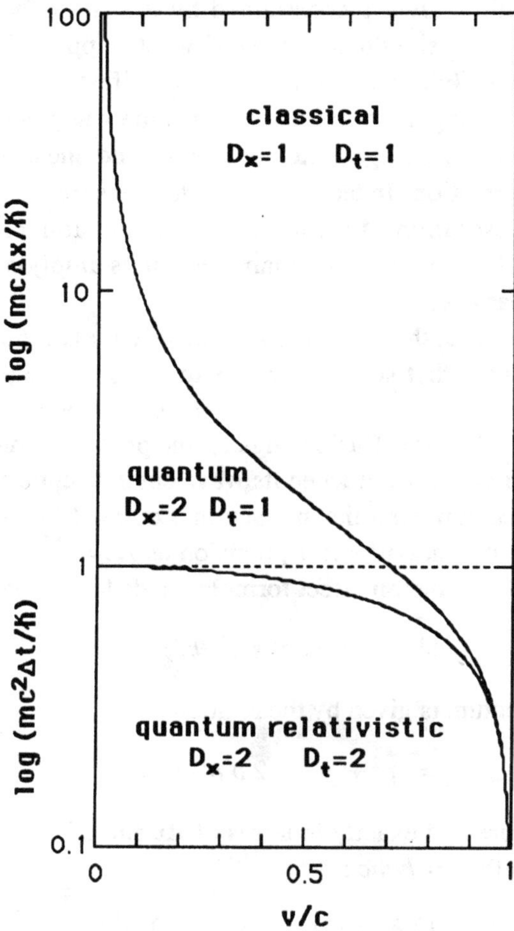

Figure 4.3. The classical / quantum and quantum / quantum-relativistic transitions in a doubly relativistic diagram (scale and motion). The first transition occurs around a resolution given by the de Broglie length $\lambda_{dB}=(\hbar/mc)(c^2/v^2-1)^{1/2}$. There, the spatial coordinate jumps from a fractal dimension D=2 to the non-fractal dimension D=1 when Δx increases. The second transition occurs around $c\tau_{dB}=(\hbar/mc)(1-v^2/c^2)^{1/2}$: it corresponds to a transition of the dimension of the temporal coordinate, that jumps from D=2 to D=1 when Δt increases.

If one now places oneself in a frame where the electron moves with energy E and velocity v, special-relativistic length contraction occurs and λ_c should be replaced by $\lambda_c(1-v^2/c^2)^{1/2}$, in agreement with the fact that now, as stated in Ref. 11, the position cannot be measured more precisely than $hc/E = (h/mc)\ (1-v^2/c^2)^{1/2}$. This is illustrated in Fig. 4.3, where the transitions from Classical to Quantum, then from Quantum to Relativistic Quantum are drawn in the plane $(ln\Delta x, v/c)$. Note that in our scale-relativistic approach, this plane is the space of states of reference systems: it sums up the lowest order information on the two fundamental states, the state of motion, through the velocity v, and the state of scale through the resolution $ln\Delta x$.

How is this new transition translated in terms of fractal dimension? From the result of (4.3.1), we briefly suggested in Ref. 12 that this loss of the notion of position should be described by a new transition to the fractal dimension 3. Cannata and Ferrari[13] have addressed this problem and come to the conclusion that the new transition occurred (i) from dimension 2 to dimension 1 and (ii) around the Compton wavelength. Such a result would neither be covariant, nor account for the additional increase of uncertainty in relativistic quantum mechanics. We proposed in Ref. 14 a new interpretation differing from these two earlier ones, which we shall recall hereafter.

Let us first demonstrate that the transition does not occur around the Compton wavelength $\lambda_c = \hbar/mc$, but instead around the length $c\tau$, where $\tau = \hbar/E$ is the de Broglie time of the particle. A simple argument seeming to yield the result that the trajectory is no more fractal in the relativistic domain is that, when the particle's velocity becomes close to the light velocity, clearly the length it travels can no more increase. Let us make this statement more specific. The velocity of a particle is given in terms of its momentum by

$$v = p\,c\,/\,(\,p^2 + m^2 c^2\,)^{1/2} \quad ,$$

so that its average absolute velocity is

$$<|v|> \;=\; \int \frac{|p|\,c\ \ |\varphi(p)|^2\ dp}{(\,p^2 + m^2 c^2\,)^{1/2}} \quad . \tag{4.3.2}$$

We remain in the one-dimensional case for simplicity. Following Refs. 4 and 13, let us describe $|\varphi(p)|^2$ by a step function: $|\varphi(p)|^2 = (2\Delta p)^{-1}$ for $|p-p_0| < \Delta p$ and 0 otherwise (but the final result does not depend on this particular choice).

*When $p_0 > \Delta p$ (i.e. $\Delta x > \lambda$, since $\Delta x.\Delta p \approx \hbar$: this is the "classical" domain), one gets

$$<|v/c|> = \left\{[(p_0+\Delta p)^2 + m^2c^2]^{1/2} - [(p_0-\Delta p)^2 + m^2c^2]^{1/2}\right\} / 2\Delta p \ ,$$

which reduces to either $\approx p_0/mc$ if $p_0 << mc$, or ≈ 1 if $p_0 > mc$. This yields in both cases the classical (respectively nonrelativistic and relativistic) non-fractal result D=1.

*When $p_0 < \Delta p$ (i.e., $\Delta x < \lambda$: "quantum" domain), one gets

$$<|v/c|> = \left\{[(p_0+\Delta p)^2+m^2c^2]^{1/2} + [(p_0-\Delta p)^2+m^2c^2]^{1/2} - 2mc\right\} / 2\Delta p \ . \quad (4.3.3)$$

Let us introduce the energy $E_0=(p_0^2c^2+m^2c^4)^{1/2}$. Equation (4.3.3) becomes

$$<|v/c|> = \{[E_0^2/c^2 + \Delta p^2 + 2\Delta p\, p_0]^{1/2} + [E_0^2/c^2 + \Delta p^2 - 2\Delta p\, p_0]^{1/2} - 2mc\} / 2\Delta p \ . \quad (4.3.4)$$

Consider first the case $\Delta p << E_0/c$ (i.e., $\Delta x >> c\tau$: quantum non-relativistic domain). Since we have $p_0 < \Delta p$, this is compatible only with $p_0 << mc$. Expanding Eq. (4.3.4) yields

$$<|v/c|> \approx (E_0 - mc^2) / c\Delta p + c\Delta p / 2E_0 \ .$$

But $(E_0-mc^2)/ c\Delta p \approx p_0^2/2mc\,\Delta p < \Delta p / 2mc \approx c\Delta p / 2E_0$, so that the dominant term is the second one, and one gets the fractal dimension 2 behaviour:

$$<|v/c|> \approx c\tau / \Delta x \quad \text{when} \quad \Delta x >> c\tau \ .$$

On the other hand, when $\Delta p > E_0/c$ (quantum relativistic domain), one gets

$$<|v/c|> \approx 1 - mc / \Delta p \approx 1 \ .$$

As announced above an apparent transition from D=2 to D=1 is indeed found, but around $c\tau$, where $\tau = \hbar/E$ is the de Broglie time, instead of around the Compton length λ_c: they are identical only in the rest frame of the particle, see Fig. 4.3. This result is now consistent with the requirement of Lorentz covariance. However we believe that this transition from fractal behaviour to a standard one in the quantum relativistic case is only apparent and should be interpreted as an artefact of another phenomenon (which we analyse in Sec. 5.3), for the following reasons.

Contrary to what is suggested by this result (from which it seems that one comes back to standard non-fractal space in the relativistic domain), both experiments and theory demonstrate that physics shows a still radically new behaviour beyond the de Broglie time. The transition found hereabove in the case of the Compton effect is an example of this increase of complexity of physics for smaller resolutions. The fundamental phenomenon of physics which occurs in the relativistic domain is the spontaneous creation-annihilation of particle-antiparticle pairs. As recalled in Ref. 11, the loss of the notion of position comes from the fact that if one wants to measure the position of a particle with a precision $\Delta x \approx \hbar/mc$ in its rest frame, the Heisenberg relation implies that the uncertainty on momentum will become of the order of the threshold for pair creation. In any rest frame, the limit of position measurement becomes[11] $\Delta x \approx \hbar c/E$, i.e., precisely the transition value we obtained above. Then a complete understanding of what really occurs around the de Broglie time can be reached only when accounting for the variable number of particles (as remarked by Cannata and Ferrari[13]), i.e., in our approach for the not only spatial but also temporal fractal structure of quantum particle paths.[12,14] Let us now address this problem.

4.4. Fractal Dimension of Time.

Nonrelativistic quantum time.

Consider first a time measurement for a nonrelativistic quantum system. Assume one measures for *fixed space intervals* δx the time interval Δt separating two measurements. The average value of Δt will be given by $\langle \Delta t \rangle \propto \delta x / \langle |p| \rangle$, and the total time measurement T will be proportional to $\langle \Delta t \rangle$. The time-energy Heisenberg relation writes $\delta E . \delta t \approx \hbar$.

Considering now that $E = p^2/2m$ for a free particle, the uncertainty relation becomes $p.\delta p \propto \hbar\ \delta t^{-1}$. But for a system nearly at "rest" ($<p>\approx 0$), we have already recalled that $<|p|>\approx\delta p$ in the fully quantum case, so that we expect $<|p|>^2 \propto \delta t^{-1}$ and we finally get

$$T \propto <\Delta t> \propto \delta x\ \delta t^{1/2} .$$

Let us suggest an interpretation for this result. A classical observer using a classical clock tends to measure time intervals with his own definition of time, which is a physical quantity of topological dimension 1. If the quantum mechanical time is actually of fractal dimension D, the measurements will yield $T \propto \delta t^{1-D}$. This might be more easily understood when remembering the two possible building processes of a fractal object.[15,16] A fractal curve may be obtained as well by adding segments as by deleting surfaces: It is characterized by an infinite length and an infinitesimal surface. Both may be effectively written in terms of Non-Standard numbers (Ref. 15 and Chapter 3). In the same way, a fractal dust is characterized by an infinite number of points, and also by an infinitesimal "length". Then it is suggested that the hereabove equation should be interpreted as demonstrating that time in nonrelativistic QM may *behave in some situations as a fractal dust of fractal dimension* 1/2, whose topological dimension is then zero.

Such a result may be related to *the quantification of energy* of bounded states and to *quantum jumps*, since energy is precisely the conservative quantity which derives from the uniformity of time: note that, indeed, in the above argument we have taken $<|p|>\approx\delta p$, which corresponds to the fully quantum case. Breaking of the uniformity of time is expected to change the behaviour of energy, and conversely the different behaviour of energy in closed or bounded systems relative to its classical behaviour is expected to lead to non-classical properties of the time coordinate itself. Let us explicitly demonstrate this connection in the case of hydrogenoid atoms (for which we recall that a similarity dimension of 2 may be obtained).

The energy spectrum of hydrogenoid atoms is given, neglecting fine structure and up to a negative constant, by

$$E = \frac{1}{n^2} \quad ,$$

which varies between 0 and 1 for the principal quantum number n varying between ∞ and 1. Consider the interval of energy E, $E+\Delta E$. We may define two integers m and p such that

$$E \approx \frac{1}{p^2} \quad , \qquad E + \Delta E \approx \frac{1}{m^2}$$

so that $p \approx E^{-1/2}$ and $m \approx E^{-1/2}(1 - \frac{1}{2}\frac{\Delta E}{E})$. Then the number of energy levels in the interval ΔE is $\mathcal{N}(E,\Delta E) = p - m$, given by

$$\mathcal{N}(E,\Delta E) \approx \frac{\Delta E}{E^{3/2}} \quad .$$

What is the *average* number on the full energy range ? Asked in such a standard way, this question has no answer, since it involves the integral

$$\int_0^1 E^{-3/2}\, dE \quad ,$$

which is divergent. So let us ask it in our perspective of a scale-dependent physics. In this frame the resolution ΔE is a smoothing ball inside which more detailed information keeps no physical meaning. If one wants to get more information, one must *change* the resolution. Then for a given resolution ΔE, pushing the integration up to 0 is meaningless: it can be stopped at ΔE itself. This yields

$$\mathcal{N}(\Delta E) \approx \Delta E \int_{\Delta E}^{1} E^{-3/2}\, dE \approx (\Delta E)^{1/2} \quad .$$

In this method we have analysed the average variation of the content of our "object" (the energy spectrum) into balls of varying sizes: this method would yield a power 1 for a line, 2 for a surface, etc..., so that we have actually found an average fractal dimension 1/2 for the energy spectrum, the same dimension as was found above for time measurements.

Before jumping to the quantum relativistic case, it should be remarked that the time-energy Heisenberg relation in the nonrelativistic case may be considered as a simple consequence of the momentum-position relation. Indeed one may calculate $\Delta E.\Delta t$ as $\Delta(p^2/2m).\Delta t = \Delta p.p\Delta t/m \approx \Delta p.\Delta x \leq \hbar$. It is also recalled that the quantification of energy in atoms is basically a consequence of a stationary phase condition bearing on momentum, $\oint p\, dr = n\, h$. Then we suggest[14] that the above result ($D_t = 1/2$) is but a consequence of the *spatial* transition to the fractal dimension 2.

Dimension of time in Quantum Electrodynamics.

Let us assume that one wants to measure the energy of an electron during a time interval Δt much smaller than its Compton time $\tau = \hbar/mc^2$. The Heisenberg relation implies that such a measurement will cause an energy fluctuation $\Delta E \approx \hbar\Delta t$, so that a number of n electron-positron pairs may be created, with $\Delta E \approx 2n\ mc^2$ and a total number of particles $2n + 1 \approx (E+\Delta E)/mc^2$. These fluctuations are supposed to occur even in the absence of any measurement, and are interpreted in present Quantum Electrodynamics via Feynman's diagrams in terms of virtual electron-positron pairs and virtual photons. However when considering a free electron, these virtual particles are considered to be a part of the nature of the electron itself, since they contribute to defining its proper self-energy.

In our present approach, we aim at representing the electron properties as the manifestations of the fractal structure of its spacetime trajectory. Along the road of the Feynman-Wheeler-Stueckelberg ideas,[17] let us assume that the virtual electrons and positrons are actually manifestations of the single real electron, more precisely of the ability of its space-time trajectory to *return into the past for very small time intervals* (a more complete argument will be presented in Chapter 5). As seen in Fig. 4.4, provided there is only one particle in the initial and final states, and *re-interpreting the particle going back in time as an antiparticle*, any continuous and self-avoiding trajectory will give rise to an apparent number of n+1 particles and n antiparticles, this interpretation being imposed by the prescription of (macroscopic) causality.

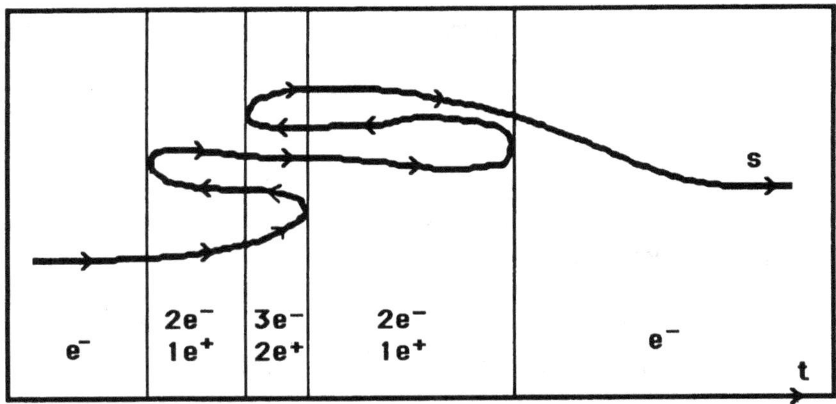

Figure 4.4. In Feynman's interpretation of the trajectory of an electron which is allowed to run back in time, the parts of the trajectory where the electron runs back in time are interpreted as antiparticles. The length travelled by the electron and the proper time elapsed is obtained by adding the lengths and times of each of the various particles and anti-particles. Note that the number of particles + antiparticles is always odd.

But if these $2n+1$ particles are actually manifestations of the same initial object, a correct calculation of the length it travelled and the proper time elapsed can only be done *by adding the lengths and times attributed to each individual particle*. If we look during a total classical time T_0 at this trajectory with a time resolution Δt, then $2n \approx \Delta E/mc^2 \approx \hbar/(mc^2 \Delta t)$ virtual particles will be counted, and the total proper time elapsed becomes

$$T = (2n+1)\, T_0 \approx [(E+\Delta E)/mc^2]\, T_0 \approx (1+ \tau/\Delta t)\, T_0 \ . \qquad (4.4.1)$$

Hence, while it is independent of resolution for $\Delta t > \tau$, it diverges as Δt^{-1} for $\Delta t < \tau$. We interpret this result as the existence of a *temporal transition from dimension* 1 (non-fractal) *to fractal dimension* 2 around the de Broglie time $\tau = \hbar/E$.

Consider now the measurement of length. It has been seen in the previous section that there is an apparent transition beyond $c\tau$ from fractal dimension 2 to dimension 1, due to the fact that particles show an ultrarelativistic behaviour ($v \approx c$) for these resolutions. At these scales spatial and temporal resolutions will be related by $\Delta x \approx c\Delta t$. But when accounting for the creation-annihilation of virtual particles reinterpreted as several aspects of the same particle, the hereabove argument made for time also holds for length, and the total length travelled by the particle is

$$L = (2n+1)\, L_0 \approx (1+\Delta E/mc^2)L_0 \approx (1+ c\tau/\Delta x)\, L_0 \ , \qquad (4.4.2)$$

so that we find again dimension 2 for $\Delta x < c\tau$, since we know from Sec. 4.3 that L_0 becomes of dimension 1 (in the uncomplete argument) precisely at the same transitional value, $\Delta x < c\tau$.

We suggest to interpret this result in the following way: There is actually no spatial transition around resolution $c\tau$, while there is a new temporal transition from standard time (D=1) to fractal time of dimension D=2 around the de Broglie time τ. The apparent spatial transition from D=2 to D=1 was only an artefact of not having properly accounted for the temporal transition. It is remarkable that under this interpretation, we get a very simple and Lorentz covariant view of the fractal character of trajectories in quantum physics:

Spatial and temporal coordinates now show the same behaviour, i.e., a transition from D=1 to D=2 around $\lambda^i = \hbar/p^i$, i=1,2,3, in the spatial case and around $\tau = \hbar/E$ in the temporal case. In other words the transitional resolutions, which are the de Broglie periods reinterpreted as structural constants of the fractal trajectories aimed at describing a free particle, are given by the inverse of the 4 components of the energy-momentum 4-vector, $\lambda^\mu = \hbar/p^\mu$, $\mu = 0$ to 3. Though λ^μ is not a vector, its transformation law in a Lorentz transform is easily derived from its definition. Confirmation of these results by different methods is presented in subsequent sections.

One of the interesting consequences of this interpretation is that it precisely accounts for the Zitterbewegung, this oscillatory motion of the center of mass of an electron resulting from the Dirac equation. This effect is known to be the result of interactions between the negative energy and positive energy solutions of the Dirac equation. Though it indeed disappears if one keeps only the positive energy solutions to describe an electron, this does not yield a satisfactory solution to the problem, since such a positive energy electron would be completely delocalized.[18] If, on the other hand, one fixes the localization assumed (or measured) for the electron, $\Delta x = c\Delta t = \hbar c/E$, one can then deduce the relative rate of positive and negative energy solutions. One gets

$$P_-/P_+ \approx [pc/(E+mc^2)]^2 = (E-mc^2)/(E+mc^2) \quad . \qquad (4.4.3)$$

Now in the fractal model, for each classical time interval we have $2n+1$ segments, $n+1$ running forward and n running backward, so that with $E=(2n+1)mc^2$, one gets $P_-/P_+ = n/(n+1) = (E-mc^2)/(E+mc^2)$, i.e., *exactly the quantum electrodynamical result.* This opens the possibility to construct fractal, localized, one-particle solutions of the Dirac equation, as we shall see in more detail in Sec. 5.9.

4 References

1. Feynman, R.P., 1948, *Rev. Mod. Phys.* **20**, 367.
2. Feynman, R.P., & Hibbs, A.R., *Quantum Mechanics and Path Integrals* (MacGraw-Hill, 1965).
3. Dirac, P.A.M., quoted in Feynman, R.P., Nobel lecture, *Physics Today,* (August 1966), p. 31.
4. Abbott, L.F., & Wise, M.B., 1981, *Am. J. Phys.* **49**, 37.
5. Le Méhauté, A., *Les Géométries Fractales* (Hermès, Paris,1990).
6. Landau, L., & Lifchitz, E., 1972, *Quantum Mechanics* (Mir, Moscow).
7. Messiah, A., *Mécanique Quantique*, (Dunod, Paris, 1959).
8. Campesino-Romeo, E., D'Olivo, J.C., Socolovsky, M., 1982, *Phys. Lett.* **89A**, 321.
9. Allen, A.D., 1983, *Speculations in Science and Technology* **6**, 165.
10. Kraemmer, A.B., Nielson, H.B., & Tze, H.C., 1974, *Nucl. Phys.* **B81**, 145.
11. Landau, L., & Lifchitz, E., 1972, *Relativistic Quantum Theory* (Mir, Moscow).
12. Nottale, L., 1988, *C. R. Acad. Sci. Paris* **306**, 341.
13. Cannata, F., & Ferrari, L.,1988, *Am. J. Phys.* **56**, 721.
14. Nottale, L., 1989, *Int. J. Mod. Phys.* **A4**, 5047.
15. Nottale, L., & Schneider, J., 1984, *J. Math. Phys.* **25**, 1296.
16. Mandelbrot, B., *The Fractal Geometry of Nature* (Freeman, San Francisco, 1982).
17. Feynman, R.P., Nobel lecture, in *Physics Today,* August 1966, p. 31.
18. Bjorken, J.D., & Drell, S.D., *Relativistic Quantum Mechanics* (MacGraw-Hill, NewYork, 1964).

Chapter 5

THE FRACTAL STRUCTURE OF QUANTUM SPACE-TIME

5.1. On Quantum Properties of Fractals.

Chapter 4 was devoted to the demonstration that quantum systems have fractal non-differentiable properties. In this new chapter, we shall attempt to push further the fractal approach (up to the demonstration that it yields Schrödinger's equation). The present section aims at introducing this subject by showing that, already at a very rough level, fractal objects own quasi-quantum properties. It is from this observation that fractal geometry may be recognized as a new hope to understand the quantum world at the level of fundamental principles. We shall only give two simple intuitive examples; more specific demonstrations will be found in subsequent sections.

Consider first the well-known fractal of generator –⌉⌊–, which we plot in Fig. 5.1 at the inclination of 45°. We know that this fractal is non-differentiable. However on any of its approximations a slope may be defined, which takes only two possible values, +1 and −1. Assume that we want to measure explicitly the slope at a given point of the curve. If this was possible to infinite precision, no slope could be obtained in the classical sense. But owing to the fact that an effective measurement is made with a non-vanishing resolution, a slope will be measured, which takes arbitrarily the values +1 and −1, with the only stable and predictable property $P(-1)=P(+1)=1/2$.

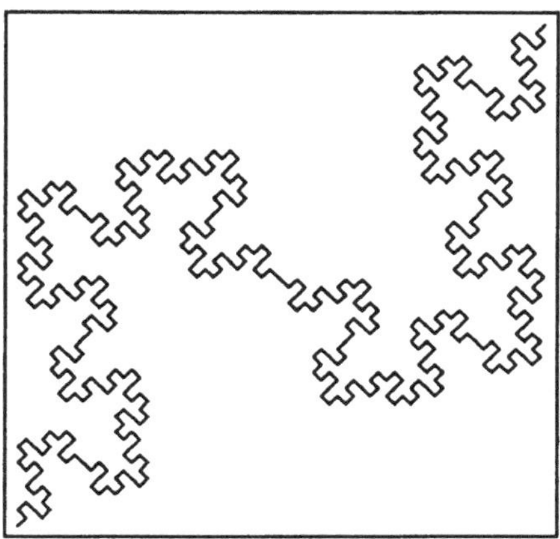

Figure 5.1. The fractal whose generator is given in Fig. 3.5, plotted at an inclination of 45° at the approximation F_3. Whatever the resolution, the slope of any segment is always either +1 or −1.

Note that even for the limiting points of the fractal which are defined by infinite sequences, a kind of generalized "slope" may be defined, given by the list of the various slopes at all resolutions, e.g. s_i ={1,-1,-1,-1,1,1, 1,-1,1,-1,1,-1, ...}, i=1,∞. There too, each of these sequences is characterized by $P(-1)=P(+1)=1/2$.

Our second example uses the Peano curve. Consider two different inclinations, 0° for case (a), and 45° for case (b), of a periodical infinite Peano curve. The probability of presence along axis Ox of a particle travelling on the fractal is given by its fractal derivative. In case (a) the distribution of probability is periodical, while it is constant in case (b), which differs from case (a) only by a rotation of 45°. The total probability is the same in both cases. Actually the probability distribution of case (b) may as well be obtained from the fractal of (a), by projection along a direction 45° from the Ox axis (see Fig. 5.2). This offers a simple model for going from an interference pattern to a "classical" pattern by a simple change of the coordinate system.

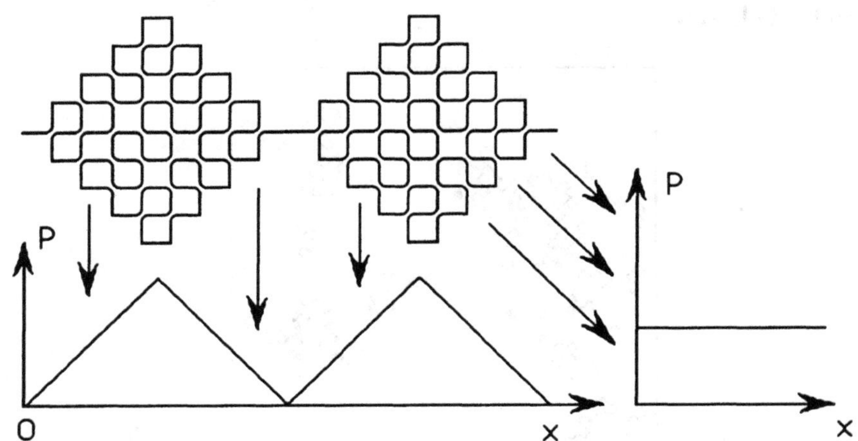

Figure 5.2. The probability of presence, $P(x)$, of a particle following a Peano curve, for two different orientations of the direction of projection.

5.2. Derivation of Heisenberg's Relations in Fractal Space-Time.

An intermediate solution to the problem of a description of the internal fractal structure of space-time consists in attributing any such internal structure to each individual coordinate separately. As already explained in Chapters 2 and 4, it is conjectured that the classical coordinate x^μ results from the smoothing with "4-balls" $\Delta x^\mu > \lambda^\mu$ of an internal fractal structure of dimension D=2 along which a fractal curvilinear coordinate X^μ is defined (see Fig. 5.3). As suggested in Ref. 3, this structure may be visualized by introducing extra-dimensions x'^μ, x''^μ, etc., though we assume that the whole physical information is contained in the properties of x^μ, X^μ, and of their relations. These properties are partly described by the concept of fractal derivatives $đX^\mu/dx^\mu$ (Chapter 3). However we shall attempt, to the best of our ability, to work in a pure 4-dimensional space-time continuum (in terms of topological dimension). A fractal structure is indeed able to provide us with non-classical quantities which may appear to us as internal (see for example the fractal model of spin presented in Sec. 5.4). Note that, if one works with *differentiable* curvilinear coordinates, one will not be able to describe a fractal trajectory by following the evolution of only one coordinate; however complicated the system of coordinate, beyond some resolution the fractal trajectory will always show extensions in the other dimensions: this means that, for each coordinate, the 3 additional space-time coordinates will play at some level the role of extra-dimensions.

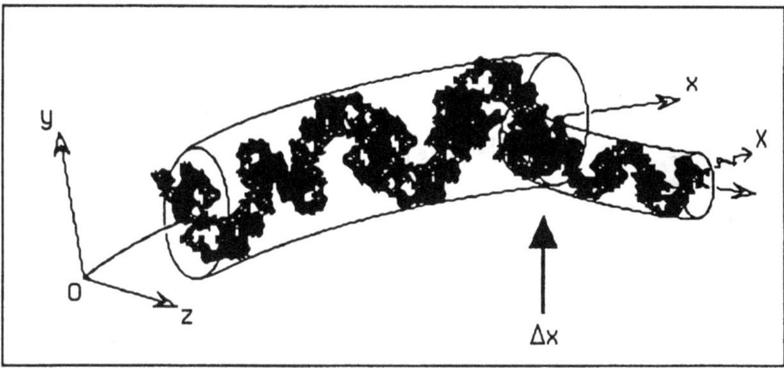

Figure 5.3. Schematic representation of classical and fractal coordinates. The classical coordinate (x) is assumed to have an internal fractal structure described by a fractal curvilinear coordinate (X). The width of the classical "tube" is of the order of the de Broglie length λ_{dB}. In the case of an effective measurement of position with a resolution $\Delta x < \lambda_{dB}$ (arrow), the internal srtucture is revealed by an "unfolding" of the initial fractal. The new trajectory is contained in a new tube of width $\approx \lambda'_{dB} \approx \Delta x$, as a consequence of Heisenberg's relation. Its new direction is unpredictable (diffraction effects) because of the non-differentiable character of the internal structure.

Fractal structure of trajectories and Heisenberg's relations.

Let us assume that a (normalized) metric invariant s can be defined along any fractal trajectory, in particular along the geodesics of the fractal space-time. It should be identified with the fractal proper time of the particle, which flows uniformly by definition. The 4 classical coordinates may now be followed in terms of s: We get 4 fractal functions $x^{\mu}(s, \Delta x^{\mu})$ (see Figs. 3.7 and 5.4) which become classical functions for $\Delta x^{\mu} > \lambda^{\mu}$. The actually observed fractal trajectory $[x(t), y(t), z(t)]$ is then assumed to be the external manifestation of one geodesic among an infinite family of geodesics, $[x(s), y(s), z(s), t(s)]_i$. The universal fractal dimension 2 found for each coordinate of a quantum particle is considered as a manifestation of a universal dimension 2 of such geodesics.

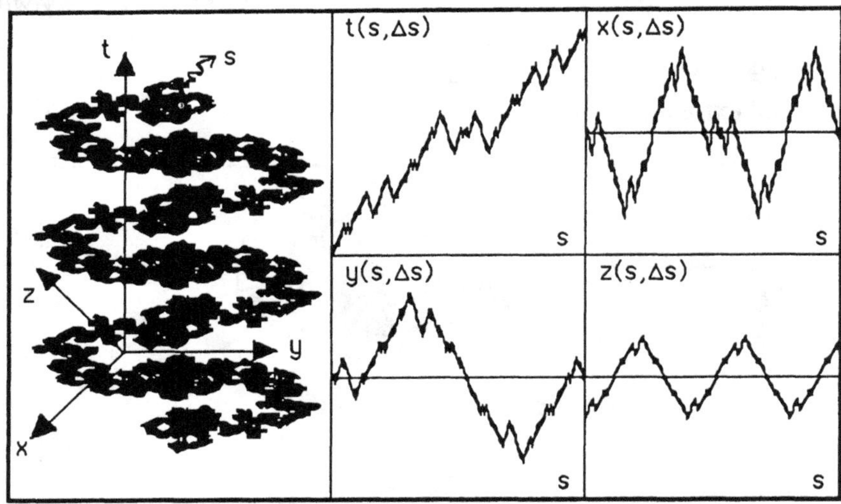

Figure 5.4. Schematic representation of a fractal curve in space-time. The evolution of its coordinates is described by four fractal functions of the normalized curvilinear coordinate s intrinsic to the fractal curve. The case shown would correspond to a particle seen in its classical rest frame ($<v>=0$).

In the previous chapter, the fractal character of particle trajectories and their dimension were deduced from the Heisenberg relations. We will now reverse the argument and show how the Heisenberg relations may be derived from the postulate of the fractal substructure of space-time. Consider the action of a system as a function of space-time coordinates, $S(x^{\mu})$, $\mu=0$ to 3. It is well known that the system momentum and energy are given by its partial derivatives, i.e., in 4-dimensional formalism, $p_{\mu} = -\partial S/\partial x^{\mu}$.[1] In classical and relativistic mechanics, the invariance of this 4-vector results from the homogeneity of space-time (i.e., homogeneity of space and uniformity of time).[1,2] This fundamental symmetry of physics is clearly broken by the assumed fractal structures beyond the transitional resolution λ^μ. Uniformity is again found only at the infinitesimal level on the fractal itself, i.e., for the (ideal) fractal coordinates X^μ and in the classical regime $\Delta x^\mu > \lambda^\mu$. In between, we expect the laws of conservation of momentum and of energy to break down. If the relation between x^μ and X^μ was monotonic and one to one, we might write $p_\mu = -\partial S/\partial x^\mu = -(\partial S/\partial X^\mu).(dX^\mu/dx^\mu)$. However, due to the essential property of the fractal to run backward with respect to the classical coordinate, there is an infinity of values $\{X_i^\mu\}$ of X^μ for a given value of x^μ (what we have called the derived fractal) so that we suggest that (dX^μ/dx^μ) should in that case be replaced by the fractal derivative $đX^\mu/dx^\mu = \Sigma_i\,(dX_i^\mu/dx^\mu)$, thus we get

$$p_\mu = -(\partial S/\partial x^\mu) = -(\partial S/\partial X^\mu) \quad (đX^\mu/dx^\mu) \quad .$$

Since X^μ is now a uniform variable, we recover $(\partial S/\partial X^\mu)$ as an invariant 4-vector:

$$(\partial S/\partial X^\mu) = -p_\mu^o \quad .$$

As seen in Chapter 3, the fractal derivative is itself a fractal function which may be decomposed into a finite part and a part diverging with decreasing resolution,

$$đX^\mu/dx^\mu = \xi^\mu(x^\mu, \Delta x^\mu) \; (\Delta x^\mu/\lambda^\mu)^{1-D} \quad ,$$

where $\xi^\mu(x^\mu, \lambda^\mu) = 1$. We now assume that several measurements of 4-momentum are made, yielding a dispersion of results σ_{p^μ}. We expect to obtain the following result:[3]

$$\sigma_{p^\mu} = p_\mu^o \quad \sigma_{\xi^\mu} \quad (\Delta x^\mu/\lambda^\mu)^{1-D} \quad . \tag{5.2.1}$$

Assume now that Δx^μ is defined in such a way that it will be related to the dispersion of position results σ_x in a series of measurements. In general the precise relation between Δx and σ_x will depend on each experimental process: Δx might be to the first approximation related to the error bars, i.e., $\Delta x = 4\ \sigma_x$ for a 2σ error, $\Delta x = 6\ \sigma_x$ for a 3σ error... We then set $\Delta x = K\ \sigma_x$ and get $\sigma_p\, \sigma_x = (p_o\, \lambda)\ (K\ \sigma_\xi)$. The momentum-position and time-energy Heisenberg relations are then finally derived in their exact form provided:[3]

$$D = 2 \ , \tag{5.2.2}$$

$$p_o^\mu \ \lambda^\mu = h \ , \tag{5.2.3}$$

$$\Delta\xi^\mu (x^\mu, \Delta x^\mu) \ \geq \ 1/4\pi \ , \tag{5.2.4}$$

where we have set $\Delta\xi = K \ \sigma_\xi$.

This confirms that fractal dimension 2 is a necessary and sufficient condition for getting the right form $\Delta p^\mu \propto (\Delta x^\mu)^{-1}$ for spatial as well as for temporal coordinates. The second of these equations carries two informations: The invariant momentum p_o^μ may be identified with the invariant classical momentum (and indeed we have remarked that the homogeneity of space-time is recovered both at the infinitesimal and classical limits); in that case the transition resolution λ^μ is identified with the de Broglie wavelength (here for a full period, corresponding to a phase change of 2π, i.e., to the periodicity of the wave function). The third equation (5.2.4) sets a constraint on the dispersion of acceptable normalised fractal derivatives in order to get the exact limit $\hbar/2$. Several trials using effective fractal curves with D=2 have verified this constraint in every case. Additional work is needed to decide whether it is generally verified for fractals of dimension 2, or whether one should define a subclass of such curves adapted to the description of the internal fractal structures of the quantum space-time.

We do not consider the above demonstration to be definitive. It still possesses a formal character, due to the remaining arbitrariness in the definition of the action and fractal internal coordinates. This is rather an intermediate step that indicates to us that a self-consistent quantum theory founded on the fractal approach appears possible.

Generalized Heisenberg relation.

Another important consequence of Eq. (5.2.1) is that it allows a generalization of the Heisenberg relation. It may indeed be written (we omit the indices for simplicity)

$$\frac{\sigma_p}{p_0} = \sigma_\xi \left(\frac{\lambda_0}{\sigma_x}\right)^\delta \ , \tag{5.2.5}$$

where δ (= $D-1$) can also be interpreted as a renormalization group anomalous dimension and $\sigma_\xi \geq 1$. The Heisenberg relation corresponds strictly to $\delta = 1$ (i.e. D=2), but we see here that any fractal space-time implies a Heisenberg-like relation with generalized power related to the fractal dimension of its geodesics. This result will be used in Chapter 6 : it will play a key role in the first developments of a *special theory of scale relativity*. Note also that (5.2.5) does *not* contradict the Heisenberg *inequality* provided either $\sigma_x < \lambda_o$ and $\delta > 1$ (we shall suggest in Chapter 6 that this is the ultrarelativistic situation), or $\sigma_x > \lambda_o$ and $\delta < 1$ (this is achieved in particular in the classical case $\delta = 0$ and during the transition to the classical case, $0<\delta<1$). Indeed the product $\sigma_p\, \sigma_x$, if larger than $p_o\, \lambda_o\, (\lambda_o/\sigma_x)^{\delta-1}$, will be automatically larger than $p_o\, \lambda_o = \hbar$ in the two above cases. Hence the universality of the Heisenberg inequality is found still reinforced; these results also help one to understand how, while assuming this universality in the quantum *and classical* domains, we were able to recover the quantum and classical situations that correspond to different fractal dimensions and the de Broglie transition between them.

Peano model.

Before ending this section, let us describe a simple 2-dimensional model based on the Peano curve, illustrating how the Heisenberg relation may be obtained from a trajectory of fractal dimension 2. Assume that a particle follows an inclined Peano curve as shown in Fig.5.5. (We disregard the multiple points: assume for example that this is the projection of a 3-dimensional curve, which could then be self-avoiding).

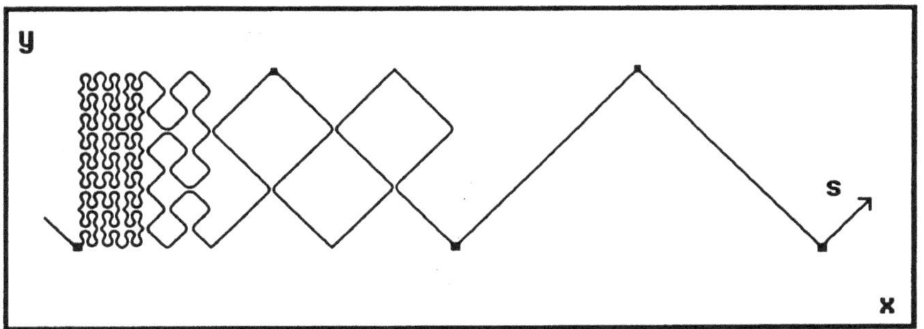

Figure 5.5. A "Peano model" of fractal trajectory, seen at various resolutions (see text).

It is immediately clear that, whatever the resolution, the velocity of the particle for a given position is now undefined and should be replaced by a probability distribution. If v is the classical velocity, the velocity at resolution $\Delta x = 3^{-n}\,\lambda$ can take only two values in this special model, $v_+ = 3^n v$ with a probability $P_+ = (1+3^{-n})/2$, and $v_- = 3^{-n} v$ with a probability $P_- = (1-3^{-n})/2$. It is easy to verify that $<v> = v$. The velocity dispersion is found to be[3]

$$\sigma_v = v\, 3^n\, [1 - 3^{-2n}]^{1/2} = (\hbar/m\Delta x)\, [1 - (\Delta x/\lambda)^2]^{1/2} \quad ,$$

so that, in the nonrelativistic case ($p = m\,v$ and $\lambda = \hbar/p$), we get for $\Delta x < \lambda$, i.e., $n \geq 1$, the quantum result

$$\Delta x\ \sigma_p \approx \hbar \quad ,$$

while for $\Delta x \geq \lambda$, i.e. $n = 0$, we obtain the classical result $v = constant$ and $\sigma_v = 0$.

5.3. Classical Quantities as Fractal Structures.

Geometrical structures of a fractal trajectory.

The ability of fractals to structure space-time offers the possibility to define the basic physical quantities describing a free particle in purely geometrical terms. Let us define an elementary particle as a perfect point whose space-time trajectory is a fractal curve of dimension 2 beyond some transitional resolutions λ, τ. (Only one spatial dimension is considered hereafter in order to simplify the argument; 3-dimensional relations are recovered by indexing λ, p and v). One does not need to attribute to this point additional physical quantities such as energy, momentum, velocity, phase velocity or mass. Indeed all of them may be defined from parameters of the fractal curves that are defined in a purely geometrical way: namely, these parameters, λ and τ, may be *defined* as the fractal / non-fractal transition, (which also corresponds to the spatial and temporal periods $2\pi\lambda$ and $2\pi\tau$):

$$E = \hbar / \tau \ , \quad p = \hbar / \lambda \ , \tag{5.3.1}$$

$$v = c^2 \tau / \lambda \ , \quad v_\phi = \lambda / \tau \ , \quad \sqrt{1 - v^2/c^2} = \sqrt{1 - c^2\tau^2/\lambda^2} \ , \tag{5.3.2}$$

$$m = \frac{\hbar}{c} \sqrt{\frac{1}{c^2\tau^2} - \frac{1}{\lambda^2}} \ . \tag{5.3.3}$$

In particular, in the particle rest frame the mass is defined by $m = \hbar/c^2\tau_0$. This definition of mass[3] is to be related to the recent equivalent proposal[4] of defining the mass from measuring a velocity and a wavelength, from de Broglie's equation, $m = (\hbar/\lambda c)\sqrt{c^2/v^2 - 1}$.

A zero-mass particle will be characterized from (5.3.3) by $\lambda = c\tau$, so that from (5.3.2) $v = v_\phi = c$. Another solution would be τ and λ infinite, but this would yield vanishing E and p, which is to be understood as meaning either non-existence, or the vacuum: this shows that the vacuum must have both very large scale (cosmological) and quantum properties. The above equations can be interpreted by describing inertial mass and energy as classical manifestations of the universal fractal character of the quantum space-time. The classical quantities may thus be related to geometrical structures that characterize quantum particle trajectories, which we conjecture to be the geodesics of such a space-time. We shall see in the next section that the quantum spin can also be obtained from fractal structures; the problem of the electric charge is more complicated and will be tackled in Sec. 5.9 and Chapter 6.

Thus we do not have to endow the point particle with mass, energy, momentum or velocity. *The particle may now be reduced to and identified with its own trajectory*. This is a complete change of view-point about the nature of elementary particles. The current model (sometimes implicit) we still have in mind in standard quantum theory for, e.g., an electron is that of a perfect point which owns intrinsic (internal) properties, namely its mass, spin and charge. This actually relies on the assumption of differentiability and rectifiability. It is assumed in this scheme that one may look at a purely infinitesimal portion of the electron trajectory and that, at this zero scale, there is no structure at all, so that mass, charge or spin are forced to be internal, i.e., *not* defined in space-time. Note that it is partly to tackle such a problem that string theories have been introduced. They attribute one-dimensional structures to particles at the Planck scale and also solve in this way the mass and charge divergence problem. It is noticeable in this respect that the trajectories of strings are then 2-dimensional (in the sense of a surface) to be compared to our *fractal* dimension 2 here.

Now in the fractal description, the trajectories are considered to be completely non-differentiable: there is no lower cutoff below which fractalization would stop. Hence there are always structures whatever the scale and one never reaches the non-structured limit that is assumed in the standard theory. This re-opens the hope that internal quantum numbers are nothing but such very internal structures of the trajectory, and that, at the end, a particle is nothing but the set of geometrical structures of its trajectory. We shall see in what follows that, indeed, such a description is possible for the quantum spin, which is known to be the simplest quantum number without any classical counterpart.

As a consequence the energy-momentum tensor,[5] and thus the right hand term of Einstein's equations, may be written in a completely geometrical form in terms of the de Broglie periods and the Planck length $\Lambda = (\hbar G/c^3)^{1/2}$ (with $X_i(\sigma_i)$ being four fractal functions defining the trajectory of particle i in terms of a curvilinear fractal proper time σ_i):[6]

$$\frac{G}{c^2} T^{\mu\nu} = \sum_i \int c\tau_i \, \frac{\Lambda}{\lambda_i^{\mu}} \frac{\Lambda}{\lambda_i^{\nu}} \, \delta^4[x - X_i(\sigma_i)] \, (dt/d\sigma_i) \, d\sigma_i \quad . \tag{5.3.4}$$

This may be seen as a first step towards Einstein's dream of a purely geometrical equation for space-time in which the field(s) and the sources would be treated on the same footing. This result is also to be related to a recent one due to Lachaud.[7] He has demonstrated that applying conformal transformations to the source-free Laplace equation $\Delta\Phi = 0$ transforms it into a Poisson equation with point sources (delta functions) at a given point: the relevance of such a result for the present approach becomes clear in the light of our initial point (Chapter 2) that the conformal group could well be the natural group-theoretical tool to be used for constructing a scale covariant physics, complementary to the geometrical (fractal) and algebraic (the renormalization group) tools that are considered here.

Dependence on scale of velocity.

Among the various variables which describe the particle trajectory, the case of velocity is worth a more complete study, in view of the particular part it plays in the derivation of the fractal dimensions, and of an interesting problem: any velocity is constrained by special relativity (of motion) not to exceed the speed of light. If one only extrapolates the nonrelativistic approach (see e.g. Feynman's result that $<v^2> \propto \delta t^{-1})$[8] one expects the velocity to diverge when scale tends to zero. But the temporal transition that we have found around the de Broglie time should be accounted for. Actually at very small scales the divergence of time begins to cancel the spatial divergence and one gets a limited ratio $\delta x/\delta t$. Hence the fractal approach sheds a new light on special relativity. Let us examine how this cancellation occurs.

Consider a free point describing a trajectory of fractal dimension 2 for $\Delta x \leq \lambda$, where λ is the de Broglie length of the particle. What is its velocity in terms of resolution ? Consider first a rough approximation. For $\Delta x > \lambda$, one gets the classical velocity v. For $\Delta x < \lambda$, the length travelled increases as $(\Delta x/\lambda)^{-1}$, so that the velocity increases as

$$V(\Delta x) = v \, \lambda/\Delta x \quad .$$

When one reaches the relativistic transition $\Delta x = c\tau$, the particle velocity becomes

$$V(c\tau) = v \, \lambda/c\tau = v \, E/pc = c \quad .$$

This result is reminiscent of the Dirac formalism in which the velocity operator owns only two proper values, $+c$ and $-c$. Let us now cross the $c\tau$ barrier. If one does not account for temporal phenomena, it has been seen in the previous section that one gets a new transition to D=1 for $\Delta x < c\tau$. The velocity would thus remain equal to c. Accounting for pair creations yields rather an interpretation in which the fractal dimension remains D=2 for spatial coordinates, while a new transition from D=1 to D=2 occurs around $\Delta t = \tau$ for the temporal coordinate. Consider a simple model for this behaviour in the frame proposed here, i.e., of a metric depending on resolutions:

$$ds^2 = c^2(1 + \tau^2/\Delta t^2)\ dt^2 - (1 + \lambda^2/\Delta x^2)\ dx^2 \ .$$

For $\Delta t >> \tau$ and $\Delta x >> \lambda$, one finds again the Minkowski metric, while for $\Delta t << \tau$ and $\Delta x << \lambda$, we introduce two divergent coordinates of fractal dimension 2, $T = t\ \tau/\Delta t$ and $X = x\ \lambda/\Delta x$. Introducing the velocity $V = dx/dt$ yields

$$ds = c\,\{\ (1 - V^2/c^2) + [\tau^2/\Delta t^2 - (\lambda^2/\Delta x^2)\,(V^2/c^2)]\ \}^{1/2}\, dt \ .$$

Assume that the temporal and spatial resolution along the trajectory are related by $\Delta x = V\Delta t$. Thanks to the relation $V/c = c\tau/\lambda$, one finds a remarkable factorization of the scale and motion factors:[3]

$$ds = c\,(1 - V^2/c^2)^{1/2}\ (1 - c^2\tau^2/\Delta x^2)^{1/2}\ dt$$

There is one and only one solution when $\Delta x < c\tau$, i.e., $V = c$, for which one gets the self-consistent isotropic result $ds = 0$. However no profound physical meaning should be given to this simple model, which we have shown here to illustrate the fact that the 2-dimensional divergence of time can cancel that of space and yield both the apparent spatial fractal dimension 1 as an artefact, and a "velocity" along the fractal trajectory always consistent with special relativity, $V \leq c$.

Let us now get a more realistic expression for the scale dependence of velocity: in a realistic model, the velocity is expected to reach the velocity of light only asymptotically, for $\delta t \to 0$. This will allow us to observe in a better way how the time divergence cancels the space divergence to finally yield the limiting velocity c, and to give an example of efficient use of the tool of fractal functions.

Let us start with two fractal functions of a fractal invariant s, $x = X(s,\Delta x)$ and $t = T(s,\Delta t)$, which define respectively the spatial and the temporal coordinate. They can be differentiated in the usual way, which we have generalized to fractal functions in Sec. 3.8. These derivatives are expressed in terms of the fractal dimension (here D=2), the fractal / non-fractal transition (here around the de Broglie length and time λ and τ), and new fractal functions $X'(s,\Delta x)$ and $T'(s,\Delta t)$ which define their "finite parts":

$$\frac{\partial x}{\partial s} = X'(s,\Delta x)\ \left(1 + \frac{\lambda^2}{\Delta x^2}\right)^{1/2} ,$$

$$\frac{\partial t}{\partial s} = T'(s,\Delta t)\ \left(1 + \frac{\tau^2}{\Delta t^2}\right)^{1/2} .$$

Let us now define a new fractal function $\xi(s,\Delta x,\Delta t)$ such that $<\xi> = 1$, and $\xi = 1$ when $\Delta x >> \lambda$ and $\Delta t >> \tau$:

$$v_0\ \xi(s,\Delta x,\Delta t) = \frac{X'(s,\Delta x)}{T'(s,\Delta t)} ,$$

where v_0 is the classical velocity of the particle. So the local velocity of the particle at resolutions Δx and Δt is given by

$$v = \frac{\partial x/\partial s}{\partial t/\partial s} = v_0\ \xi(s,\Delta x,\Delta t)\ \frac{\lambda}{\tau}\ \frac{\Delta t}{\Delta x}\ \left(\frac{1 + \Delta x^2/\lambda^2}{1 + \Delta t^2/\tau^2}\right)^{1/2} . \quad (5.3.5)$$

We shall now specialize and consider two possible choices of the relations between the position and time resolutions, namely $\Delta x/\Delta t = c$ and $\Delta x/\Delta t = v$. Both give, as they have to, $\Delta x/\Delta t = c$ when $\Delta x \to 0$ and $\Delta t \to 0$. In the first case we obtain (remembering that $\lambda/\tau = c^2/v_0$)

$$v = c\ \xi(s,\Delta x,\Delta t)\ \left(\frac{1 + \Delta x^2/\lambda^2}{1 + \Delta x^2/c^2\tau^2}\right)^{1/2} . \quad (5.3.6)$$

In the second case, v appears in both the right- and the left-hand sides of (5.3.5), so we get a second order equation which is solved to give

$$\left(\frac{v}{c}\right)^2 = \frac{\frac{1}{2}\xi^2\frac{c^2\Delta t^2}{\lambda^2} + |\xi|\left[1 + \frac{\Delta t^2}{\tau^2} + \frac{1}{4}\xi^2\frac{c^4\Delta t^4}{\lambda^4}\right]^{1/2}}{1 + \frac{\Delta t^2}{\tau^2}} . \quad (5.3.7)$$

The scale dependence obtained in (5.3.6) and (5.3.7) is illustrated in the following Fig. 5.6 (in which we take $\xi=1$, disregarding the additional fluctuation), for various values of the classical velocity. The velocity asymptotically reaches v_0 at large scale and c at small scale, beyond the two transitions λ and $c\tau$ (which nearly coincide when $v_0 \to c$).

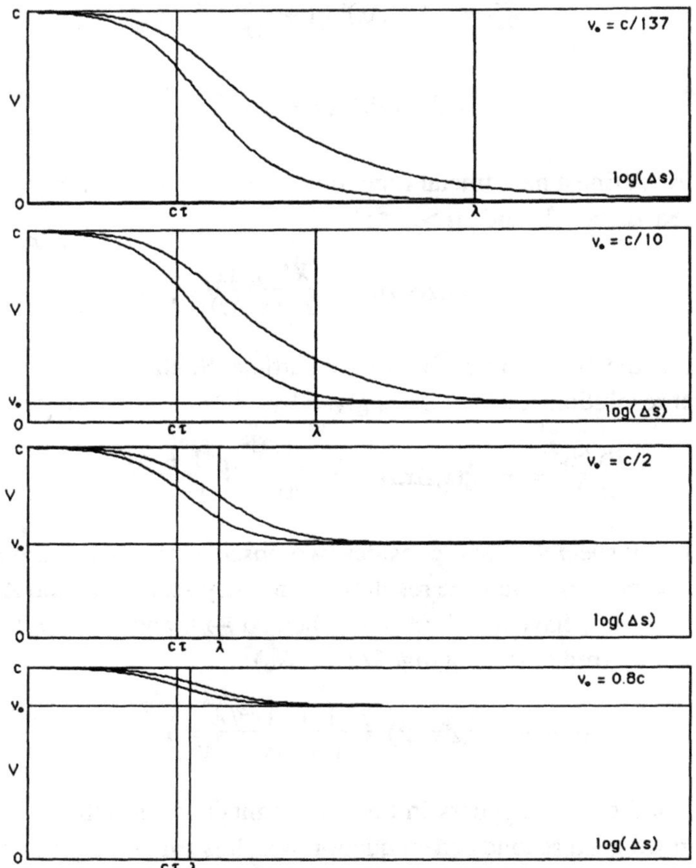

Figure 5.6. Variation with scale of the velocity of a "particle" on its fractal trajectory for various values of the classical velocity, v_0. The upper curves correspond to the choice $\Delta x/\Delta t=v$ and the lower curves to $\Delta x/\Delta t=c$. The two (spatial and temporal) de Broglie transitions are shown. For $\Delta x>>\lambda_{dB}$, the classical velocity is recovered, while for $\Delta x<<c\tau_{dB}$ the velocity becomes c, in agreement with the solutions of the Dirac equation.

5.4. Emergence of Spin from Fractalization.

The spin of the electron is considered as an essentially quantum-mechanical property without any classical analog. First introduced by

Goudsmit and Uhlenbeck[9,10] as a proper angular momentum of the particle, this mechanical picture was quickly proved to be untenable since the speed of rotation of its surface (assuming an extension of the electron equal to its classical radius) would have been in excess of the speed of light. It was recently recalled[11] that spin may be described as a wave property of the electron. Indeed it was demonstated by Belinfante[12] that spin may be regarded as the result of a circulating flow of energy in the particle wave field. This allows one to understand the mechanism which produces spin in terms of the wave nature of the electron. However it leaves open the problem of understanding the spin in terms of its particle nature, while it is this last aspect of the nature of the electron which is revealed in actual detection.

We demonstrate hereafter that spin may be obtained[3] as a specific property of perfect point masses following trajectories of fractal dimension 2. Let us first present a simple calculation which reveals the profound relation between the emergence of a finite intrinsic angular momentum (neither null nor infinite) and the precise value D=2 of the fractal dimension. In cylindrical coordinates, the z component of the angular momentum of a point mass m is:

$$M_z = m\, r^2 \dot{\varphi} \quad ,$$

where $\dot{}=d/dt$. Consider a fractal trajectory in 3-space built from a generator made of p segments of length $1/q$ (see Fig. 3.12). Assume that, seen at a given resolution $\Delta x = q^{-n}$, the trajectory F_n is such that an average angular momentum $M_n = mr^2\dot{\varphi}$ is found with respect to the lowest order approximation of the trajectory F_{n-1}. Consider now the next stage of the fractalization and compute the angular momentum of F_{n+1} with respect to F_n. Every length will be divided by q while the total number of segments is multiplied by p; so the number of turns in the unit of time, i.e. $\dot{\varphi}$, will also be multiplied by p. Finally one gets $M_{n+1} = m\,(r/q)^2\,(p\,\dot{\varphi}) = (p.q^{-2})\,M_n$. By recurrence we infer that the limiting value on the fractal will vanish if $(p.q^{-2}) < 1$, i.e. $D<2$, diverge if $(p.q^{-2}) > 1$, i.e., $D>2$, while it may be finite and non-vanishing only for the particular value $(p.q^{-2}) = 1$, i.e., $D=2$. The same argument will hold if one calculates the average angular momentum of F_{n+m} with respect to F_n, m being fixed and n variable.

This finding relates the existence of spin to a result (D=2) which was obtained on a ground where spin was not at all assumed. In imaged words, we may express this property of fractal 2-dimensional trajectories by saying that spin emerges as the result of the product ($0 \times \infty$): the particle is indeed assumed to be elementary (totally pointlike: $r = 0$), and as such can classically possess no proper angular momentum in its rest frame, but it rotates with an infinite angular velocity ($\dot{\varphi} = \infty$), the final proper angular momentum being finite.

Let us now give a more complete calculation of the emergence of spin, which yields its precise value in terms of the fractal structures and also accounts for its size scale $\approx\hbar$. The angular momentum vector is $\boldsymbol{M}=\sum \boldsymbol{r}\boldsymbol{x}\boldsymbol{p}$. We consider first a nonrelativistic calculation (corresponding, e.g., to the electron rest frame). We place ourselves in standard space (x,y,z) and consider the 2-dimensional fractal trajectory of a particle, $[x(s),y(s),z(s)]$. The angular momentum of a particle of mass m with respect to its classical trajectory (defined as the axis Oz) writes

$$\sigma = (m/N) \sum_{i=1}^{N} \left(x_i \frac{y_{i+1}-y_i}{t_{i+1}-t_i} - y_i \frac{x_{i+1}-x_i}{t_{i+1}-t_i}\right) , \qquad (5.4.1)$$

where N is the total number of segments: in a NSA formulation, $N=p^{\omega}$ on the non-standard fractal curve and σ is given by the standard part of the hereabove expression, which is defined precisely for D=2. Summing over a de Broglie full period $2\pi\lambda = h/mv$, identified with a fractal period, we may write $N(t_{i+1}-t_i) = T$ with $T= v/\lambda$. Introducing the dimensionless lengths $X= x/\lambda$ and $Y= y/\lambda$, we get[3]

$$\sigma = \hbar \sum_{i=1}^{p^{\omega}} (X_i\, Y_{i+1} - Y_i\, X_{i+1}) \quad . \qquad (5.4.2)$$

Successive approximations of this formula may be found by computing the sum on F_n for increasing values of n. Let us comment on this result.

The term under the summation sign is now a *purely geometrical term* which depends only on the fractal structure. Examples of fractal curves which have a spin of exactly $\hbar/2$ are given in Fig. 3.15 a,b,c. But some other curves of fractal dimension 2 may be built with spins which are not multiples of $\hbar/2$ (for example, 3.15 d, $\sigma \approx 0.18\,\hbar$ and e, $\sigma \approx 0.42\,\hbar$). This is an additional indication that the quantum space-time is not describable by any fractal space-time whose geodesics are of dimension 2, but possibly only by a sub-class verifying Eq. (5.2.4) and the relation $\sum_i (X_i Y_{i+1} - Y_i X_{i+1}) = k.\hbar/2$ with k integer.

The second point to be noticed is that σ in Eq. (5.4.2) actually behaves as an internal angular momentum, since it does not depend on the origin of the axis Oz. Indeed changing this origin would introduce terms like $a\Sigma(Y_{i+1}-Y_i)-b\Sigma(X_{i+1}-X_i)$ which vanish thanks to the fractal periodicity.

Last but not least, if one assumes self-similarity for the fractal trajectory, the value of the spin obtained around F_0 ($=Oz$, identified with the classical trajectory), will also be found exactly around any segment of any approximation F_n of the fractal curve (with sign + or – depending on the orientation), as a consequence of the fractal dimension 2 (i.e. $pq^{-2}=1$). While the direction of such a segment may have any orientation thanks to the folding of the fractal curve (see Fig. 3.15), once this direction is chosen the spin is clearly undefined on any other axis, in agreement with standard quantum mechanics. Equation (5.4.2) holds equally for an infinitesimal segment, which means that the spin is finally defined locally. In that case a preliminary relativistic calculation (i.e., introducing the Lorentz factors in λ and p) yields the same result when $\Delta x \to 0$, though a complete treatment of the problem needs additional work, since it should integrate both the assumed fractal structure of time and our proposal that mass and momentum are derived precisely from the fractal structure.

5.5. Geodesical Interpretation of Wave-Particle Duality.

Introduction.

In the present approach, our working hypothesis is that some, if not all, of the quantum properties of the microphysical world are characteristic of properties of the geodesics of a fractal space-time. A complete demons-

-tration of this conjecture, provided it happens to be correct, requires substantial additionnal work. We shall only hereafter illustrate this point with some examples, aiming at a demonstration that such an interpretation is indeed possible and could solve some of the well-known quantum paradoxes. It is recalled at the end of this Section that a congruence of geodesics in the present archetype for space-time theories, general relativity, is indeed described by a Schrödinger-like equation.

The geodesics of a fractal space-time are expected to be characterized by a mixing of individual and collective properties, to which should be added the effects of projection in standard space-time (see Figs. 3.5, 3.15). In such an interpretation, there is a clear difference between the properties of a prediction (or equivalently of retroprediction) which is based upon the wave function (more generally on the probability amplitude) and thus on the wave nature of the "particle", and the properties of an actual measurement (e.g., impact of a particle on a screen) which reveals its particle nature. In a space-time theory, the answer to the question "what is the trajectory of a free particle?", is that it follows a space-time geodesic. However the non-differentiability, the infinity of obstacles at all scales and the projection effects of a fractal space-time of the kind considered here imply that an infinity of geodesics will exist between any two classical points, all of them by definition equiprobable (see Figs. 3.16-22). We may thus consider that the particle indeed followed one particular geodesic among the family of possible ones (and thus get the well defined particle behaviour given by a single position measurement), and at the same time that the prediction of which particular geodesic has been actually followed is definitively impossible, all geodesics being equiprobable; then predictions (or equivalently retropredictions) can only be of a statistical nature, in agreement with the current interpretation of quantum mechanics.

The fractal approach, rather than bringing back determinism in the classical sense, definitely excludes it for purely mathematical reasons. But the introduction of the fractal or "zoom-space-time", a concept which is absent from present quantum physics, sets the problem in a completely different way: We stress the fact that this property of infinite multiplication of geodesics holds in a *completely defined* fractal space-time. One does not have to work with random fractals, since statistical and non-classical properties may be found for physical objects defined on a fractal space-time owning by itself no statistical properties. Hence one may hope for a general relativistic type of approach in which the matter and energy content determines the space-time fractal structures, while the quantum, undeterministic, statistical behaviour would appear as a property of geodesics.

Let us now consider some of the basic quantum experiments and show how their paradoxical properties are illuminated by the fractal geodesics interpretation.

Quantum mechanics and Gödel's theorem.

Consider a two slit (or Young's) experiment. This is certainly one of the most fundamental quantum experiments, since it shows both the wave-particle nature of quantum objects and the phenomenon of complementarity. Let us first comment on the so-called Copenhagen interpretation of this experiment and of quantum mechanics. The fact that the observation of the interference pattern makes it impossible to know by which hole the particle did pass has been interpreted as indicating that there was no content of truth (or of reality) in the assertion that it indeed passes through one or the other hole. By extension this leads to the idea that no reality should be attributed to the particle between the measurements, and finally that the measurement results are determined by the measurement apparatus and just at the time of measurement (rather than revealing a reality which was pre-existing to the measurement). This interpretation assumes that there can be no element of truth attributed to something which is either non-measurable or non-predictible.

Concerning at least the domain of predictions, which comes under mathematical physics, one may remark that, as such, it should be dependent on Gödel's theorem.[13] We recall that this theorem states that in any self-consistent system based on a set of axioms which includes number theory, there exists non-demonstrable true statements. Its main consequence is that it has definitively demonstrated *the logical difference between truth and demonstrability.* We have suggested[6] that some of the paradoxes of quantum mechanics reveal the first intrusion of this limitation into mathematical physics. What is after all the status of a prediction in theoretical physics ? A physical theory is an ensemble of principles which are translated into definitions and equations, i.e. from the mathematical view point it starts with a set of axioms. That number theory is indeed contained in any physical axiomatic results from the very nature of physics: indeed physical theories try to describe the result of *measurements* and relations between them, while these results are, ultimately, always described by numbers. So a physical prediction or statement is nothing but a theorem in the frame of an axiomatic which should come under Gödel's theorem. As a consequence it should not be unexpected that physics be one day confronted with undecidable statements, which could thus be considered as true but non-demonstrable (i.e. non-predictable by the physical theory), not due to our own limitations (insufficient theory or measurement apparatus) but to profound *logical* (in Gödel's sense) reasons. We claim that this situation already happened in physics, namely in quantum mechanics.

Applied to Young's double-slit problem, this means that our inability to predict or retropredict the hole through which the particle pass *does not* imply that it is not true that it passed through one of the holes. The statement "the particle indeed passed through one of the holes" might be true but undemonstrable in the framework of the axioms of the theory (and indeed, when making a *measurement* just behind the holes, it is always found that the particle does indeed exist and had passed through one of the holes, but not both of them). In other words, the strong complementarity principle à la Bohr, stating that the particle is either a corpuscle (when measured), or a wave (between measurements), but never both, is logically unnecessary. The logical reformulation of quantum mechanics by Omnes[14] may be understood as leading to similar conclusions. The "particle" may coexist with the wave, provided an unescapable unpredictability of some of its parameters is ensured (we shall see in the following how, in the geodesical interpretation, the wave nature and the particle nature correspond to *different* aspects of a same reality).

Fractal geodesics and Young's holes.

Let us illustrate this analysis by a more detailed description of the two-slits problem. A first paradox is the completely different results of the distributions of measurement positions in the one and two holes cases. This is indeed expected of a space-time approach, since the geometrical structures of the zoom-space-time are assumed to be determined by the whole distribution of matter and energy (in a way which remains to be understood), including the holes themselves and any detector aimed at deciding through which hole the particle did pass. The probability of presence is determined by the effective set of geodesics coming from the source and arriving at the point where the particle was measured. Closing a hole suppresses all the geodesics which passed through that hole; detecting the particle behind a given hole implies that one should consider now only the geodesics connecting the source, the screen and the point of detection, which is close to the one hole case (see Fig. C8).

A second and stronger paradox might at first sight appear to work against the geodesical interpretation. Consider a point of the screen corresponding to destructive interferences. How could it be so, while the opening of both holes implies that the total number of geodesics connecting this point to the source has been *increased* with respect to the one-hole case? Once more the fact that the number of geodesics is infinite is decisive in answering this question. The number of geodesics may both be increased from the one-hole to the two-hole case, and be infinitesimal with respect to those arriving at points of the screen where constructive interference occurs, in the same experimental disposition (see Fig. C8).

In fact this virtual set of fractal geodesics fills space like a fluid, so that it becomes easy to endow it with wave properties. When an experiment is actually run, either the number of particles is large (sometimes even undefined), and the structures of the geodesics are immediately visualized, or particles are cast one by one, and the interference pattern will little by little be established in a probabilistic way, since each particle will randomly "choose" one of the geodesics of the family.

The Einstein-Podolsky-Rosen paradox.

The EPR paradox, with its explicit violation of Bell's inequalities,[15] plays a central role in our understanding of quantum mechanics. Consider the anticorrelated emission along two separated beams of two particles in a mixed state of some quantum number s which may take two values, say +1 and -1. The state in which the two particles are emitted is such that $\{P_1(+1) = 1/2, P_1(-1) = 1/2\}$, $\{P_2(+1) = 1/2, P_2(-1) = 1/2\}$ and $s_1+s_2=0$. There is no way to specify in the absence of measurement whether particle 1 (or 2) has $s = -1$ or $+1$, but any measurement of s_1 alone allows one to immediately infer that $s_2 = -s_1$. In the current standard interpretation of quantum mechanics, the state mixing is a property *which is attributed to the particle itself*. In these conditions, nothing more than $\{P_1(+1) = 1/2, P_1(-1) = 1/2\}$ can be said of the particle prior to a measurement of s, so that the value of s given by the measurement *does not pre-exist* to the measurement. The determination of s_2 being simultaneous to that of s_1, we thus get the usual paradoxical result, either of strong nonlocality, or of propagation of information faster than the velocity of light. Let us propose another interpretation. As recalled hereabove, in the geodesical interpretation, the properties of quantum systems will consist of a *mixture of individual properties*, those of individual geodesics, and *collective properties*, those of the families of equivalent geodesics from which predictions can be drawn. Let us then assume that a mixed state is made up of a beam containing a fractal distribution of geodesics, each of which is characterized by some value of s. Assuming that the distribution is fractal means that, in whatever domain, either of space, or of phase space, or of any other physical quantity considered to whatever resolution, there will always be found a mixing of an infinite number of $\{+1\}$ geodesics and $\{-1\}$ geodesics. This is a (non-differentiable) model in which *there exists no hidden parameter* which could allow one to tell the value of s prior to a measurement, but in which we may also admit that *one particular geodesic has actually been followed by the particle*. Following our Gödel-like interpretation, we assume that such a geodesic *exists*, but that it is *undetermined* (i.e., there is no equation or parameter which would allow one to distinguish between it and a $\{-s\}$ equivalent geodesic). A mixed state is then a fractal mixing of geodesics, each geodesic corresponding to a pure state. In this interpretation a measurement of the kind described above does not determine the measured quantity, but reveals its preexisting value, even though we should admit our definitive (and fundamental) inability to predict it.

We think that the intervention of Gödel's theorem sheds some light on the EPR experiment, but we do not claim that the fractal model outlined above has exhausted all its paradoxical properties. In particular, we have not considered spin measurements on various axes: this case could require to take into account the internal structure of the fractal trajectory. We conjecture that a better understanding of the EPR paradox will be obtained by a more thorough description of quantum trajectories, not only as fractal curves, but mainly as geodesics of a fractal non-differentiable space-time.

Indiscernability.

The indiscernability of equivalent particles is a straighforward consequence of the identification of "particles" with the geometric structures of fractal trajectories themselves. The particles being nothing but these structures, there is clearly no way to distinguish between two equivalent structures. When two indiscernable particles share the same region of space, there is no way to distinguish between the geodesics of one or the other particle. We are left with a new family of geodesics which is characteristic of the *pair* of particles as such, and this is a new object certainly different from the sum of geodesics of individual particles.

Light beams in General Relativity.

Let us now attempt to push further the geodesical interpretation. Our conjecture is that the wave function should be identified with properties of a family of geodesics. General relativity offers a well-developed model for this approach. A light beam in a Riemannian space-time is described as a congruence of null geodesics, whose equations (to the geometric optics approximation) have been written by Sachs.[16] These "optical scalar equations" are widely used in observational cosmology, in particular for the relativistic approach to the problem of gravitational lensing.[17-23]

The various possible deformations of a light beam during its propagation may be reduced to 3 infinitesimal effects, *expansion*, *rotation* and *shear*. Let A be the beam cross-sectional area, ω an affine parameter along the beam, $k^a = dx^a/d\omega$ the wave 4-vector, t^a a complex null vector orthogonal to k^a, t^{*a} its complex conjugate, R_{aibj} the Riemann tensor and R_{ab} the Ricci tensor. Between ω and $\omega + d\omega$, a circular beam is subjected to an expansion $\theta\, d\omega = dA/2A = d\sqrt{A}/\sqrt{A}$, a rotation $\Omega\, d\omega = dW$ and is deformed to an ellipsoidal shape, the length of its axes differing from the initial diameter by $\alpha\, d\omega$ and $\beta\, d\omega$; defining a phase ϕ from the large axis position angle, a complex shear may then be built as $\sigma=|\sigma|\, e^{i\phi}$ with $|\sigma| = (\alpha+\beta)\,/\,2 = \xi/A$.

Note that one may use indifferently as variables, θ or A, Ω or W, and $|\sigma|$ or ξ. The optical scalar equations write[16,17]

$$\frac{d(\theta + i\,\Omega)}{d\omega} + (\theta + i\,\Omega)^2 + |\sigma|^2 = \frac{1}{2}\, R_{ij}\, k^i\, k^j \ , \qquad (5.5.1)$$

$$A^{-1}\ \frac{d(\xi . e^{i\phi})}{d\omega} = R_{aibj}\, k^a\, k^b\ t^{*i}\, t^{*j} \ . \qquad (5.5.2)$$

When applied to problems of observational cosmology, these equations are solved for plane or spherical waves, for which the rotation term vanishes, while the expansion and shear terms give rise to the several effects of gravitational lensing.[23] On the contrary, let us consider here the shear-free case with rotation. We are left with the first equation (5.5.1). Let us introduce a new complex variable

$$\Psi = e^{\int(\theta+i\Omega)\, d\omega} = \sqrt{A}\ e^{iW} \ .$$

Denoting by $\boldsymbol{R}$ the driving term $\boldsymbol{R} = \frac{1}{2}\, R_{ij}\, k^i\, k^j$, the optical scalar equation now reads

$$\frac{d^2\Psi}{d\omega^2} - \boldsymbol{R}\ \Psi = 0 \ .$$

This has exactly the form of the one-dimensional time-independent Schrödinger equation written in terms of a wave function $\psi = \sqrt{P}\ e^{iS}$. In this analogy, the *probability of presence* of quantum mechanics is identified with the *cross-sectional area of a beam of geodesics*, while the *quantum phase* is identified with the total *beam rotation*.

In general relativity the driving term may be expressed in terms of the energy-momentum tensor T_{ab}: $\boldsymbol{R} = -4\pi G c^{-2}\, T_{ij}\, k^i\, k^j$. For example in Friedmann cosmology, it becomes proportional to the density of the Universe, and more generally to the average density of matter and energy crossed by the light beam.[23] Thus it plays a role quite similar to that of the equivalent term in the Schrödinger equation, i.e., the energy term $2m(E-V)/\hbar^2$.

The aim of this analogy was mainly to demonstrate that, when considering a *family* of geodesics, rather than a unique geodesical line, even a space-time as simple as that of general relativity (still differentiable) is able to yield the equivalent of the *complex* probability amplitude: this is particularly remarkable owing to the fact that the optical scalar equations assume the hypothesis of geometrical optics. Note also that the shear term is expected, from this point of view, to correspond to polarization, i.e., spin.

However a still better connection of the fractal approach to quantum mechanics may be constructed in the framework of *stochastic quantum mechanics*. This is the subject of the next section.

5.6. Non-Differentiable Space and Stochastic Quantum Mechanics.

This section is devoted to one of the essential points to be clarified in our approach. We have demonstrated in the previous sections that the basic features of the quantum behaviour can be understood in the frame of a fractal non-differentiable approach. But if we want these ideas to be developed one day into a full theory, one must demonstrate that quantum mechanics itself would indeed be recovered as an approximation or a limit of such a theory. By quantum mechanics, we mean here not only its basic laws like the de Broglie and Heisenberg relations, but the full theory itself: the concepts of *complex* probability amplitude and wave function, the operator description, and finally the Schrödinger equation itself (in the nonrelativistic case).

The question that is asked here is a fundamental one. Underlying is the problem of the completeness of quantum mechanics and the understanding of the origin of the inescapable probabilistic description in microphysics. Let us recall again Einstein's far reaching position on this question (often misunderstood in our opinion).

Einstein's main criticism on the quantum theory was that, in this theory, *the probability* (more precisely the probability amplitude concept, completed by Born's statistical interpretation) *is set as a founding concept*, rather than *deduced from more fundamental principles*. In other words, contrary to what is often pretended, Einstein was perfectly ready to accept undeterminism as a limitation of our capability to make predictions,[24] but not undeterminism of the *fundamental* physical laws. Following Einstein's prescriptions for a real understanding of microphysics, the future theory must be able to *describe the processes affecting individual systems*, and it is *from such a description that the probabilistic description should naturally emerge*.[25] The confusion came from the fact that for most physicists, it seemed logical to admit that perfectly determined elementary physical laws are unable to yield undeterministic predictions. We shall try to demonstrate in the following that the concept of a fractal non-differentiable space-time actually leads to such a situation, in which one may get both an elementary description in terms of individual space-time events and non-predictability of particle trajectories.

In this respect, we remark that the various attempts that have been made to interpret quantum mechanics while keeping classical concepts are unsatisfactory, since none of them has proved to be able to satisfy to the hereabove "Einstein prescriptions". Three of these attempts are particularly interesting in the present context and have received considerable attention in recent years: the *quantum potential* approach[26,27] of de Broglie and Bohm, *stochastic quantum mechanics*,[29,31] mainly developed by Nelson, and *geometric quantum mechanics*,[32,34] as proposed by Santamato.

In the quantum potential approach, the trajectories are deterministic and the statistical behaviour is artificially introduced into the formalism by *assuming* that the initial conditions are at random; the physical origin of the "quantum force" that produces quantum effects remains mysterious.

In geometric quantum mechanics, the statistical description is obtained by the same postulate of random initial conditions; then "the theory does not describe the motion of an individual particle; rather it describes the statistical behaviour of an ensemble of identical particles".[32] Though not fulfilling Einstein's prescriptions, this theory is however very interesting, since the quantum behaviour is found to be a consequence of the underlying Weyl geometry. As demonstrated by Castro,[34] the Bohm quantum potential equation is recovered (rather than set) from a least-action principle acting on the Weyl gauge potential. Such an approach may be partly related to ours, since Weyl's geometry is closely related to conformal transformations, which include dilatations (i.e., scale transformations).

Stochastic quantum mechanics is a less "classical" approach than the two previously cited. Indeed the trajectories in the quantum potential approach and in geometric quantum mechanics remain deterministic: as such these theories are some kind of “hidden parameter” theories and so should be disproved by Bell's theorem. In stochastic quantum mechanics one assumes that an underlying Brownian motion, of unknown origin, is at work on every particle. This Brownian force induces a Wiener-like process which is at the origin of the quantum behaviour. In this theory the trajectories are *continuous, non-differentiable and non-deterministic*, as prescribed by Feynman's analysis of quantum trajectories. The Schrödinger equation can be recovered as a transcription of Newton's equation and of the Fokker-Planck equations for the diffusion process. We shall see in the following how our own fractal approach may be connected to Nelson's: the main link is apparent in the now well-known fact that Brownian motion is of fractal dimension 2,[35] while this is exactly the dimension which has been computed for quantum particle trajectories (Chapter 4). Let us only remark here that this theory does not satisfy Einstein's prescription either. The basic tool is a stochastic process, so that, as in standard quantum mechanics, the statistical description is set as a fundamental unexplained principle. Nothing is said, either on individual phenomena, or on the physical origin of the underlying stochastic process. Subsequent attempts have tried to ascribe quantum fluctuations to some subquantum medium, second-quantized into hypothetical particles.[36,37] But the physical nature of such particles remains unclear: there is presently no place for them in the standard model of elementary particles. (See also Rosen[36] for a critical point of view on stochastic quantum mechanics by one of its originators).

Before jumping to the fractal approach, let us say a few additional words on Feynman's path integral reformulation of quantum mechanics.[8] As was perhaps realized at that time by Wheeler (see Ref. 39) this interpretation of quantum mechanics is far more "realistic" (in Einstein's meaning) than the Copenhagen one. Recall that one demonstrates that the probability amplitude for a particle to go from a point a to another point b is given by the sum over all possible trajectories:

$$K(a,b) = \int_a^b exp\{\frac{i}{\hbar} S(a,b)\} \mathcal{D}x(t) \quad ,$$

where $S(a,b)$ is the classical action for the path considered and $\mathcal{D}x$ is a differential element coined for this special integral. The integration should be performed over all possible continuous paths connecting a and b, however distant or complicated may they be. Actually the paths which are too distant from the classical trajectory are very unprobable, due to destructive interferences between the $exp(iS/\hbar)$ terms. One finds, for a free particle of de Broglie wavelength λ travelling between two points a and b separated by a distance ℓ, that the most probable paths are included in an ellipsoïd of revolution with thickness at half-way of order $\sqrt{\lambda \ell}$. Indeed the paths which deviate from the a-b axis by more than y, with $y^2 = (\lambda \ell/2)\,[1-(2x/\ell)^2]$, (where x is the coordinate on this axis, the origin being at the middle of the a-b segment) become increasingly destroyed by destructive interferences.

It is intuitively clear that in such a volume, the number of non-differentiable paths greatly exceeds those which are differentiable. Indeed, as recalled in Sec. 4.1, Feynman finds the mean quadratic velocity to scale as $<v^2> \approx \delta t^{-1}$, corresponding to fractal dimension 2 (see Secs. 3.8 and 4.1).

Hence in Feynman's approach, one can use the concept of particle trajectory, while this was forbidden in the Copenhagen interpretation of quantum mechanics. This is not a return to determinism, since a Feynman path is only one possibility among an infinity: the various possible paths are precisely characterized as being equiprobable (they have equal probability, but *not* equal probability amplitudes). In Bohr's intepretation of quantum mechanics, following his principle of complementarity, only the wave nature of particles is actualized in the absence of measurement, and no information finer than that given by the wave function has physical sense. On the contrary in Feynman's perspective, one may explicitly consider individual possible trajectories, then analyse them and describe their structure. This does not result in trivial statements, since even though the number of possible trajectories is infinite, they share common properties, among which their non-differentiability and $D=2$ fractal character.

However if one wants to reach "Einstein's realism," one must go one step further. One should not only consider individual trajectories but must also look at the structure of all the individual points which constitute them. This means looking at the very structure of space-time, and more specifically, asking oneself what are the elementary properties of space-time which drive the particles into such complicated paths. We thus fall back to our original proposal: trajectories are fractal because space-time itself is fractal and non-differentiable and because they follow its geodesical lines.

Let us now make an attempt at explicitely looking at one point of a fractal space-time. Useful intuitive models of what happens may be found in Sec. 3.6 about fractal surfaces, and in the generalization to fractal spaces in Sec. 3.10. Actually every point of a fractal space happens to be singular, and any trajectory passing through such a singular point is expected to be broken, in agreement with its non-differentiability. Hence whatever detailed model of singular space is used, the minimal prescription in the description of what happens at a given point P of the fractal space implies two main features:

(1) Even if one assumes the incoming trajectory to be well-defined, the outcoming trajectory will be defined only in a probabilistic way. This is a strongly chaotic situation, in which *infinitesimal* differences in the initial conditions lead to completely different trajectories (see Fig. 5.7). For example, in a simple conic model, the local properties of each point of a singular space can be described by the value of the ratio of the circonference of a circle over its diameter, π_X, different from the Euclidean value π. In the 2-dimensional case (singular surface), two outcoming trajectories correspond to each incoming trajectory. The point P

behaves as if it was attractive when $\pi_X < \pi$ ("spherical") and repulsive when $\pi_X > \pi$ ("hyperbolic"). We conjecture that a generalization to 3-space will yield outcoming trajectories making up a full cone, the opening angle of which is related to the value of π_X at point *P*.

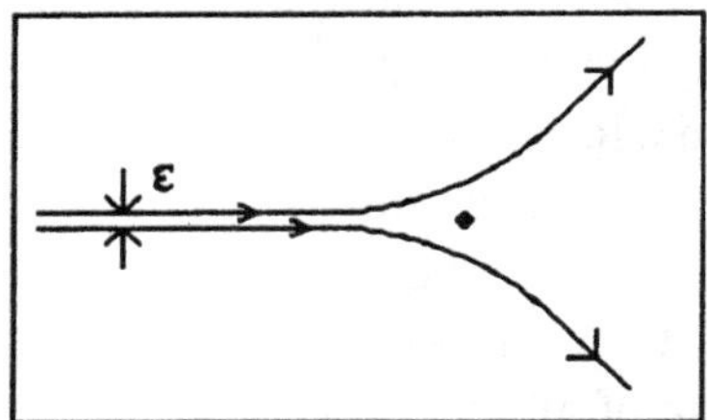

Figure 5.7. Two trajectories initially infinitely close may diverge in a non-differentiable space.

Now, reversing the arrow of time, one finds that to a given outcoming trajectory, there also correspond several incoming trajectories. The general situation at point *P* is finally that several trajectories with broken slopes are possible (an infinity in 3-space).

(2) Once one of the possible incoming and outcoming trajectories is chosen, the trajectory at point *P* is characterized by *two velocities* instead of one classically, a velocity v_- before the point and a velocity v_+ after the point (see Fig. 5.8). So the two velocity vectors v_+ and v_- define a plane which passes through the point *P* and which has no classical counterpart. We shall come back later to the physical meaning of this plane. From these two vectors, one may now define in the same plane

$$\boldsymbol{u} = \frac{v_+ - v_-}{2} \quad , \quad v = \frac{v_+ + v_-}{2} \quad .$$

It is straighforward that the classical differentiable case is recovered for $v_+ = v_-$, so that v is the generalization of the classical velocity ($v = v_+ = v_-$), while $\boldsymbol{u}$ is a new quantity which vanishes in the classical (differentiable) approximation.

What is the meaning of this new velocity $\boldsymbol{u}$? Being built from a *difference* of velocities, one may expect it to be related to an acceleration. However the acceleration at point *P* is actually given by $\Gamma = (v_+ - v_-)/2\delta t$, with $\delta t \rightarrow 0$, so that it is undefined from the view point of standard methods:

$\Gamma = \infty$. So let us attack the problem from the fractal point of view. In Sec. 3.8 we have introduced the concept of fractal functions $f(x,\varepsilon)$.

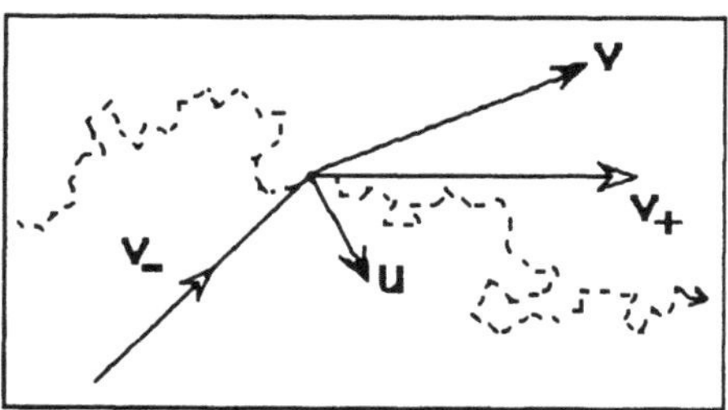

Figure 5.8. The backward and forward velocity on a trajectory whose slope is broken at each of its points, and their combination v, which generalizes the classical velocity, and $\boldsymbol{u}$, which vanishes in the differentiable case.

Start with the position vector of a particle. In the fractal approach this is a finite fractal function $\boldsymbol{x}(t,\varepsilon)$, where, depending on the experiment performed, ε is either a spatial resolution ($\varepsilon=\delta x$) or a temporal one ($\varepsilon=\delta t$). Then the three components of the velocity of the particle are divergent fractal functions:

$$\frac{d\boldsymbol{x}}{dt}(t,\varepsilon) = \boldsymbol{w}(t,\varepsilon)\left(\frac{\tau}{\varepsilon}\right)^{\beta} ,$$

where we have assumed the resolution to be a temporal one, i.e., $\varepsilon=\delta t$, so that standard quantum mechanics yields $\beta=1/2$. The finite fractal vector $\boldsymbol{w}(t,\varepsilon)$ is another representation for the hereabove v_+ and v_- velocities. The fractal acceleration will itself be a fractal function, now independent of the value of β, from the theorem of Sec. 3.8:

$$\Gamma(t,\varepsilon) = \frac{d\boldsymbol{w}}{dt}(t,\varepsilon) = \gamma(t,\varepsilon)\left(\frac{\tau}{\varepsilon}\right) . \tag{5.6.1}$$

We may now compute the value of the hereabove new velocity $\boldsymbol{u}$ to any resolution ε and eventually take the limit $\varepsilon\to 0$. One may write

$$\boldsymbol{u} = \frac{\boldsymbol{w}(t+\varepsilon,\varepsilon) - \boldsymbol{w}(t-\varepsilon,\varepsilon)}{2} .$$

But we may also compute Γ as

$$\Gamma(t,\varepsilon) = \frac{w(t+\varepsilon,\varepsilon) - w(t-\varepsilon,\varepsilon)}{2\,\varepsilon} \quad ,$$

so that we get $\boldsymbol{u} = \varepsilon\, \Gamma(t,\varepsilon)$, and from Eq. (5.6.1) we finally obtain the remarkable result

$$\boldsymbol{u} = \gamma(t,\varepsilon)\ \tau \quad .$$

This formula states that *the new "velocity"* $\boldsymbol{u}$ *is nothing but the finite part of the fractal acceleration* (to the constant multiplicative factor τ, which is fixed by the *classical* state of motion of the particle). The infinite acceleration is naturally renormalized thanks to the fractal structures, and this renormalized acceleration is identified with $\boldsymbol{u}/\tau$. We shall see in the following that this quantity has a profound physical sense (the probability of presence is deduced from its average): this demonstrates that the introduction of the finite part of fractal functions was indeed endowed with a physical meaning. Consider in this respect some of the difficulties encountered by the various models which have been proposed for Brownian motion (see also hereafter). In the Einstein-Smoluchowski and Wiener methods, there is a well-defined position but the velocity is undefined. In the Ornstein-Uhlenbeck method, one assumes a defined velocity, but then the acceleration is undefined, being always infinite. Our method of separation of finite and infinite parts of fractal functions sheds some light on the meaning of these models: the Ornstein-Uhlenbeck case corresponds to taking the finite part $\boldsymbol{w}$ of the fractal velocity, while the Wiener process starts with the position variable $\boldsymbol{x}$, which is itself a finite fractal function. Finally we have shown that the acceleration itself may be defined after our fractal renormalization is performed.

Let us proceed further. We have seen that the non-differentiable fractal structure of space implies emergence of a fundamental and elementary probabilistic behaviour of trajectories at any point of this space. (We shall not explicitely consider in this section the space-time case: let us simply say that the generalization of the concept of non-differentiable and probabilistic trajectories to space-time is related to the probabilistic creation and annihilation of particle-antiparticle pairs: see a preliminary account of the relativistic case in Sec. 5.8).

This probabilistic behaviour is now a consequence of a more fundamental postulate, the postulate that microphysics space-time is fractal, which is itself a consequence of the principle of relativity of motion, in its extended version ("the laws of nature should apply to systems of coordinates whatever their state of motion, even non differentiable"); or equivalently of the principle of general covariance, itself extended ("the equations of physics should be covariant under general continuous coordinate transformations, not only differentiable ones"). In this respect we hope that, if a future well-developed theory is to be built upon these principles, it will satisfy "Einstein's prescriptions" for a realistic theory of microphysics: one may have, at least in principle, a *completely determined continuous space-time*, the geodesics of which are *undeterministic* because of its *non-differentiability*.

The elementary probability introduced at the infinitesimal level *implies a statistical treatment* of physical laws. We shall see that the setting of such a statistical approach for a fractal motion *of fractal dimension* 2, as imposed by Heisenberg's relations, is fully equivalent to stochastic quantum mechanics. Since the Schrödinger equation is obtained in stochastic quantum mechanics as a consequence of a Wiener process, we may now be certain that a fractal theory of microphysics is expected to include quantum mechanics as an approximation.

Let us now recall how the stochastic approach is able to yield Schrödinger's equation. Though the essential information is already contained in Nelson's work,[31,40] we present here a new derivation (with occasional different notations) in which we attempt at fully using the new "bi-velocity" structure $(\boldsymbol{u},v)$: this allows us to better grasp the physical origin of the complex probability amplitude and of the complex operators, and to *demonstrate* two of Nelson's main assumptions rather than postulate them.

Let us place ourselves at point P at some given instant t. The basic hypothesis in the stochastic approach is that, even if the processes considered are no more differentiable in the usual sense, one may still define[31] a mean forward derivative d_+/dt and a mean backward derivative d_-/dt:

$$\frac{d_+}{dt}y(t) = \lim_{\Delta t \to 0+} \left\langle \frac{y(t+\Delta t) - y(t)}{\Delta t} \right\rangle ,$$

$$\frac{d_-}{dt}y(t) = \lim_{\Delta t \to 0+} \left\langle \frac{y(t) - y(t-\Delta t)}{\Delta t} \right\rangle .$$

Once applied to the position vector $\boldsymbol{x}$, they yield *forward and backward mean velocities*:

$$\frac{d_+}{dt}\boldsymbol{x}(t) = \boldsymbol{b}_+ \quad , \quad \frac{d_-}{dt}\boldsymbol{x}(t) = \boldsymbol{b}_- \quad .$$

In our approach, these velocities are defined as the average at point P and time t of the respective velocities of the outcoming and incoming fractal trajectories; in stochastic quantum mechanics, this corresponds to an average on the quantum state.

It is clear that, in the fractal approach, these slopes, averaged over families of virtual trajectories, are expected to be related to the fractal space-time itself. However the hypothesis that one can define mean forward and backward derivatives may not be fulfilled *in the general case*: for most of its points (those which are given by an infinite number of digits of their curvilinear coordinate, see Sec. 3) even this generalized derivative may be undefined on a fractal space. However the "physical" points are those which correspond to a finite resolution (arbitrarily small but never zero), while we have shown that these points, corresponding to a finite number of digits, may indeed be characterized by a "left derivative" and a "right derivative" for some classes of fractals: those whose generator has a null slope at the origin (see Fig. 3.9).

In the following, we shall disregard these difficulties and consider Nelson's assumption as the strongest simplifying assumption one may make about fractals, i.e., in other words, as the simplest possible breaking of differentiability. We shall see that this yields Schrödinger's equation in its simplest form, without spin or charge. One may then expect new structures to emerge when relaxing this hypothesis (this will not be considered in the present book).

The position vector $x(t)$ of the particle is thus assimilated to a stochastic process which satisfies, respectively after ($dt > 0$) and before ($dt < 0$) the instant t,

$$dx(t) = b_+[x(t)]\, dt + d\xi_+(t) = b_-[x(t)]\, dt + d\xi_-(t) ,$$

where $\xi(t)$ is a Wiener process.[41] It is in the description of ξ that the D=2 fractal character of trajectories is input. Indeed, that ξ is a Wiener process means that the $d\xi$'s are assumed to be Gaussian with mean 0, mutually independent and such that

$$<d\xi_{+i}(t)\, d\xi_{+j}(t)> = 2\, \mathcal{D}\, \delta_{ij}\, dt , \qquad (5.6.2a)$$

$$<d\xi_{-i}(t)\, d\xi_{-j}(t)> = -2\, \mathcal{D}\, \delta_{ij}\, dt , \qquad (5.6.2b)$$

where <> denotes averaging, and where $\mathcal{D}$ is the diffusion coefficient.

Nelson's postulate is that[31] $\mathcal{D} = \hbar/2m$. This value can be easily justified. Indeed the diffusion coefficient is expected to be given by the product L^2T^{-1} of the characteristic length and time period of the system, i.e., in an equivalent way, $2\mathcal{D} = Lv$. In the quantum non-relativistic case, $L = \lambda_{dB} = \hbar/p = \hbar/mv$, so that we finally obtain $\mathcal{D} = \hbar/2m$. Note however that it has been demonstrated by Davidson[42] that any value of $\mathcal{D}$ may lead to quantum mechanics, and by Shucker[43] that in the limit $\mathcal{D} \to 0$, the theory becomes equivalent to Bohm's[27] deterministic quantum potential approach. We shall in the following work essentially with the value $\mathcal{D} = \hbar/2m$.

That Eq. (5.6.2) is indeed a consequence of a fractal dimension 2 of trajectories is straighforward: it may be written $<d\xi^2>/dt^2 \approx dt^{-1}$, i.e. precisely Feynman's result $<v^2>^{1/2} \approx \delta t^{-1/2}$. We have demonstrated (Sec. 3.8) that a fractal dimension D leads to the relation $\mathcal{L} \approx <v^2>^{1/2} \approx \delta x^{1-D}$ when the measurement resolution is spatial, and $\mathcal{L} \approx \delta t^{1/D-1}$ when it is temporal. Feynman's power $-1/2$ is thus translated into $D = 2$. In Nelson's approach, this choice for the stochastic basic process is attributed to the hypothesis that quantum particles are subject to a Brownian motion of unknown origin. The connection with fractals is clear: Brownian motion is now known to be of fractal dimension 2.[35] However the fractal conjecture is far more general than the Brownian motion hypothesis: the fractal hypothesis is a general geometrical description which supersedes any of its possible physical cause, the second is a particular choice of a $D = 2$ physical phenomenon which has at present no experimental support, the existence of a diffusing particle being up to now at variance with the standard model of elementary particles.

Let us now examine the implications of this Wiener process for the mean derivatives. Start from any function f of $\boldsymbol{x}$ and t, and expand f in a Taylor series up to order 2. Then take the average and use the properties of the Wiener process ξ (Eq. 5.6.2), one gets[31]

$$d_+f/dt = (\partial/\partial t + \boldsymbol{b}_+ . \nabla + \mathcal{D}\Delta)f \ , \tag{5.6.3a}$$

$$d_-f/dt = (\partial/\partial t + \boldsymbol{b}_- . \nabla - \mathcal{D}\Delta)f \ . \tag{5.6.3b}$$

Let $\rho(\boldsymbol{x},t)$ be the probability density of $\boldsymbol{x}(t)$. It has been demonstrated by Kolmogorov[44] that for any Markov process (the Wiener process is indeed a particular case of a Markov process[41]), the probability density satisfies a forward equation:

$$\partial\rho/\partial t + div(\rho\boldsymbol{b}_+) = \mathcal{D}\Delta\rho \ ,$$

and a backward equation:

$$\partial\rho/\partial t + div(\rho\boldsymbol{b}_-) = -\mathcal{D}\Delta\rho \ .$$

These two equations are often called forward and backward Fokker-Planck equations. We may now define two new *average* velocities:

$$\boldsymbol{V} = \frac{\boldsymbol{b}_+ + \boldsymbol{b}_-}{2} \quad ; \quad \boldsymbol{U} = \frac{\boldsymbol{b}_+ - \boldsymbol{b}_-}{2}$$

i.e., in the space-time approach, they are the statistical averages of the individual velocities, $\boldsymbol{V} = \langle \boldsymbol{v} \rangle$ and $\boldsymbol{U} = \langle \boldsymbol{u} \rangle$. Adding the two Fokker-Planck equations yields

$$\partial\rho/\partial t + div(\rho\boldsymbol{V}) = 0$$

which is nothing but the well-known equation of continuity. This confirms $\boldsymbol{V}$ as a generalization of the classical velocity. Subtracting the two Fokker-Planck equations yields

$$div(\rho U) - \mathcal{D}\, \Delta \rho \;=\; 0 \quad ,$$

which may be written

$$div\{\rho\ [U - \mathcal{D}\ \nabla ln\rho\,]\,\} = 0 \quad .$$

One may actually demonstrate[31,40] by using the properties of (5.6.3) that the term under the *div* operator is itself null, so that U is a gradient:

$$U \;=\; \mathcal{D}\ \nabla ln\rho \quad .$$

We shall now introduce new notations aimed at fully using the bi-vector new structure (U,V). From now on, the derivation of the Schrödinger equation which is presented here is original (to the best of our knowledge). We place ourselves in the (U,V) plane and introduce a new complex velocity

$$\mathcal{V} = \ V - i\ U \ .$$

Consider now the backward and forward mean derivatives. In the same way as U and V have been defined, we may set

$$\frac{d_v}{dt} = \frac{1}{2}\ \frac{d_+ + d_-}{dt} \qquad , \qquad \frac{d_u}{dt} = \frac{1}{2}\ \frac{d_+ - d_-}{dt} \quad .$$

By combining Eqs. (5.6.3 a) and (5.6.3 b), these mean derivatives write

$$\frac{d_v}{dt} = \frac{\partial}{\partial t} + V \,.\, \nabla \qquad , \qquad \frac{d_u}{dt} = \mathcal{D}\ \Delta + U \,.\, \nabla \quad . \tag{5.6.4}$$

From these two new operators, we now define a complex operator

$$\frac{d'}{dt} = \frac{d_v}{dt} - i\ \frac{d_u}{dt} \quad ,$$

which, from (5.6.4), is finally given by

$$\frac{d'}{dt} = \left(\ \frac{\partial}{\partial t} - i\ \mathcal{D}\Delta\ \right) + \mathcal{V} \,.\, \nabla \quad , \tag{5.6.5}$$

since $V - i\ U = \mathcal{V}$ by definition.

We shall now postulate that the passage from classical (differentiable) mechanics to the new nondifferentiable processes that are considered here can be implemented by a unique prescription: *Replace the standard time derivative d/dt by the new complex operator* $\mathcal{d}/dt$. Let us indicate the main steps by which one may generalize classical mechanics using this prescription.

We assume that any mechanical system can be characterized by a Lagrange function $\mathcal{L}(x_i, \mathcal{V}_i, t)$, from which a *mean* action $\mathcal{S}$ is defined:

$$\mathcal{S} = \left\langle \int_{t_1}^{t_2} \mathcal{L}(x, \mathcal{V}, t)\, dt \right\rangle \quad . \tag{5.6.6}$$

The least-action principle, applied on this new action with both ends of the above integral fixed, leads to generalized Euler-Lagrange equations

$$\frac{\mathcal{d}}{dt} \frac{\partial \mathcal{L}}{\partial \mathcal{V}_i} = \frac{\partial \mathcal{L}}{\partial x_i} \quad , \tag{5.6.7}$$

in agreement with the correspondence ($d/dt \rightarrow \mathcal{d}/dt$). Other fundamental results of classical mechanics are also generalized in the same way. In particular, assuming homogeneity of space *in the mean* leads to defining a complex momentum

$$\mathcal{P}_i = \frac{\partial \mathcal{L}}{\partial \mathcal{V}_i} \quad .$$

If one now considers the action as a functional of the upper limit of integration in (5.6.6), the variation of the action from a trajectory to another nearby trajectory, when combined with Eq. (5.6.7), yields a generalization of another well-known result, namely, that the complex momentum is the gradient of the complex action:

$$\mathcal{P} = \nabla \mathcal{S} \quad . \tag{5.6.8}$$

We shall now specialize and consider Newtonian mechanics. The Lagrange function of a closed system, $L(\boldsymbol{x},v,t)=\frac{1}{2}mv^2-\mathcal{U}$, is generalized as $\mathcal{L}(\boldsymbol{x},\mathcal{V},t)=\frac{1}{2}m\mathcal{V}^2-\mathcal{U}$, so that the Euler-Lagrange equation keeps the form of Newton's fundamental equation of dynamics:

$$-\nabla\mathcal{U} = m\,\frac{d\!\!\!/}{dt}\,\mathcal{V}\ , \tag{5.6.9}$$

which is now written in terms of complex variables and operator.

Let us separate the real and imaginary parts of the complex acceleration $\gamma = d\!\!\!/\,\mathcal{V}/dt$. We find

$$d\!\!\!/\,\mathcal{V} = (d_v - i\, d_u)(V - i\,U) = (d_v V - d_u U) - i\,(d_u V + d_v U)\ .$$

The force $F=-\nabla\mathcal{U}$ being real, the imaginary part of the complex acceleration vanishes. It is given by

$$\frac{d_u}{dt}V + \frac{d_v}{dt}U = \partial U/\partial t + U.\nabla V + V.\nabla U + \mathcal{D}\Delta V = 0\ ,$$

from which $\partial U/\partial t$ may be obtained. Differentiating the expression $U = \mathcal{D}\,\nabla\ln\rho$ and using the equation of continuity yields another expression for $\partial U/\partial t$:

$$\frac{\partial U}{\partial t} = -\,\mathcal{D}\,\nabla(divV) - \nabla(V.\,U)\ .$$

The comparison of these two relations yields $\nabla(divV) = \Delta V - U\wedge curlV$, where the term in $curlU$ vanishes since U is already known to be a gradient. But in the Newtonian case now considered, the complex momentum becomes $\mathcal{P} = m\mathcal{V}$, so that Eq. (5.6.8) implies that *$\mathcal{V}$ is a gradient.* This demonstrates that the "classical" velocity V is a gradient (while this was postulated in Nelson's original paper). We can now introduce a generalization of the classical action S (in dimensionless units) by the relation

$$V = 2\mathcal{D}\,\nabla S\ .$$

Combining this relation with the expression for U, we find the complex action to be given by $\mathcal{S} = 2\,m\,\mathcal{D}\,(S - i\ \ln\rho^{1/2})$, i.e., its imaginary part is the logarithm of the probability density.

Note that Nelson[31] arbitrarily defines the acceleration as

$$d_N^2 x/dt^2 = \frac{1}{2}\,\frac{d_+ d_- + d_- d_+}{dt^2}\,x$$

(it could *a priori* have been any second order combination of d_+ and d_-; however see Nelson[40]). It is easy to show that Nelson's acceleration is nothing but the real part of our complex acceleration $\mathit{d}\mathcal{V}/dt$. It is also noticeable that a stochastic least-action principle was introduced by Guerra and Morato,[99] based on the *real* Lagrange function $L = \frac{1}{2}m(V^2 - U^2) - \mathcal{U}$, which is nothing but the real part of our complex Lagrange function $\mathcal{L}$.

We shall see that the way to Schrödinger's equation is now remarkably short. We now introduce the complex function $\psi = e^{i\mathcal{S}}$, that is, in terms of probability density and real part of action:

$$\psi = \sqrt{\rho}\ e^{iS}\ ,$$

and the complex velocity is now related to this new function by

$$\mathcal{V} = -2\,i\ \mathcal{D}\ \nabla(\ln\psi)\ . \qquad (5.6.10)$$

Let us stop one moment on this result. In terms of our *complex momentum* $\mathcal{P} = m\mathcal{V}$, it writes (when $\mathcal{D} = \hbar/2m$) $\mathcal{P}\psi = -i\hbar\nabla\psi$, i.e., in operator terms

$$\mathcal{P} = -i\,\hbar\nabla\ .$$

Hence one of the most mysterious "recipes" (or postulates) of quantum mechanics, the correspondence rule $p \to -i\hbar\nabla$, finds a natural interpretation once the complex "bi-velocity" is introduced.

Let us now introduce the wave function ψ in the equation of motion (5.6.9), which generalizes Newton's equation to nondifferentiable space. It takes the new form

$$\nabla\mathcal{U} = 2\,i\ \mathcal{D}m\,\frac{\mathit{d}}{dt}\,(\nabla\ln\psi)\ .$$

Being aware that d' and ∇ do not commute, we replace d'/dt by its expression (5.6.5):

$$\nabla \mathcal{U} = 2i\ \mathcal{D} m \left[\frac{\partial}{\partial t}\ \nabla \ln\psi - i\ \mathcal{D} \Delta(\nabla \ln\psi) - 2i\ \mathcal{D}(\nabla \ln\psi \,.\, \nabla)(\nabla \ln\psi) \right]$$

This expression may be simplified thanks to the three following identities, which may be established by straightforward calculation:

$$\nabla \Delta = \Delta \nabla$$

$$(\nabla f \,.\, \nabla)(\nabla f) = \frac{1}{2}\ \nabla(\nabla f)^2$$

$$\frac{\Delta f}{f} = \Delta \ln f + (\nabla \ln f)^2$$

This implies

$$\frac{1}{2} \Delta(\nabla \ln\psi) + (\nabla \ln\psi \,.\, \nabla)(\nabla \ln\psi) = \frac{1}{2}\ \nabla \frac{\Delta\psi}{\psi}\ ,$$

and we finally obtain

$$\frac{d'}{dt}\ \mathcal{V} = -\nabla \mathcal{U} / m = -2\ \mathcal{D} \nabla\ \left\{ i\ \frac{\partial}{\partial t} \ln\psi + \mathcal{D} \frac{\Delta\psi}{\psi} \right\}.$$

Integrating this equation finally yields

$$\mathcal{D}^2 \Delta\psi + i\ \mathcal{D} \frac{\partial}{\partial t} \psi - \frac{\mathcal{U}}{2m} \psi = 0\ , \qquad (5.6.11)$$

up to an arbitrary phase factor $\alpha(t)$ which may be set to zero by a suitable choice of the phase S. Replacing $\mathcal{D}$ by $\hbar/2m$, we get Schrödinger's equation

$$\mathcal{U} = \frac{i\hbar}{\psi} \frac{\partial}{\partial t} \psi + \frac{\hbar^2}{2m} \frac{\Delta\psi}{\psi}\ .$$

This is, in our opinion, a demonstration that an eventual future theory based on the principle of scale relativity and the fractal space-time conjecture will give back quantum mechanics as an approximation. The very direct route from Newton's equation to Schrödinger's equation suggests to us the following interpretation of quantum mechanics: *quantum mechanics is mechanics in a non-differentiable space.* We shall also suggest in Sec 7.2 that the above formalism may be used in a different situation, namely that it may help solve the problem of structures arising from chaos.

We think that the above formalism solves the problem of the physical origin of the "complex plane" of quantum mechanics. One of the most mysterious feature of this theory was the complex nature of the probability amplitude: being irreducible to classical laws, it led to the belief that it is impossible to find its origin in space-time and to the interpretation of quantum mechanics in terms of an abstract space. Here we have shown, by extending Nelson's formalism, that in a non-differentiable space-time one may attach to each point of space-time a plane (U,V) (this plane is a coordinate-dependent "field"), and that the wave function is directly linked to this plane:

$$\nabla \ln\psi = \frac{1}{2\mathcal{D}} \; (U + i\,V) \; .$$

This returns the probability amplitude into 4-dimensional space-time, but into a space-time which is basically and irreducibly non-differentiable.

Paradoxically, if one accepts the above interpretation of quantum mechanics, one of the main difficulties is to understand why this "approximation" is so good. Recent theoretical and experimental work on nonlinear perturbations on the Schrödinger equation[45,46] give a first clue of an answer to this question: in the domain of low energy processes (spectra of atomic transitions and so on), nonlinear corrections are expected to be extremely faint.

In our own approach, the most relevant comparison concerning possible new predictions is in the relations between Newton's and Einstein's theories. The Newtonian theory of gravitation is indeed so precise in its predictions that it was thought for centuries that it was definitive. The relativistic theory, special and general, nevertheless embeds the Newtonian theory, and brings "corrections" to it in well-defined situations: specifically, when a variable or a relevant physical quantity becomes very large. The theory of relativity should be used when describing or using very high velocities ($v \approx c$), large distances (the case of cosmology) [but in special relativity already, even for $v << c$, the time transformation writes $t' = t + vx/c^2$, and there is a non-negligible relativistic correction for $x \gtrsim c^2/v$], high densities (neutron stars) or strong gravitational fields (black holes).

We suggest that the same is true of quantum mechanics. Most current experiments testing quantum mechanics to a very high degree of precision correspond to weak field and/or relatively large length scale. So possible deviations from quantum mechanical predictions are to be searched in two situations:

**Strong field*: we have suggested[3] that the unexplained anomalous positron and electron lines observed at Darmstadt in heavy ion collisions are precisely a first occurrence of such a breaking of quantum mechanics in strong electromagnetic field. We shall show in Sec. 5.10 that these spectra have indeed fractal properties whose main features are accounted for by fractal derivatives of the kind described in Sec. 3.7.

**Very small scale*, i.e., *very high energies* : we shall address more fully this case in Chapter 6, in which a theory of special scale relativity will be developed. It leads to a new relation between the length-time scale and the mass-energy-momentum scale and to several new predictions, among which that of the value of two fundamental scales in particle physics, namely the scale of "grand unification", and the electroweak symmetry breaking scale.

5.7. Classical-Quantum Transformation.

We have seen in the previous section that, starting from Newton's classical law of dynamics, $F = m\gamma$, its transformation by the way of a non-differentiable Wiener process leads to Schrödinger's equation, i.e., to quantum mechanics. Conversely, for the last ten years, there has been extensive research activity on the transition from quantum to classical.[47-54]

This work has led to the conclusion that this transition was the result of an interaction with the environment. The dominant effect of this environment, described e.g. by a scalar field interacting with the particle under consideration, *is a Brownian motion* which destroys quantum coherence and yields classical states. Such a result then leads to an extraordinary paradox: the same effect, a Brownian motion, i.e. a diffusion process that is often considered as an increase of disorder, allows one to go from quantum to classical *and* from classical to quantum. This clearly deserves additional analysis, which is the aim of the present section.

The problem addressed by Zurek[55] and other workers in this field is a long standing one. We know that the elementary laws of physics are quantum laws, while we also know from everyday experience that in most situations the macroscopic world obeys classical laws. Thus there must be some classical-quantum transition. The problem is that this transition does not seem to be apparent concerning the basic quantum mechanical equations, neither Schrödinger's equation, nor Heisenberg's relations, which are expected to be universal. Moreover, attributing the quantum laws to microphysics and the classical laws to macrophysics, and thus looking for an absolute scale of transition is unacceptable: we know that there exists macroscopic quantum systems such as superconductors. If such a transition does exist (some authors claim that it does not), it must be *relative*, i.e., a function of some physical characteristics of the system (mass, velocity, energy, temperature...).

There is a well-known solution to this problem, already described at length in most text-books. The conditions of validity of the classical approximation are those of geometric optics, i.e. the quantum-classical transition is nothing but the de Broglie length, $\lambda = \hbar / m v$ for a free particle, generalized to $\lambda = \hbar / \sqrt{2m(E-V)}$ for a particle in a potential. However what remained unknown or badly known was the precise way the transition occurred. In particular no one was able to describe in detail the way by which specific quantum phenomena as delocalization or coherence (large scale correlations) were destroyed in the classical domain (e.g., the well-known Schrödinger's cat problem). Moreover, applying naively the hereabove de Broglie formula to macroscopic systems leads to a paradox which was hardly solved: consider indeed two bodies set on a firm ground.

Their relative velocity may be expected to be vanishing, so that, whatever high their mass, the de Broglie wavelength of one of the bodies in the other body's rest frame tends to infinity. Should we admit that states where there is non-zero probability for the body to be in various places relative to the other are possible ?

Unruh and Zurek's solution,[54] which continues the work by Zeh,[47] Zurek,[48,49] Caldeira and Leggett,[51,52] and others, is as follows. They first remark that *quantum systems are seldom isolated but rather interact with their environment.* Note that this premise is exactly the same as that set as basis for construction of stochasic quantum mechanics. Hence Nelson[40] writes in the introduction of his book *Quantum Fluctuations*: "No physical system of finitely many degrees of freedom is truly isolated: it is always in interaction with a background field". Describing the environment as a scalar field, Zurek and collegues find, as stated above, that its coupling to a particle induces a Brownian-like motion which destroys quantum coherence.

Start with the time evolution equation of the density matrix which describes the system under consideration

$$\dot{\rho} = L\,\rho \ , \tag{5.7.1}$$

where L is an evolution operator taking into account the system and its interaction with the environment. This equation is known as a 'master equation', and is found to take the general form:[54,55]

$$\dot{\rho} = -\frac{i}{\hbar}\ [H, \rho] - \gamma\,(x - x')\left(\frac{d\rho}{dx} - \frac{d\rho}{dx'}\right) - \frac{2m\gamma k_B T}{\hbar^2}\ (x - x')^2\,\rho\ . \tag{5.7.2}$$

The interaction with the environment gives rise to the two last terms. The second term causes dissipation, but it is the last term which destroys quantum coherence: it describes a Markov diffusion process due to random kicks by the heat bath of the interacting scalar field. Anyway Unruh and Zurek show that in the Wigner representation, the master equation takes the Fokker-Planck form (but with a time dependent diffusion coefficient).[54]

There is another important remark one may make about (5.7.2) in the light of our own approach. Let us quote Unruh and Zurek:[54] "we have discovered that the calculation becomes more tractable when the density matrix ρ is given not in the position representation but rather in a new $(k,\Delta x)$ representation; here Δx measures the distance from the diagonal in the position representation". Hence Δx is nothing but a resolution, $\Delta x = x - x'$, and it is quite apparent in Eq. (5.7.2) that it is the relevant variable. We shall see that it is precisely in terms of the ratio of this variable over the de Broglie scale that the quantum classical transition is elucidated in this approach. So the master equation may be viewed as the first example (to my knowledge) of an *explicitly scale-dependent equation* in the domain of low energies (at high energies, nearly every fundamental equations become scale dependent through the renormalization group).

Let us proceed further and answer the main question: where is the classical-quantum transition? Unruh and Zurek consider a coherent superposition of two Gaussians separated by a distance Δx much larger than their width. The corresponding density matrix has four peaks (see Fig. 5.9 a). For example, such a situation occurs in a two slit experiment with both slits open and no detector behind the slits: the quantum probability is then $P = (\Psi_1 + \Psi_2)^2 = \Psi_1^2 + \Psi_2^2 + \Psi_1\Psi_2 + \Psi_2\Psi_1$. It is given not only by the two classical terms $P_1 = \Psi_1^2$ and $P_2 = \Psi_2^2$, but also by two additional off-diagonal terms. They find that the interaction with the surrounding medium results in a decay of the off-diagonal peaks, thus leaving only the two diagonal peaks: this corresponds to the classical law of addition of probabilities, $P = P_1 + P_2$ (see Fig. 5.9 b, in which decoherence is nearly achieved). But the really clarifying result is their calculation of the decoherence time scale. They find that the off-diagonal peaks decay as $P \approx e^{-t/\tau_D}$ with the decoherence time scale τ_D given by

$$\tau_D = \tau_R \left(\frac{\lambda_T}{\Delta x}\right)^2 ,$$

where τ_R is the relaxation time of the system and λ_T is the *thermal* de Broglie wavelength

$$\lambda_T = \frac{\hbar}{\sqrt{2mk_B T}} .$$

This confirms that the classical-quantum transition is nothing but the de

Broglie scale, as was already found in other approaches, in particular the fractal one which we try to develop here.

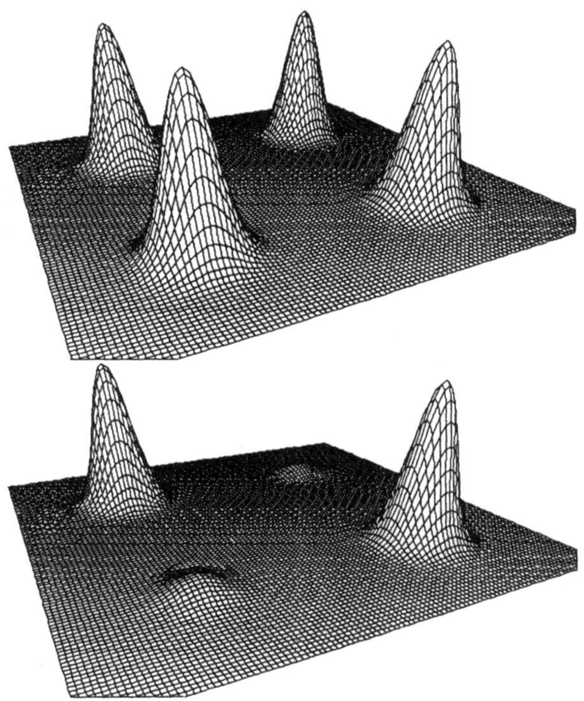

Figure 5.9. (Adapted from Fig. 4 of Ref. 55). Density matrix of a particle including off-diagonal terms which describe quantum coherence (up). Partially decohered density matrix (down).

Let us comment on this result. My first comment is that it definitively solves one of the worse puzzles of quantum mechanics: why does a macroscopic body *at rest* always have a well-defined position, while its de Broglie length $\hbar/mv$ is formally infinite for $v = 0$? The solution is that, for an extended object, the de Broglie length should be expressed in terms of *temperature*. Take e.g. a body weighing 1 g at room temperature (≈300 K). One finds the de Broglie transition to be $\lambda_T \approx 10^{-20}$ cm. At a resolution Δx = 1 cm, one should reach a temperature $T \approx 10^{-38}$ K in order to observe macroscopic quantum effects at such a scale: this is to be compared with the present limit of $\approx 10^{-6}$ K in laser trapped atom experiments.[56] This means that interaction with the 3 K cosmic microwave background photons are enough to decohere most macroscopic systems.[50] Moreover, as remarked by Zurek,[55] for such a system (m = 1 g, Δx = 1 cm, T = 300 K) prepared in a quantum coherent state, the decoherence time scale is 10^{-40} times the relaxation time, so that, even if τ_R was of the order of the age of the Universe ($\approx 10^{17}$ s), it would remain as small as 10^{-23} s (that is, of the order of the de Broglie time of an electron). This result is also in good agreement with the fact that macroscopic quantum effects, such as superconductivity, are observed in most cases at cryogenic temperature.

My second comment concerns the quantum-classical dualism. Let us quote Woo:[53] "Zurek's aim seems to ultimately deduce the finality of a classical record by first obtaining effective superselection rules; whereas we think that the implication goes in the opposite direction [...] – once one accepts the classical-quantal dualism and properly classifies information into the classical and the quantal, the superselection rules become understandable". Said in other words, Zurek's position seems to be that the really fundamental laws are the quantum laws, while the classical laws are only approximations of them (he indeed writes:[55] "I suggest that this idealization is responsable for our confidence in classical mechanics, and, more generally, our belief in classical reality"), while Woo and others claim that there is a fundamental and irreducible classical-quantum dualism, i.e., that a *reduction* of classical laws to approximate quantum laws is impossible.

I would say that the recent results by Unruh and Zurek actually make the two points of view converge. Indeed, starting from the hope to find classical laws as a degenerated case of quantum laws, Zurek finds that the classical-quantum transition *does exist,* that it is *inevitable*, that it is *nearly instantaneous*, and overall that there is *an exponential decay of non-classical terms*: this is the key point, which shows that Zurek has, in the end, demonstrated the classical-quantum dualism. Indeed in the above example (m = 1 g, Δx = 1 cm, T = 300 K, $\tau_D = 10^{-23}$ s), after 1 second the contribution of non-classical terms is $P = e^{-10^{23}}$, which is absurdly small and merely *indicates that the system has become totally classical.* Thus the classical laws are *not* approximations of quantum laws: the classical sum of probability is, by all standards of meaningful precision (see the hereabove number), an *exact* law; the passage from quantum to classical is an actual (but *relative*) transition, similar to phase transitions in statistical physics.

My third comment concerns the claim that the transition to classical comes solely from interaction with the environment. Let us come back on the fractal approach. Consider an *isolated* free particle. It verifies Heisenberg's relation, $\Delta p\, \Delta x \approx \hbar$, assumed universal, i.e., at the classical approximation, $\Delta v\, \Delta x \approx \hbar/m$. Recall that the fractal approach consists in introducing a new quantity, the integrated scale dependent length $\mathcal{L}(\Delta x)$ which serves as generalization in the quantum case of classical curvilinear coordinates. We stress that, even though $\mathcal{L}(\Delta x)$ may be obtained by actual measurements of positions, it may also be defined and calculated independently of any measurement: we postulate in that case that it gives us information about the virtual internal structure of the particle "trajectory", and that this is precisely such a structure which is revealed by an actual (single) measurement.

The key point is that $\mathcal{L}(\Delta x) \propto <|v|>$: it is *this precise form*, which comes from the definite choice to come back to a geometric description of physics and to define a coordinate system that may follow the particle in its quantum motion, which gives rise to the appearance of the de Brogrie transition. If one rather attempts to characterize the trajectory by using what could seem a more "natural" quantity such as $<v^2>^{1/2}$, as was done by Feynman in his pionneering work[8] (see Secs. 4.1 and 5.6), one finds that Heisenberg's relation is translated into a *universal* scaling law $<v^2>^{1/2} \propto (\Delta x)^{-1}$, *without any transition to classical.*

Let us recall the fractal result once again, owing to its importance in the present context. The fact that $\mathcal{L}(\Delta x)$ involves $<|v|>$ while the Heisenberg relation is written in terms of Δv implies the need to connect these two quantities. Two cases may happen (see Sec. 4.2 and Figures therein):

*$<v> \gg \Delta v$: in that case it is clear that $<|v|> \approx <v>$ and that $\mathcal{L}$ is scale independent. The de Broglie length is naturally introduced without any additional hypothesis: using the Heisenberg relation $\Delta v \Delta x \geq \hbar/m$, the condition $<v> \gg \Delta v$ writes $\Delta x \gg \hbar/<v> = \lambda_{dB}$.

*$<v> \ll \Delta v$, i.e., from the above remark, $\Delta x \ll \hbar/<v>$: in that case $<|v|>$

$\approx \Delta v$, and the Heisenberg relation yields the quantum behaviour $\mathcal{L}(\Delta x) \propto (\Delta x)^{-1}$, which corresponds to fractal dimension 2.

Thus when characterized by the fractal test (quantum is fractal while classical is not), the classical-quantum transition is obtained without invoking any interaction with the environment. Note that the Heisenberg relation remains universally valid, even in the classical domain where nevertheless the classical laws are now found to hold exactly. This shows that the classical-quantum transition is in some sense intrinsical to physical laws. This result is to be related to a recent work by Balian[57] who recalls that the *reduction of the wave packet*, i.e. *the projection hypothesis*, does not need to be postulated, but may be deduced from the other postulates of quantum mechanics: it is actually implicitly contained in the principle of repeatability, which states that two ideal measurements of the same quantity A performed in *immediate* succession on the *same* system provide the same answer a_a. This principle (which was explicitly added by Von Neuman) is often forgotten in textbooks on quantum mechanics, maybe because it is at variance with the current interpretation that the results of measurement are determined by the measurement itself (see Chapter 2 on this point). Anyway it is made necessary by experience itself: it is well known that if we place a polaroid in front of a light beam and then measure again the polarization with a new polaroid, it is found to have been unchanged between the two polaroids.

However we are conscious that the fractal argument, in its present version, is incomplete and badly adapted to an extended system, and more generally to real systems: in particular the de Broglie wavelength is a Lorentz covariant quantity, and one may place oneself in a reference system where $\langle v \rangle = 0$; the transition to classical would then be rejected to infinity. In fact the de Broglie length in its simplest form, $\lambda = \hbar/p$, corresponds to a plane wave, while in actual experiments, such as laser trapped atoms,[56] the effective quantum-classical transition is rather a quantum phase correlation length, i.e. a thermal wavelength. This kind of paradox is clearly solved in the environmental approach. Moreover the transition anyway keeps its intrinsic character, since the whole system {object + environment} is treated quantum mechanically : it is at the end an authentic, though complicated, quantum system that eventually gives rise to classical behaviour. Concerning the fractal approach, this means that extensive work is still needed if one wants to demonstrate in a general way that the quantum-classical transition can be identified with a fractal / non-fractal transition.

Let us now come to the point which mainly motivates the present section. Start from quantum laws, add Brownian motion, and you get classical laws. But we have seen in the previous section that the reverse is also true: start from classical, add Brownian motion, and you get Schrödinger's equation, i.e. quantum laws. This is an extremely mysterious and paradoxical result. A Markov or Wiener process is usually considered as a method for describing disorder, and, in a naive view, one may expect that applying "disorder" twice would yield even more disorder. On the contrary, the combination of the results of stochastic quantum mechanics and the environment-induced decoherence seems to imply rather that the Wiener process transformation is "reversible".

In order to understand such a paradoxical result, one must come back to the notion of order and disorder. The first reactions of physicists who were confronted with the quantum behaviour of microphysics at the beginning of the century were that the quantum world was far more disordered than the classical one: loss of simultaneously well-defined position and momentum, uncertainty, unavoidable statistical and probabilistic description, undeterminism and delocalization were the signature of the new realm. Then a natural tendency was to hope to understand such a "complicated" behaviour by the Brownian motion-like influence of a background which would render messy the "simple" classical laws.

Since then physicists have become accustomed with the principles of quantum mechanics and have studied the quantum behaviour in itself, without reference to classical laws. Some of them have started to consider them as the really fundamental laws of nature. Once admitted that the quantum fundamental "object" is a probability amplitude, one notes that its equation of evolution is extremely simple, linear and deterministic. The quantum world is correlated, highly coherent, and far less subjected to chaos than is the classical. In this view the question which arises becomes: how can such a messy world as the classical one exist in Nature ? The only solution is to add "disorder", by the way of the Brownian-like effect of the environment.

We see only one solution out of this dilemna. We must give up the concepts of order and disorder as physically meaningful concepts and admit that both domains, classical and quantum, are *organized*, but differently. Then the D=2 Wiener process $\mathcal{W}$ becomes a reversible transformation (i.e., $\mathcal{W}^2 = 1$) from one type of organization to the other. While delocalizing, $\mathcal{W}$ brings coherence, and while decohering, $\mathcal{W}$ relocalizes.

To summarize this section, the combination of the results of Nelson *et al.* and Zurek *et al.* leads to the conclusion that the quantum-classical dualism is a fundamental and inevitable property of Nature, that the classical and quantum domains are two "orthogonal" types of organization which transform "reversibly" one into the other by way of a Wiener process, their transition occurring around the thermal de Broglie wavelength:

$$\begin{bmatrix} \text{CLASSICAL} \\ \textit{localized} \\ \textit{uncoherent} \end{bmatrix} \leftarrow \begin{bmatrix} \text{Wiener process } \mathcal{W} \\ \mathcal{W}^2 = 1 \\ \lambda_{\text{dB}} \end{bmatrix} \rightarrow \begin{bmatrix} \text{QUANTUM} \\ \textit{delocalized} \\ \textit{coherent} \end{bmatrix} .$$

5.8. Microscopic Reversibility of Fractal Time.

The introduction of a fractal time running backward for time intervals smaller than $\tau \approx \hbar/E$ allows one to bring new insights into the question of the arrow of time. Consider a curve of fractal dimension 2 aimed at describing an elementary particle in its rest frame. We assume the curve to be drawn in a space described with coordinates t, t', t''... , where the coordinate t is to be identified with classical time and t' and t'' are extra-dimensions aimed at describing the "thickness" of the fractal curve. The precise meaning of these extra-coordinates has no importance in the demonstration to follow (they may actually be the spatial coordinates themselves).

We assume that the proper time of the particle, i.e. by definition the quantity which flows uniformly, is the curvilinear coordinate on the fractal curve, while smoothing the fractal with balls $\Delta t > \tau$ reduces the curve to the t axis (classical trajectory). Consider now the fractal curve as being made of p^{ω} non-standard segments of length $q^{-\omega}$. Each of these segments is a vector of components t_i, t'_i, t''_i, ... Denoting with a + subscript the components of plus sign and with a – subscript the (absolute values of the) components of minus sign, the hereabove choice leads one to write for one period the following sums over the p^{ω} segments:

$$\Sigma t'_+ = \Sigma t'_- \,, \quad \Sigma t''_+ = \Sigma t''_- \,, \ldots$$

$$\Sigma t_+ - \Sigma t_- = \tau \,,$$

$$\Sigma t_+ + \Sigma t_- = K \; \tau q^{\omega} \,,$$

where K is a finite number $0 < K < 1$ (e.g. for the Peano curve, K=1/2; in Euclidean 3-space, isotropy on the fractal would yield K=1/3). Coming back to the finite case by replacing as usual q^{ω} by $(\tau/\Delta t)$, we get the following probabilities, P_- for the particle to run backward in time (i.e., to be viewed as an antiparticle) and P_+ to follow the arrow of classical time:

$$P_+ = \frac{1}{2} \; [K(\Delta t) + \frac{\Delta t}{\tau}] \,,$$

$$P_- = \frac{1}{2} \; [K(\Delta t) - \frac{\Delta t}{\tau}] \,,$$

where $K(\Delta t) \to K$ when $\Delta t \to 0$. As a consequence one gets complete reversibility on the limiting fractal, since $P_+ = P_-$ when $\Delta t \to 0$, while the classical time t was *a priori* oriented in one direction. This result allows one to set the problem of the relations between microscopic reversibility and macroscopic irreversibility in a renewed manner: As demonstrated hereabove, in the fractal model of time, both of them naturally coexist.

5.9. Fractals and Quantum Electrodynamics.

One of the ways by which we arrived in Chapter 4 at the concept of a fractal dimension 2 for the proper time of elementary particles was a reinterpretation of the virtual electron-positron pairs occurring in quantum electrodynamics (QED) as a manifestation of the fractal "trajectory" of the particle itself. It is now conjectured that the same is also feasible for the virtual photons.

The result that the fractal dimension of a particle trajectory is 2, combined with the ansatz that it should be self-avoiding (in fractal coordinates) implies that such a curve (if embeddable in Euclidean space) cannot be represented in $\mathbb{R}^2$, but at least in $\mathbb{R}^3$. Consider the fractal 2-dimensional orthogonal generator of Fig. 5.10b (note that it is a kind of 3-dimensional version of the Peano curve). We assume that the variable t should be identified with the classical proper time of the particle, i.e., with the Minkowskian metric invariant. Hence displacements along the two extra-axes t' and t'' correspond to $dt = 0$. (We recall once more that these two extra-dimensions are defined only for the need of representation, while we assume that the only physical variables are the classical time t, the curvilinear coordinate on the fractal curve s, and the three space variables). The generator is made of 9 segments of length $\tau/3 = \hbar/3mc^2$. Let us follow the classical proper time from $t = 0$ to $t = \tau$. At resolution $\Delta t = \tau$, one sees a single electron of mass $m_e=\hbar/\tau c^2$. At resolution $\Delta t = \tau/3$, one "sees" the same electron between $t = 0$ and $t = \tau/3$; then at the instant $t = \tau/3$ what is seen is a neutral pair of fermion-antifermion with $dt = 0$, i.e., a spin 1 vectorial particle which will be interpreted as a virtual photon. During the following $\Delta t = \tau/3$ interval, one sees an electron plus an electron-positron pair, then again a virtual photon, then the single initial electron. The fractal trajectory at resolution $\tau/3$ would be finally identified in terms of virtual photons and electron-positron pairs with the 4-vertices Feynman graph of Fig. 5.10b, which is one of the fourth order contributions to the self-energy of the electron, corresponding to the emission of a virtual photon, decaying into an electron-positron pair, which annihilates into a photon finally reabsorbed by the electron.

In Fig. 5.10a is given another generator which we suggest is to be identified with the second order diagram (first order in terms of the fine structure constant α) with one virtual photon emitted, then reabsorbed by the electron. Note that the "trajectory" projected on the plane (t', t''), describes a closed loop, which is characteristic of a given virtual photon in this representation. The two antiparallel segments at $t = \tau/3$ are to be identified with the emitted (or absorbed) photon, and the two antiparallel segments at $t = 2\tau/3$ with the absorbed (respectively emitted) photon, while the two antiparallel segments at $t = \tau/3$ and $2\tau/3$ are assumed to correspond to the same transmitted virtual photon. This may be generalized, as seen in Figs. 5.10c and 5.10d, to the cases of 4-vertices Feynman graphs with two virtual photons (with generators made of 16 segments of length $\hbar/4\tau c^2$). Note that in all these new representations, various states of the electron correspond to different coordinates in the plane (t', t''). The vertices correspond to the ends of "electronic" segments $(dt \neq 0)$, i.e., to rotations from the t axis to the t' or t'' axis.

If one considers now higher orders of the fractal construction, the generator structure will be embedded into individual segments, possibly after a rotation of $\pi/2$. In such a rotation, an electron-positron pair (antiparallel segments in the planes t,t' or t,t'') may become a photon (antiparallel segments in the plane t', t'') and the reciprocal is true (see Fig. 5.11e). This is an essential result for the *relativistic* point of view which we attempt to adopt here.

Even in standard QED the fractal character of physical objects is apparent in the decomposition of their self-energy (and more generally of any process) into Feynman diagrams of increasing order. It is this structure which allows the introduction of quantities like the polarisation operator for the photon, the mass operator for the electron, and which leads to their relations to propagators and to the Dyson equation (see e.g. Ref 58). The various levels of the decomposition of the self-energy of the electron are successive corrections to the internal lines, but also to the vertices of the only irreducible second order diagram with one virtual photon[58] (Fig. 5.10a). This decomposition is made by using two fundamental building blocks, virtual photons and virtual electron-positron pairs, with the additional recipe that the second one appears two orders (in terms of charge e) after the first one. As a consequence this fractal structure is not a simple self-similar one of the kind we most of the time considered up to now, but a more complicated construction involving at each level these two "stones" and their insertion, in internal lines and also around vertices.

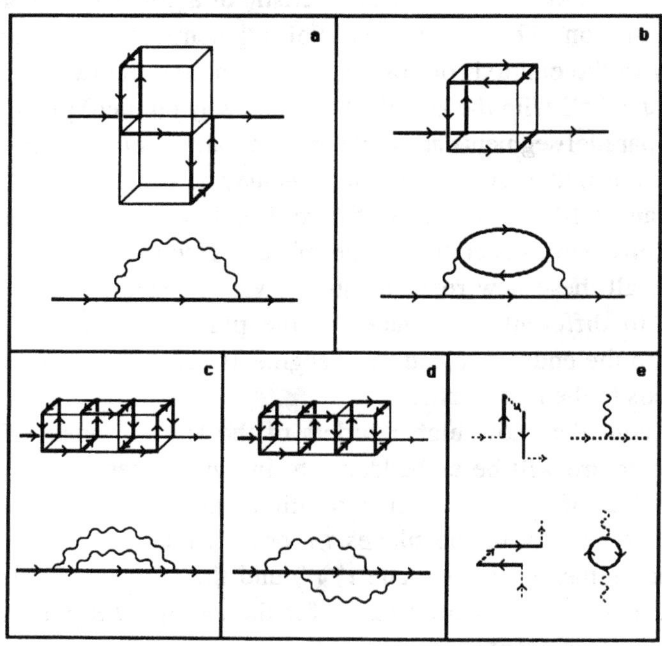

Figure 5.10. The analogy between Feynman diagrams and the generators of orthogonal fractals of dimension 2 (see text).

In fact, as already stated in Chapter 3, the digitalized resolution $\Delta t = t_0.q^{-n}$ should be finally replaced by a continuous one, and the discretized generators by infinitesimal ones. In the case of the fractal of Fig. 5.10a, this may be realized by taking a high order version of its construction, then smoothing it with balls of continuously increasing radius. But an infinity of different generators may then lead to the same final fractal. For example a self-similar fractal is equally obtained by taking as generator any of its approximations F_n. This remark leads to reinterpreting an orthogonal generator, like the one in Fig. 5.10a, as the limit for pair creation. Indeed its smoothing with balls $\Delta t > t_o/3$ provides no backward segments (only virtual photons will be "seen"), while electron-positron pairs will be revealed in this case with an increasing probability only when Δt becomes smaller than $t_o/3$.

More generally, in a continuous description of the new "zoom" or scale dimension, orthogonality cannot be conserved, so that we suggest that the virtual pairs / virtual photons elements of the QED approach should finally be identified to the *components* of a vector $(t,t',t'',...)$, respectively (t) to electrons and positrons, this being justified by $\sum t = \tau$ (even number of segments, see Figures); and the orthogonal vector $(t',t'',...)$ to virtual photons, this being justified by $dt = 0$ and $\sum t' = \sum t'' ... = 0$.

Disregarding the fact that the lack of strict self-similarity should be accounted for in higher order Feynman diagrams, we will now show that, despite the absence of a detailed theory, a general class of fractal structures emerges if one tries to give a fractal description of elementary phenomena, even in the first approximation. We have remarked that the resolution corresponding to the first occurrence of fractal invariant time elements orthogonal to the classical time axis should also correspond to the limit for the occurrence of virtual electron-positron pairs. The Heisenberg relation $\Delta E \, \Delta t \geq \hbar/2$ suggests this limit to be $\Delta t = \hbar/4mc^2$, while the electron Compton time (i.e. its rest frame de Broglie time) is $\lambda_c = \hbar/mc^2$.

A more detailed argument may be made for the effective occurrence of such a scale $\approx \lambda_c/4$ in connection with the electric charge. The variation of the fine structure constant due to vacuum polarization is given to lowest order, for scales $r << \lambda_c$ in space representation, by the relation (see Section 6.2):[58,59]

$$\alpha(r) = \alpha \left\{ 1 + \frac{2\alpha}{3\pi} \left[\ln \frac{\lambda_c}{r} - (\gamma + \frac{5}{6}) \right] \right\},$$

where $\gamma = 0.577\ 215...$ is Euler's constant. This relation may also be written in the form:

$$\alpha(r) = \alpha \left\{ 1 + \frac{2\alpha}{3\pi} \ln \frac{\lambda_c/Q}{r} \right\},$$

where the number Q is indeed very close to 4: $Q = e^{\gamma+5/6} = 4.098\ 204...$.

This leads us to tentatively consider fractals built up from orthogonal generators based on a structural constant $q = 4$ (see Chapter 3). Some examples of such generators made of 16 segments of length 1/4 are given in Figs. 3.12 a,b,c. The generator of Fig. 3.12a has been built from a symmetrization of the $q = 3$ generator of Fig. 5.11a [thanks to the identity $4^2=2(3^2-1)$]. Note that one obtains its symmetric version by time reversal, in agreement with the particle-antiparticle symmetry.

The interesting point here is that the fractal structures which appear in the fractal derivative of the fractal of Figs. 3.12 a,b,c, in particular the distribution of peaks and holes, are not specific to a particular model, but to a general class of symmetric fractal curves whose orthogonal generators are characterized by $q = 2$ or $q = 4$. This universality of structures is related to the loss of information implied in the projection process from the fractal curve to its fractal derivative. For example in Ref. 60 the same kind of structures was found from a fractal curve of dimension $D = 3/2$. But the $D = 3/2$ generator we used (which is studied in Chapter 3) was precisely a projection of the $D = 2$ generators of Fig. 3.12 a or b.

Consider indeed a $q \approx 4$ orthogonal generator without back-running segments. It will be made of 4 horizontal segments and 4^D-4 orthogonal segments situated at $t = 0.1$, 0.2 and 0.3 in the counting base 4. This will correspond to a first order fractal derivative with a central peak (.2) and two symmetrical secondary peaks (.1) and (.3), as seen in Figs. 5.11 and 3.5. To second order these peaks will show substructures, while tertiary peaks (of coordinates .02, .12, .22, .32) appear at the center of the inter-peak holes, giving to these holes a characteristic doublet structure, since the second order equivalent of the secondary peaks are embedded in the wings of the first order peaks. The final expected distribution of holes is given in Eq. (5.10.1). This similarity of structures of $q = 4$ fractal derivatives is illustrated in Fig. 5.12. The same result also holds for a large class of fractals with $q = 2$, of which the hereabove $q = 4$ generators would be the second order approximation F_2.

Clearly the hereabove approach proceeds from mere analogy, so that only experiment may for the moment decide whether it is endowed with physical meaning. It will indeed be shown in the next section (5.10) that the hereabove fractal derivative structures is able to describe new resonances which have been recently found in strong field QED experiments.

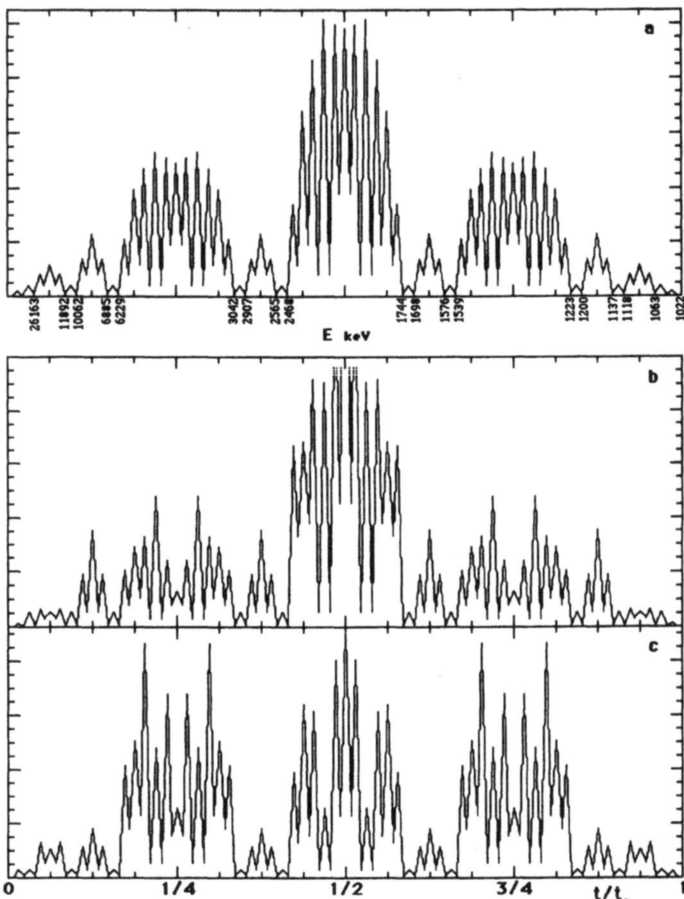

Figure 5.11. Comparison between the fractal derivatives of curves of fractal dimension 2 in $\mathbb{R}^3$ based on orthogonal generators with q=4 (level of fractalization $\xi_{3,5}$). Figures **a**, **b**, **c** correspond respectively to the fractals of Figures 3.12 a, b, c. (Note that the central peak of Fig. **b** has been truncated). While the peak structures differ from one fractal to the other, the structure of minima remains the same, as explained in text (compare also to Fig. 3.5). Identifying the main period with $E_0=2m_ec^2=1022$ keV yields energies of minima given by $E_m=E_0.t_0/t_m=\hbar/t_m$, as plotted below in Fig. 5.11**a**.

Let us conclude this section by an additional illustration of the way fractal models allow one multiple interpretation of QED phenomena.

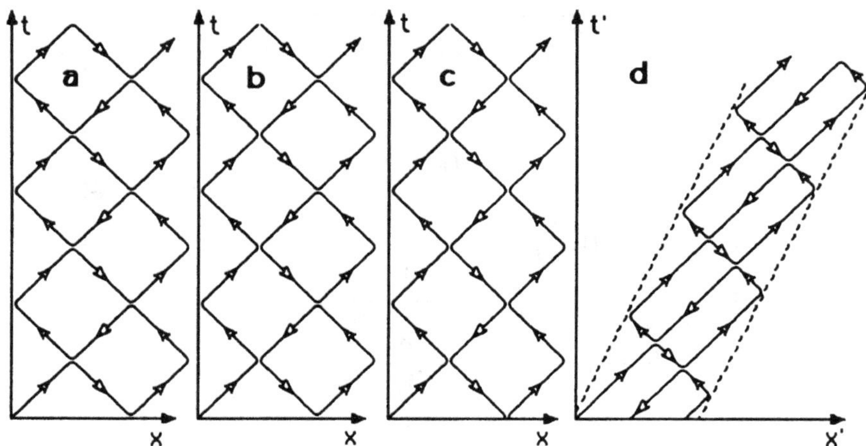

Figure. 5.12. Various interpretations of a fractal trajectory in space-time (see text). Figure **d** is a Lorentz transformation of Fig. **a.**

In Fig. 5.12 **a,** a fractal trajectory in space-time of a particle in its classical rest-frame is described by a Peano curve (approximation F_1) containing segments running back in time. The same trajectory may be interpreted (Fig. **b**) as one particle following an oscillatory motion (Zitterbewegung), plus particle-antiparticle loops; or (Fig. **c**) as two particles plus one antiparticle. The velocity of each individual segment is the velocity of light, in agreement with the solutions to the Dirac equation.[58,59] In Fig. 5.12 **d,** the effect of a Lorentz transformation is shown. The particle gets an average classical velocity $v < c$, while the velocity on the fractal remains $\pm c$: the difference with the rest frame is only in the length of the individual segments depending on their orientation. All these properties are conserved for the higher order approximations of the fractal curve.

Consider a general free-particle solution of the Dirac equation including both positive and negative energy terms[59]

$$\psi(x,t) = \int \frac{d^3p}{(2\pi\hbar)^{3/2}} \sqrt{\frac{mc^2}{E}} \; [\, b(p)\, u(p)\, e^{-\theta} + d^*(p)\, v(p)\, e^{\theta} \,] , \quad (5.9.1)$$

where $\theta = ip^{\mu}x_{\mu}/\hbar$, and $u(p,s)$ {resp. $v(p,s)$} denotes the spinor which is a positive energy solution (resp. negative) of the Dirac equation with momentum p^{μ} and spin s^{μ}. In the above integral, we have omitted for simplicity the sum over the spin states. We have already demonstrated in Sec. 4.4 that the rate of positive and negative energy solutions for a given localization of the particle is precisely the rate of the number of segments which go respectively forward and backward in time on a D=2 fractal trajectory. So, while the current interpretation of (5.9.1) corresponds to Fig. 5.12 **c**, it is now immediately clear that it can also be interpreted in terms of Fig. 5.12 **a**. In other words, the Dirac equation allows *localized, one-particle solutions*, provided they are *dimension* 2 *fractals*, which manifest themselves by the Zitterbewegung.

5.10. Possible Obvervable Implications.

On some possible consequences of a fractal structure of space-time.

Though the theoretical attempt which has been outlined hereabove is still in a very early state, one may already wonder about its possible observable consequences. However the very high precision with which quantum physics is now experimentally tested, in particular in quantum electrodynamics (as demonstrated, e.g., by the Lamb shift or the anomalous magnetic moment of the electron), implies that any new attempt should reduce to the present quantum theory to a high order of approximation in every domain in which QED makes definite predictions. This is not excluded by the hereabove fractal approach: In particular we stress once more the fact that we pass from fractal to standard classical coordinates by a projection effect, by which the underlying fractal structure may completely disappear (apart from its manifestation through the uncertainty relations). A good example of this is the Peano curve, whose fractal derivative is a continuous, differentiable non-fractal function.

In this connection, some recent works[61,62] have addressed the problem of the dimension of space-time and obtained $|D-4|<10^{-6}$. This result clearly does not apply to our attempt, since this determination relies precisely on the assumption that any deviation from quadri-dimensional space-time is to be searched for in possible experimental discrepancies with respect to quantum electrodynamics, while we tentatively postulate here that the whole quantum behaviour might come from the fractal geometry of microphysical space-time.

But we may also expect the new approach to predict in some cases new structures with respect to present quantum physics. The most interesting domains are clearly the transition zones around which the fractal dimension rapidly changes from 1 to 2, i.e. the de Broglie periods. We leave to future work the question of spatial structures around the de Broglie wavelength.

We will consider here only the purely temporal case, which is directly connected with the problem of energy structures, i.e. of energy and mass spectra. Indeed, placing ourselves in the rest frame of the particle or system under investigation allows us to set $dx=dy=dz=0$ (strictly speaking this is a comoving rest frame in terms of curvilinear coordinates which follows the fractal motion), so that we are left with a relation involving only the temporal coordinate and the proper time s_o:

$$ds_o = \frac{ds}{dt}\, dt = \xi(t,\Delta t)\,(\tau/\Delta t)\, dt \ .$$

(Here s_o is the generalization of the standard classical proper time while s is the fractal proper time: they are related by $ds_o=\Sigma_i ds_i$ between t and $t+dt$).

This situation is reminiscent of the redshift effect in general relativity, that depends only on the g_{oo} term of the metric and may be inferred from the principle of equivalence and Newton's theory.[63] In the rest frame of a particle of mass m, the standard/fractal transition occurs around the time interval $\tau=\hbar/mc^2$. The smallest mass presently known for an elementary particle is the *electron* mass, so we are motivated to expect that, if new structures are to be found, one should preferentially search around the rest energy of an electron (511 keV) and of electron-positron pairs (1.022 MeV). One may also remark that vacuum energy fluctuations for time intervals smaller than $\tau_0=\hbar/4mc^2$ become larger than $\Delta E = (\hbar/2).\tau_0^{-1} = 2mc^2$, so that electron-positron pairs are spontaneously created. This is a basic phenomenon of physics which already plays a central part in QED, and the same feature is expected to remain true in the present attempt (see Sec. 6).

It was suggested in 1972 that actual observation of vacuum polarization effects was possible in strong Coulomb fields.[64,65] If the charge of a nucleus becomes larger than the critical value Z=173, the binding energy of the K shell exceeds $2m_ec^2$. The neutral vacuum becomes unstable and is expected to decay via the creation of a real electron-positron pair. Since ≈1976, a large number of experiments of heavy ion collisions have been carried out mainly at GSI (Darmstadt), allowing one to study the new domain of strong field QED ($Z\alpha > 1$). These experiments have yield unexpected new structures in the spectra of positrons, electrons and photons emitted during collisions. We suggest hereafter an explanation of these spectral lines in terms of fractal space-time structures and show that they may be related to the muon and the tau masses.

The anomalous positron peak from heavy ion collisions: a reminder.

Let us first recall the present status of the GSI Darmstadt experiments (see e.g. the review in Ref. 66) and of some related effects. First suspected in ~1980,[67] the existence of a positron energy peak was definitely established in 1983-85 by Schweppe et al.,[68] Clemente et al.,[69] Cowan et al.,[70] and Tsertos et al.[71] When observing collisions of heavy atoms or ions, these authors found, in addition to the QED prediction, a significant peak in the energy distribution of positrons arising from the collision process. In Ref. 69 the mean energy of the peak for U+U and U+Th collisions was found to be E_p≈300 keV and its width to be W≈70 keV; nearly twice the number of positrons predicted by QED were found in the peak ($R=N/N_{QED}\leq 2$), while the energy resolution was ΔE≈35 keV. In Ref. 68, U+Cm collisions were studied and a peak at E_p≈320 keV was found with W≈80 keV, R=2.3 while ΔE=20 keV. Several additional experiments with resolution 20 keV are reported by Cowan et al.[70]. From their Fig.1 one gets peak energies from 330 to 350 keV and ratios R from 1.8 to 2.6. The mean peak energy is E_p=340±6 keV.

Additional structures have been suspected,[67,72] and become now statistically confirmed.[73,76] Absolute maxima at 310 and 370 keV are often reported, as noticed by Cowan *et al.*[72] in Th+Th collisions (see Fig. 5.13). This is especially meaningful because a well-defined peak at 340 keV was found earlier[70] in Th+Th collisions at the same projectile energy. It thus seems that rather large fluctuations of the peak energy and profile might be found in various experiments *under the same experimental conditions*, and that *additional structures appear when the resolution is improved.* This was recently confirmed in Refs. 74, 75, which report on the observation of peaks whose energies cluster around 337±6 keV and 396±5 keV, plus a grouping around 255 keV for some subcritical (Z<172) collisions. In Ref. 76, *three* significant peaks at center of mass energies 238, 352 and 409 keV are reported in the *same* experiment with 12 keV resolution.

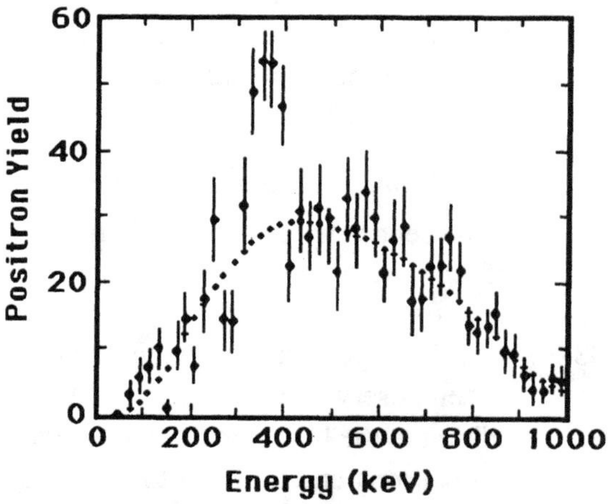

Figure 5.13. Example of a positron energy spectrum obtained at Darmstadt from heavy ion collision (adapted from Fig. 1 of Ref. 70: Th + Cm, 6.02 MeV/u, Z_u=186). The small crosses represent the QED expectation, normalized to the data outside the peak.

A particularly important fact concerning the understanding of the Darmstadt peaks is the discovery[72,75] that electrons at the same energy are emitted in correlation with the positrons in the peaks. Cowan et al.[72] report an electron peak at the same energy for U+Th collisions and an enhancement of the peaks when keeping only correlated pairs, which reach

R=2.6 times the QED prediction. They obtained similar results, albeit of lower statistical significance, in the Th+Th and Th+Cm collision systems. Kienle[75] reports on results of a new coincidence experiment with improved statistics on U+Th collisions. Very sharp peaks are found in the summed energies of positrons and electrons at energies ≈2x300 and 2x400 keV for some selected kinematical conditions; the counts in the peaks are $R \approx 3$ times the QED prediction for a resolution 10 keV.

No completely convincing and self-consistent theoretical explanation is presently known for the anomalous peaks. The suggestion of the existence of a transitional 10^{-19}s system was excluded by the observed independency on Z_u (since, in this hypothesis, a proportionality law of the peak energy $\propto Z^{20}$ was predicted). Note however that it has been recently established[74,75] that the production cross section of the peak rises as (Z^{20}), which strongly suggests that the peak production is *connected with processes induced by the strong time-changing Coulomb field.*

The only remaining possibility in the standard picture seems to be the existence[70,75] of a new neutral particle or system of mass ≈1.8 Mev/c^2. It would in particular clearly account for the observed behaviour of correlated electrons and positrons in the same narrow peaks. But several facts are to be questioned. The first is that such a relatively low energy particle does not fit into present elementary particle theories, and that one may wonder why it would not have been observed in other low energy (≈MeV) experiments (the answer may involve presence of strong electromagnetic field). Second, fluctuations of the e^+ energy peak from one experiment to another imply the existence of several different states, which seem to appear at random and which have not been related to one another up to now.

The discovery[77] of correlated equal energy photons from U+Th collisions seemed at first to confirm the neutral system hypothesis, though it would still increase its complexity. Indeed their sum energy is 1062±1 keV (with a narrow intrinsic width of less than 2.5 keV), while the various masses obtained from positrons and electron-positron pairs are found, from the survey of Ref. 74, to be grouped around 1498, 1538±8, 1579, 1648±3, 1686±4, 1723±2, 1779±3 and 1836 keV (see Figs.5.14 and 5.15). However it has been shown by the same group that the $\gamma\gamma$ line in U+Th can be explained by a cascade of Coulomb-excited transitions starting at high spin (32^+) in ^{238}U projectiles.[78] They suggest that the 1062 keV line results from a cascade $32^+ \to 30^+ \to 28^+$, and that another narrow line that they observed at 1043 keV would result from another cascade $32^+ \to 30^+$ and $28^+ \to 26^+$. Most authors have then considered the question of the $\gamma\gamma$ lines to be solved. There is yet a shortcoming in this proposal: it relies on the *hypothesis* that the energy of the transition $32^+ \to 30^+$ is 543 keV, while there is still no published measurement of transition energies from states higher than 30^+ in ^{238}U.[78]

Even more recently El-Nadi and Badawy[79] reported on the production of 13 electron-positron pairs from ≈700 collisions with emulsion nuclei of secondary fragments (of Z=2 to 10) from primary ^{12}C and ^{22}Ne projectiles. In addition they identified 7 pairs in a study of 1600 primary interactions of α particles under the same conditions. These data were reanalysed by de Boer and Dantzig[80] who found that they were consistent with old still-misunderstood events[81] and with the production and decay of neutral bosons with masses around $1.14^{+0.18}_{-0.07}$, 2.1±0.4 and 9.2±1.4 MeV and lifetimes of the order of 10^{-15} - 10^{-16} sec. This result was obtained from a plot of the full data of events ("Cairo"[79]+"Bristol"[81]) in the (mass versus lifetime) plane : the events are found to cluster into three groups, respectively A, B and C (see Fig. 5.16). Cluster A is consistent with the 1062 narrow $\gamma\gamma$ peak and cluster B with the Darmstadt peaks.

Similar resonances were also searched for in e^+e^- Bhabha scattering. A detection at an energy corresponding to a mass of 1830 keV has been recently claimed[82] but higher resolution experiments did not confirm this result.[83] A marginal 2.2σ effect was also found,[84] corresponding to a neutral system mass of 1730 keV.

We suggest hereafter[3] that the Darmstadt structures are related to the fractal nature of the microphysical vacuum around the Compton length of the electron in a strong electromagnetic field.

Fractal time and new energy peaks.

Let us try to analyse the possible implications of the assumed fractal structure of time.[3] The electron mass has been identified with a periodic fractal structure of the "zoom-space-time". This structure is revealed in the standard world (classical coordinates) through the fractal derivative $\xi(t,\Delta t)$. Though in most complex many-body situations this quantity is expected to become a quasi-random one, we assume as a working hypothesis that to such elementary physical objects as a free electron or an electron-positron pair, there corresponds regular and completely defined fractal structures of rest frame periods given respectively by the Compton times $\hbar/mc^2$ and $\hbar/2mc^2$. But we also expect the fractal derivative to show substructures, and in particular subperiodicities. Having identified the main absolute minima with the $2mc^2$ periodicity (see Fig. 5.11), we now suggest to identify the secondary minima with new masses of unstable resonances which would finally decay in e^+e^- or $\gamma\gamma$ pairs (Fig. 5.11). Basing ourselves on the increasing experimental evidence for an interpretation of the data in terms of a new particle, we have given up the method of Ref. 60 and finally tried to understand in terms of particles the newly discovered neutral system.[3]

Consider a field neutral vacuum with energy density high enough to allow the creation of an electron-positron pair. Let us assume that the proper time of such a system is to be represented by the general class of fractal curves with q=4 which has been described hereabove: As shown in Sec. 5.9, all members of this class are characterized by the same fractal structures of minima and maxima (see Fig. 5.11), even if they may differ from one particular model to the other in the details. We identify the main period with the time $\tau=\hbar/2m_ec^2$, i.e., to the energy E_0=1022 keV.

At low resolution (ξ_2), two holes appear corresponding to time intervals $\Delta t=(5/8)\tau$ and $\Delta t=(3/8)\tau$ elapsed from the origin of time. The hole at $(5/8)\tau$, which lies just behind the main peak in the fractal derivative, is expected to be the main one, being far more contrasted with respect to the preceding structures. The corresponding energy of this hole is 1635 keV; a system having such a rest mass would then decay into 307 keV kinetic energy positrons and electrons. This result corresponds to an energy resolution of about 80 keV. But seen at a better resolution ξ_3, (i.e. $\Delta E\approx$20 keV), the hole is decomposed into two substructures ($19\tau/32$ and $21\tau/32$) giving two lines at 1721 keV and 1557 keV (i.e., kinetic energies of 349 and 267 keV), the first one being expected to be the main structure of the two. These themselves are decomposed at the level ξ_4 ($\Delta E\approx$5 keV) into lines at 1744, 1698, 1576 and 1539 keV, i.e. 361, 338, 277, and 258 keV for the kinetic energies of the individual electrons and positrons. These values are quite close to the observed Darmstadt peaks. For example we compare in Fig. 5.14 the expected energies to those observed in the various collisions, and in particular to the twin peaks found in U-Au + Pb-Pb subcritical collisions (from Ref. 74, Table IV and Figs. 12 and 14).

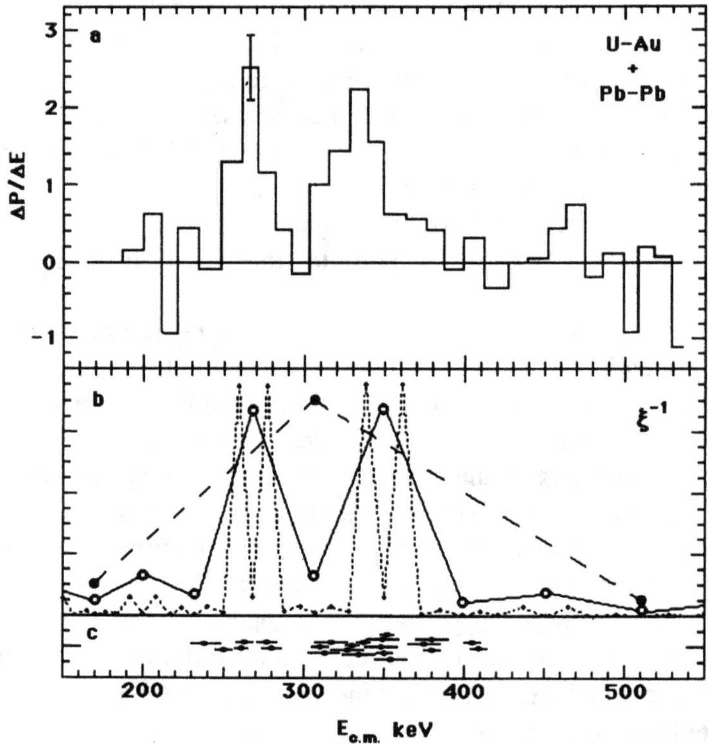

Figure 5.14. Comparison between the observed energies of the Darmstadt positron peaks and the minima of the fractal derivatives of Fig. 5.11. In (**a**) is reproduced from Fig. 12 of Ref. 74 the sum of positron spectra from U+Au and Pb+Pb collisions measured under similar kinematical constraints, (after subtraction of the smooth contributions of nuclear and dynamically induced positrons), in terms of center of mass energy. A typical error bar is shown. In (**b**) we have drawn the inverse of the fractal derivative of Fig. 5.11a (so that minima appear as peaks) for three values of the resolution, $\xi_{1.5}$ (black dots), $\xi_{2.5}$ (open circles and full lines) and $\xi_{3.5}$ (dotted line). Note the expected non-standard behaviour: when the resolution is improved, we predict that the peaks should be resolved into doublet structures, as observed in (**a**). In (**c**) we have plotted the distribution of line energies observed at Darmstadt in various spectra, from Table IV of Ref. 74.

The expected doublet structure is evident in the observations (the large 267, 349 keV doublet, but also possibly the substructure 258, 277 keV).[3] When $\Delta t \to 0$, the internal and external minima of this structure tend to masses of $3m_e$ and $(24/7)m_e$, the center being at $(16/5)m_e$. It is also remarkable that this principal structure self-reproduces itself for smaller time intervals. We get a spectrum of masses m_R in the range $3.4^n\ m_e$ to $(24/7).4^n\ m_e$ with n integer, and the symmetrical structures m'_R such that[3] $(1/m_R)+(1/m'_R) = (1/2m_e)$. Among these structures (see Fig. 5.11), there is a remarkable one (n=2) at $\Delta t = (123/128)\tau$, corresponding to E =1063 keV, in excellent agreement with the 1062±1 keV $\gamma\gamma$ line[77] (see Fig. 5.15).

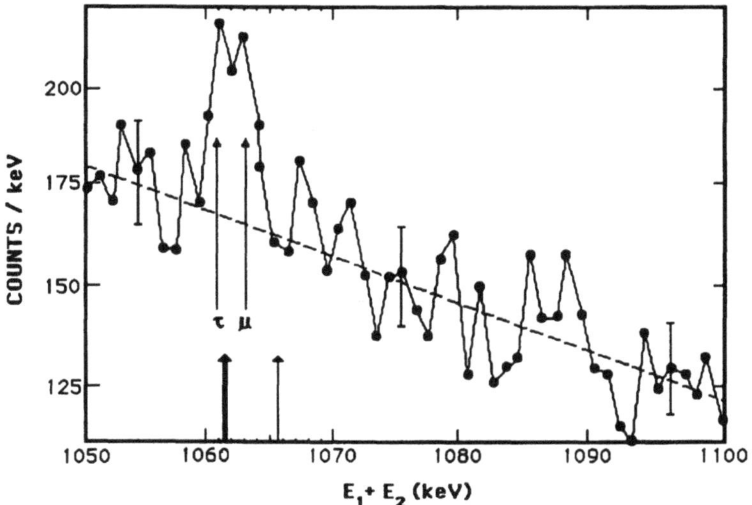

Figure 5.15. Comparison of the observed peak of correlated equal-energy photons from U+Th collisions[77] to the energies corresponding to minima in the fractal derivatives of Fig. 5.11. A line at 1063.5 keV is expected to split with improved resolution into a doublet at 1061.4 keV and 1065.7 keV (lower arrows). Also shown are the energies deduced from the μ and τ lepton masses, 1060.9 and 1063.1 keV (see text). Note however that an explanation in terms of atomic process has been proposed for the γ–γ lines, which could then be unrelated to the electron-positron pairs seen at higher energy.

We also get a good agreement with the Cairo[79] + Bristol[80] data.[3] Indeed the holes of the fractal derivative closely reproduce the cluster structure identified in the lifetime versus mass diagram (see Fig. 5.16).

Cluster A is compatible with the 1109-1234 keV holes (and possibly also the 1063 keV one); they are expected to split at higher resolution into two energy bands, 1109-1147 and 1189-1234 keV. Cluster B corresponds to two systems: the first of "full width at half minimum" 1486 to 1842 keV, in close agreement with the Darmstadt peaks (the observed values range from 1498±20 keV to 1840±10 keV)[71], the second ranges from ≈2300 to 3200 keV. Cluster C agrees with another double domain, 5946-7568 keV and 9344-13082 keV. Finally the highest energy object at 26.8±0.6 MeV, which was not separated from cluster C in Ref. 80 falls into a hole of energy 26.163 MeV and full extent 24.5-28 MeV. This structure is the symmetrical one of the 1062 keV line, and indeed this symmetry is verified in the experimental values, since $2m_e(1/m_1+1/m_2) = 1.022\ [(1.062\pm0.001)^{-1} + (26.8\pm0.6)^{-1}] = 1.0005\pm0.0011$, in excellent agreement with the expected value 1.

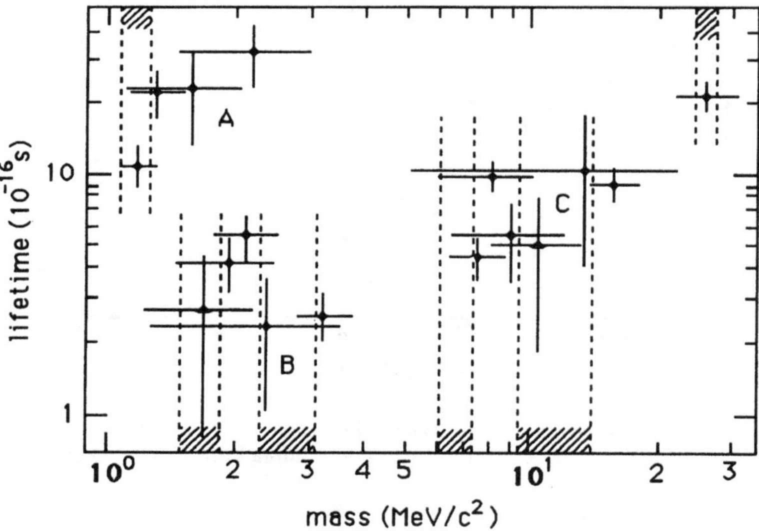

Figure 5.16. The "Cairo[77]" + "Bristol[81]" events, plotted in a lifetime versus mass diagram, from Fig. 1 of Ref. 80 , are compared with the expected energies corresponding to the minima of the fractal derivatives of Fig. 5.11. The three clusters noticed by de Boer and van Danzig[80] are well reproduced by the fractal structure. Clusters B and C are expected to split with improved resolution. The event at 26.8 MeV is considered as independent from cluster C, since it falls into an expected structure which is the symmetric part of the structure accounting for the 1062 keV $\gamma\gamma$ line.

Let us assume now that the same kind of structure is also appropriate for the description of the proper time of the electron, at least as a first approximation. The two other charged leptons, the muon and the tauon, are precisely known to be identical with the electron except for the mass and lifetime (it is the leptonic universality). The electron Compton time $\tau_e=\hbar/m_e c^2$ is twice the electron-positron period. The main structures m_R, which are situated just behind the main peaks (and give rise to the Darmstadt peaks in our interpretation), have been shown to correspond to masses in the range $(16/5).4^n.m_e$ to $(24/7).4^n.m_e$. For n=3, we get an interval [204.8-219.4] times the electron mass, in which the muon mass [206.77 m_e] is contained. For n=5 we get [3277-3511] m_e, while the tauon mass is 3491 m_e. These are encouraging results, if one takes into account the fact that the present model was not expected to hold in detail for high energies.

This fractal approach to the nature of leptons will be developed elsewhere. Here we shall only try to use their measured masses as a more precise estimate of the m_R and m'_R structures. We get for the Darmstadt energies (n=0): $m_\mu/(2.4^3)$=1651 keV, i.e., a positron kinetic energy 314 keV, in very good agreement with the observation of at least 4 lines between 310 and 316 keV[72,74] (see Fig. 5.14); and $m_\tau/(2.4^5)$=1742 keV, i.e., a kinetic energy 360 keV, closer than 2σ to at least 6 observed lines. We get for n=2: $m_\mu/8$=26.4 MeV, and $m_\tau/(2.4^2)$=27.9 MeV, to be compared with the Cairo resonance at 26.8±6 MeV. For the symmetric structure (n=2) we get $2m_e/(1-8m_e/m_\mu)$ = 1063.13 keV and $2m_e/(1-128m_e/m_\tau)$ = 1060.9 keV, in excellent agreement with the observed 1062±1 keV $\gamma\gamma$ line (see Fig. 5.15).

Several new predictions arise from this method of minima in the fractal derivative. The mass spectrum which has been described above and in Fig. 5.11 is a fractal dust of fractal dimension 1/2, given by the following law (written in the counting base 4):

$$M/2m_e = \frac{4}{x_1\, 4^{k-1} + x_2\, 4^{k-2} + \ldots + x_k} \qquad (5.10.1)$$

with x_k=2 and

*either x_1=0,1,2,3 and (x_2 to x_{k-1})=1 or 2,

*or (x_{i+1} to x_{k-1})=1 and either (x_1 to x_i)=0 or (x_1 to x_i)=3, with i=2 to k–2.

The prediction includes in addition to the already observed lines, among other details (see Fig. 5.11), new structures in the 1109-1234 keV range, splitting at better resolution of the 1062 keV $\gamma\gamma$ line (this prediction could allow one to separate our explanation from the atomic process explanation) and of the Cairo B and C clusters, and the occurrence at larger energies of a 43.6 MeV resonance and at lower energies of a 1046 keV line[3]. This last prediction may already have been fulfilled by the observation of a 1043 keV line,[78] as reminded above, provided the cascade interpretation fails in the end. However we remain conscious that this is only a model, not a fully developed theory. We do not expect such a model to remain correct for too large an amelioration of resolution, although the splitting in the Darmstadt data may have already been observed on two orders when passing from 80 keV to ≈<10 keV energy resolution.

Even more recently, new experiments with a resolution better than 10 keV have been reported, that show at least *four peaks* in the same spectra.[85] We consider this as a fair confirmation of our suggestion of the fractal nature of the spectrum.[3,60] But at the same time these recent observations show the limit of our model since the energies reported are ≈1.54, ≈1.66, ≈1.72, ≈1.83, and possibly a new line at 1.92 MeV/c^2, to be compared with our prediction (see Fig. 5.11), 1.54, 1.58, 1.70 and 1.74 MeV/c^2.

Let us finally remark that our proposal (even if it is done in a non-standard framework) is not so different from the most promising theoretical attempts to understand the Darmstadt structures.[86-90] They contemplate the possibility of the existence of a new phase of QED, the transition to which would be induced by the strong magnetic field.

5.11. Link to General Relativity.

The relation of the fractal approach to general relativity is a far reaching problem and will only be touched on here. Let us attempt to analyse how we can deal with it. General relativity, even in its purely geometric interpretation, is most of the time considered as a relativistic theory of a particular field, gravitation.

But, as remarked by Einstein,[91] "gravitation occupies an exceptional position with regard to other forces". This is partly because the "gravitational charge" is not only mass, as believed by Newton, but more generally energy-momentum. This means that everything that exists in Nature has gravitational properties. While particles or physical systems may have zero electric charge, we admit that to strictly zero energy-momentum (for sources *and* fields) corresponds non-existence.

The structure of Einstein's equations allows one to go even further. Indeed the vanishing of active matter and radiation ($T_{\mu\nu} = 0$) does not imply the vanishing of curvature (annulation of the Riemann tensor). Curvature thus appears to stand as an even more general property of the world than matter and radiation do. In this sense the theory has superseded Einstein's initial aims at describing gravitation as a space-time effect of the curvature *produced* by matter and radiation. The existence of curved empty solutions of Einstein's equations implies existence of pure gravitational fields[92] (i.e., there is no input energy-momentum, but there is indeed energy and momentum related to the field, as described by the gravitational energy momentum pseudo-tensor: see e.g. Ref. 2). Penrose and Rindler[93] go even further when they remark that the non-locality of the "energy" associated with pure gravitational fields is such that it may still exist in local regions where the space-time is completely free of curvature.

Hence a radical interpretation of the nature of gravitation is to state that it is nothing but the manifestation of a universal property of space-time: curvature.[92] Under this form it gains the same status of "superlaw" of physics which was recognized in special relativity,[94] and which, as proposed here, should also be attributed to the universal quantum properties, such as the de Broglie and Heisenberg relations (particularly when they are themselves interpreted as a manifestation of *scale relativity*, see Chapter 6). In the end, the working postulate followed in the present approach is that these various properties of Nature are but multiple aspects of relativity.

Let us analyse the present state of general relativity in the light of these comments. Three masses have been introduced in physics: the active gravitational mass, the passive gravitational mass and the inertial mass.[5] Concerning the last two masses, the general theory of relativity offers a profound explanation of their identity: particles follow the geodesics of the curved space-time and inertia is retrieved (and in the end defined) as a local property of a freely falling reference system. Thus there is no need to introduce a passive gravitational mass: the "passive gravitational content" of a body is nothing but its "inertial content".

The present status of the active mass is different. The principle of equivalence states that it is equal to the inertial mass, and indeed this is demanded by the action-reaction law at the Newtonian approximation, and more generally this is a necessity for the internal self-consistency of the theory.[5] If this equivalence is indeed to be an identity, as already achieved, concerning the two other masses, *the action of gravitation is to be attributed to inertia itself.* Then a still unanswered fundamental question is asked: Why does the *inertial* content of matter (and energy) curve space-time?

In other words, while gravitation may be completely understood as a manifestation of curvature, the same cannot be said of inertia. It should be recalled, indeed, that although the question of inertia was one of the main premises for Einstein's foundation of the general theory of relativity,[91] the problem finally remained partly unsolved. Let us recall the argument. Following Mach's analysis, Einstein considers two distant, macroscopic, fluid, self-gravitating bodies which rotate, one relatively to the other, with constant angular velocity about the line which joins them. However, one of them is a perfect sphere while the other is flattened into an ellipsoid of revolution along their common axis. While the motion of the two bodies is purely relative, their geometry is demonstrable from purely local measurements, independent of the other body. Newton's interpretation of such a system would have been that the spherical body is at rest with respect to the absolute space, while the other body is flattened precisely because of the appearance of a centrifugal force due to its rotation with respect to the absolute space.

However, Einstein remarks that Newton's solution is not satisfactory. Newton's absolute space is defined precisely from the very experiment that it claims to explain. The cause for this different behaviour of the two bodies cannot lie in the system consisting of them alone, since the motions of each body with respect to the rest frame of the other are perfectly symmetrical. The Mach/Einstein conclusion then was that inertia should be in a large part conditioned by distant masses which were not at first included in the system under consideration. This was not achieved by general relativity: This theory allows solution like Schwarzschild's (which is moreover the basis for most of its experimental tests) describing the space-time of a unique and isolated body of gravitational and *inertial* mass M, while it contains no mass at infinity.[95] The full problem of inertia is still with us.

If the answer lies neither in a closed macroscopic system nor in the general relativistic space-time, it should be searched for still elsewhere. One possibility is the global topology of the Universe, for which no physical equation is known at present.

A second possibility is to consider the internal microstructure of macroscopic systems, and more generally the additional scale dimensions which we propose to introduce here. In this "super-relativistic" picture, the nature of inertia is to be searched not at a given scale, but for all scales simultaneously, including the relations between different scales (expressed in terms of geometric "zoom" structures). Indeed in the macroscopic domain, the scale dependence of motions is a well-known fact of observational cosmology: the Earth orbits around the Sun at v= 30 km/s, the Solar System runs towards nearby stars at ≈20 km/s, our local group of stars is dragged along in the rotation of our Galaxy at ≈250 km/s, our Galaxy and its local group fall towards the Virgo cluster at a velocity of at least 200 km/s, and both fall towards a large distant mass (the "Great Attractor") at a velocity of ≈600 km/s, galaxies may have proper velocities as large as 3000 km/s in rich clusters of galaxies; to even larger scales, the velocity field becomes dominated by the expansion of the Universe. Note that such an "all scales of length approach" is akin to the global topology (i.e. in the end Machian) ideas, since it should include in its analysis the Universe as a whole. In Sec. 7.1, we shall tentatively make first steps in an attempt to connect microphysics and cosmology using the principle of scale relativity.

In the microscopic world, the quantum behaviour clearly leads to asking the question of inertia in a completely new way. While the "natural motion" of a free macroscopic body is uniform translation, a quantum body follows a continuous but non-differentiable, probabilistically described trajectory. While an experiment like the hereabove two-body system may be easily thought of in the macroscopic case (this means that one should consider an observer alternatively at rest with respect to each body), this is no longer the case in the quantum domain, since e.g. one cannot cancel out the spin of an electron, whatever the reference system chosen. Indeed in the hereabove fractal interpretation of spin (Sec. 5.4) it corresponds to an infinite velocity of rotation.

Is the fractal space-time approach able to bring new insights into these problems? Several arguments seem to open such a hope. As we have seen, it offers a new view of what inertia may be. The various physical properties attributed to a free particle, mass, energy, momentum, velocity, angular momentum, have been reinterpreted as the manifestations in classical reference frames of a universal property of space-time, its fractal character. The rest inertial mass is identified with a property of time: to be of fractal dimension 2 for time intervals smaller than $\tau_0=\hbar c^{-2}/m$. This ability of fractals to structure the "internal" part of objects in a geometric way leads to the idea that gravitation, i.e. the manifestation of curvature, might be described as a residual effect for macroscopic distances and masses of these structures.

Indeed fractalization, as recalled hereabove, is characterized by an "hypercurvature", infinite everywhere but flipping in a fractal way from one sign to the other. The result of averaging this hypercurvature at large scale (i.e. $+ \infty - \infty$) may yield flatness, but more generally any finite curvature. To be more specific, the interpretation of a general relativistic space-time as the limit of a fractal space-time is a direct consequence of our description of fractal space-time as a family of Riemannian space-times, $R_{ijkl}(\varepsilon)\rightarrow R_{ijkl}$, that becomes independent of ε when $\varepsilon\rightarrow\infty$. The form obtained in Eq. (5.3.4) for the right-hand side of Einstein's equations is also interesting in this connection: it is symmetrical in G and $\hbar$, and may be read as totally geometric in nature. The manifestation of the fractal structures of space-time are expected to vanish at resolutions larger than the de Broglie length and time, but one should keep in mind that gravitation is both a very faint effect ($Gm_e^2/\hbar c\approx10^{-44}$) and a collective effect ($\approx10^{60}$ particles in stars). This gives the hope to account for gravitation (which may be identified with the universal Riemannian nature of the macroscopic space-time) as a residual "field", resulting from a large scale smoothing of quantum non-local structures (which we propose here to identify with the universal fractal structures of the microscopic space-time).

Let us now develop a general argument which demonstrates that the present structure of physical theories (general relativity as well as quantum mechanics) needs revision, possibly not only at the Planck scale. The wave nature of any physical system is at present assumed to be universal. However the wave properties of a system have physical sense only if one may, at least in a "gedanken experiment", measure them (i.e., by a diffraction or interference experiment). But when the total mass m becomes larger than about the Planck mass $m_{\mathbb{P}}=(\hbar c/G)^{1/2}\approx 2\ 10^{-5}$ g, its Compton length becomes smaller than its Schwarzschild radius $r_S=2Gm/c^2$, thus becoming unmeasurable, not only for technological limitations, but mainly for a profound physical limitation, since it enters into a black hole horizon.[3]

Indeed let us try to describe more precisely such an experiment: in order to evidence an interference pattern in, e.g., a two-slit experiment, the hole width should be smaller than or at most of the order of the wavelength of the object which we want to make interfere. So it is first clear that we cannot get interferences with objects much larger than their de Broglie wavelength. To deal with this problem and accounting for the fact that we are interested here only in the mass dependence of the wave behaviour, independently from the nature of the object, let us assume that we have reduced the mass to a point. Even in this case, when the mass becomes larger than the Planck mass, it is anyway surrounded by its Schwarzschild sphere: if the experiment was to be actually made, the black hole horizon being larger than the holes would break the apparatus, thus preventing us from pursuing the experiment.

Strictly speaking, it is the Compton length which becomes smaller than the Schwarzschild radius, and one may argue that with a low enough velocity one would get a large enough de Broglie length. There are (at least) two answers to this objection: first in such a (gedanken) experiment, the resolution of the measurement apparatus should be of the order of the Planck length; then the Heisenberg relation implies $p\gtrsim m_{\mathbb{P}}c$, and since we deal with a mass of the order of the Planck mass, its velocity is close to that of light; second the de Broglie length $\hbar/p$ derives from the de Broglie time in rest frame $\hbar/mc^2=\lambda_c/c$ through a Lorentz transformation, so that if the Compton length loses its physical meaning, so will the case for the de Broglie length: the result should not depend on the arbitrary choice of the reference system.

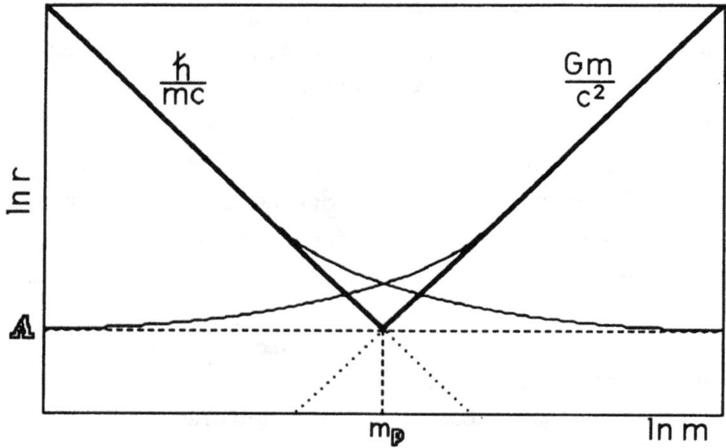

Figure 5.17. The two fundamental mass-length relations in a log-log diagram. The *classical* Schwarzschild relation (up to a factor of 2) is a direct proportionality law, while the *quantum* Compton relation is an inverse law: their crossing defines the Planck scale, $m_{\mathbb{P}} = (\hbar c/G)^{1/2}$ and $\Lambda = (\hbar G/c^3)^{1/2}$. As demonstrated in the text, the standard expression for the Compton length loses its physical meaning for $m>m_{\mathbb{P}}$, while the same is true of the Schwarzschild radius for $m<m_{\mathbb{P}}$. In Chapter 6, we argue that the principle of scale relativity implies that the Planck length is a lower, unpassable, universal length in Nature which is invariant under dilatations. The above paradox is then solved by a generalization of the Compton and Schwarzschild relations agreeing with this new principle, as schematized by the two curves reaching Λ asymptotically.

The reverse is also true: for $m<m_{\mathbb{P}}$ the Schwarzschild radius becomes smaller than the Compton wavelength. But for resolution smaller than the Compton length, we enter into the quantum relativistic domain inside which the concept of precise position itself loses its usual physical meaning[58] because of particle-antiparticle creation. This means that one cannot localize a particle with a resolution better than its Compton length.

However a black hole state can be reached only provided the whole mass m is confined into its Schwarzschild radius. So it appears[6] that the concept of Schwarzschild radius loses its physical meaning for $m<m_{\mathbb{P}}$ (see Fig. 5.17).

Combining the two results, we reach the remarkable conclusion that the Compton length loses its physical meaning for $m>m_{\mathbb{P}}$ and the Schwarzschild radius loses its physical meaning for $m<m_{\mathbb{P}}$, while they become equal for $m=m_{\mathbb{P}}$ (Fig. 5.17). Confronted with such a puzzle, we proposed in Refs. 3, 6 and 96, that there was a continuity between the two concepts, the Compton length and Schwarzschild radius reversing their roles in micro- and macro-physics. Such a suggestion would imply a breaking of Newton's law for active gravitational masses smaller than the Planck mass.[3,96]

However, the first developments of a special theory of scale relativity, as we shall see in Chapters 6 and 7, provide us with a better solution of the problem, since in this framework a completely new role is attributed to the Planck length scale: it becomes a limiting, unexceedable length in Nature that plays for scaling laws the same part as played by the velocity of light for motion laws, and possesses all the previous properties of the perfect zero point. This implies generalization of the Compton and Schwarzschild relations as schematically shown in Fig. 5.17 (see Secs. 6.9 and 7.1).

Nevertheless, the idea to test experimentally Newton's law of gravitation at the Planck mass scale keeps all its interest. Indeed the Planck mass (2×10^{-5} g) is clearly a transitional mass between microscopic and macroscopic masses: for example the gravitational force between two Planck masses is the same as the Coulomb force between two bodies each of them carrying ≈ 12 elementary electric charges. One may design a space experiment (since it is the active gravitational mass that is here in question) for studying the attractive force of two dust grains on each other. The Newtonian falling time of two particles of equal masses m and radius r separated at rest by a distance R is the solution of the equation $x^2(d^2x/dt^2) = -Gm$. It is given by[97]

$$T = \sqrt{\frac{R\,r\,(R-2r)}{Gm}} + \sqrt{\frac{R^3}{2Gm}}\,Arctg\sqrt{\frac{R}{2r}-1} \quad .$$

This formula supersedes the approximate solution given in Ref. 3. For example grains of 1 Planck mass with a high density[98] d=18 have a radius of 0.064 mm. Initially separated by 0.24 mm, their Newtonian falling time is of the order of 10 mn. Such an experiment may be at the limit of the present possibilities in space: The microgravity would be smaller or of the order of the vessel gravity gradient, while the electromagnetic perturbations should be controlled with high precision.

5 References

1. Landau, L. & Lifchitz, E., 1969, *Mechanics* (Mir, Moscow, 1969).
2. Landau, L. & Lifchitz, E., *The classical theory of fields* (Mir, Moscow, 1970).
3. Nottale, L., 1989, *Int. J. Mod. Phys.* **A4**, 5047.
4. Wignal, J.W.G., 1992, *Phys. Rev. Lett.* **68**, 5.
5. Weinberg, S., *Gravitation and Cosmology* (John Wiley & Sons, New York, 1972).
6. Nottale, L., 1991, in *Applying fractals in Astronomy*, Heck & Perdang, Eds. (Springer-Verlag), p.181.
7. Lachaud, Y., 1992, submitted for publication.
8. Feynman, R.P., & Hibbs, A.R., *Quantum Mechanics and Path Integrals* (MacGraw-Hill, 1965).
9. Goudsmit, S.A., 1976, *Physics Today* **29**, 40.
10. Uhlenbeck, G.E., 1976, *Physics Today* **29**, 43.
11. Ohanian, H.C., 1986, *Am. J. Phys.* **54**, 500.
12. Belinfante, F. J., 1939, *Physica* **6** , 887.
13. Gödel, K., 1931, in *Le Théorème de Gödel* (Seuil, 1989).
14. Omnes, R., 1988, *J. Stat. Phys.* (USA) **53**, 893, 933, 957.
15. Aspect, A., Dalibar, J., Roger, G., 1982, *Phys. Rev. Lett.* **49**, 1804.
16. Sachs, P. K., 1961, *Proc. Roy. Soc. London* **A264**, 309.
17. Kantowski, R., 1969, *Astrophys. J.* **155**, 89.
18. Dyer, C.C., Roeder, R.C., 1972, *Astrophys. J. Lett.*, **174**, L115.
19. Nottale, L., 1982, *Astron. Astrophys.*, **110**, 9; **114**, 261; **118**, 85.
20. Nottale, L., 1984, *Monthly Notices Roy. Astron. Soc.*, **206**, 713.
21. Nottale, L., Hammer, F., 1984, *Astron. Astrophys.*, **141**, 144.
22. Nottale, L., Chauvineau, B., 1986, *Astron. Astrophys.*, **162**, 1.

23. Nottale, L., 1988, *Ann. Phys. Fr.* **13**, 223.

24. Einstein, A., in *Albert Einstein, Oeuvres choisies*, Vol. 1 (Seuil-CNRS, Paris, 1990).

25. Pais, A., *'Subtle is the Lord', the Science and Life of Albert Einstein* (Oxford University Press, 1982).

26. de Broglie, L., 1926, *C.R. Acad. Sci. Paris* **183**, 447; **185**, 380.

27. Bohm, D., 1952, *Phys. Rev.* **85**, 166, 180.

28. Dewdney, C., Holland, P.R., Kyprianidis, A., & Vigier, J.P., 1988, *Nature* **336**, 536.

29. Fényes, I., 1952, *Z. Physik* **132**, 81.

30. Kershaw, D., 1964, *Phys. Rev.* **B 136**, 1850.

31. Nelson, E., 1966, *Phys. Rev.* **150**, 1079.

32. Santamato, E., 1984, *Phys. Rev.* **D29**, 216; **D32**, 2615.

33. Santamato, E., 1988, *Phys. Lett.* **A 130**, 199.

34. Castro, C., 1990, *J. Math. Phys.* **31**, 2633.

35. Mandelbrot, B., *The Fractal Geometry of Nature* (Freeman, San Francisco, 1982).

36. Bohm, D., & Vigier, J.P., 1954, *Phys. Rev.* **96**, 208.

37. Weizel, W., 1954, *Z. Physik* **134**, 264; **135**, 270; **136**, 582.

38. Rosen, N., 1984, *Found. Phys.* **14**, 579.

39. Wheeler, J.A., 1989, in *Richard Feynman, Physics Today* (special issue, February), p.24.

40. Nelson, E., *Quantum Fluctuations* (Princeton Univ. Press, 1985).

41. Welsh, D.J.A., in *Mathematics Applied to Physics* (Springer,1970), p.465.

42. Davidson, M., 1979, *Physica*, **96A**, 465.

43. Shuker, D., *Letters in Math. Phys.* **4**, 61.

44. Kolmogorov, A.N., 1931, *Math. Annalen* **104**, 415

45. Weinberg, S., 1989, *Phys. Rev. Lett.* **62**, 485.

46. Bollinger, J.J., *et al.*, 1989, *Phys. Rev. Lett.* **63**, 1031.

47. Zeh, H.D., 1970, *Found. Phys.* **1**, 69.

48. Zurek, W.H., 1981, *Phys. Rev.* D **24**, 1516.

49. Zurek, W.H., 1982, *Phys. Rev.* D **26**, 1862.

50. Joos, E., & Zeh, H.D., 1985, *Z. Phys.*. B **59**, 223.

51. Caldeira, A.O., Leggett, A.J., 1983, *Physica* **A121**, 587.

52. Caldeira, A.O., Leggett, A.J., 1985, *Phys. Rev.* A **31**, 1057.

53. Woo, C.H., 1986, *Am. J. Phys.* **54**, 923.

54. Unruh, W.G., Zurek, W.H., 1989, *Phys. Rev.* D **40**, 1071.

55. Zurek, W.H., 1991, *Physics Today* (October), p. 36.

56. Aspect, A., Arimondo, E., Kaiser, R., Vansteenkiste, N., & Cohen-Tannoudji, C., 1988, *Phys. Rev. Lett.* **61**, 826.

57. Balian, R., 1989, *Am. J. Phys.* **57**, 1019.

58. Landau, L., & Lifchitz, E., *Relativistic Quantum Theory* (Mir, Moscow, 1972).

59. Itzykson, C., & Zuber, J.B., *Quantum Field Theory* (McGraw-Hill, 1980).

60. Nottale, L., 1988, *C. R. Acad. Sci. Paris* **306**, 341.

61. Svozil, K., & Zeilinger, A., 1985, *Int. J. Mod. Phys.* **A1**, 971.

62. Müller, B., & Schäfer,A., 1986, *Phys. Rev. Lett.* **56**, 1215.

63. Einstein, A., 1911, *Annalen der Physik* **35** , 898. English translation in "*The principle of relativity*", (Dover publications), p. 99.

64. Müller, B., Rafelski, J., & Greiner, W., 1972, *Z. Physik* **257**, 62.

65. Zeldovich, Ya.B., & Popov, V.S., 1972, *Sov. Phys. Usp.*, **14**, 673.

66. Kienle, P., 1986, *Ann. Rev. Nucl. Part. Sci.* **36**, 605.

67. Reinhardt, J., Müller, B., & Greiner, W., 1981, *Phys. Rev.* **A24**, 103.

68. Schweppe, J., *et al* ., 1983, *Phys. Rev. Lett.* **51**, 2261.

69. Clemente, M., *et al.*, 1984, *Phys. Lett.* **137B**, 41.

70. Cowan, T., *et al.*, 1985, *Phys. Rev. Lett.* **54**, 1761.

71. Tsertos, H., *et al.*, 1985, *Phys. Lett.* **162B**, 273.

72. Cowan, T., *et al.*, 1986, *Phys. Rev. Lett.* **56**, 444.

73. Kienle, P., in *Atomic Physics* **10**, 105 (Elsevier, 1987).

74. Koenig, W.*et al.*, 1987, *Z. Phys.* **A 328**, 129.

75. Kienle, P., in *7th Gen. Conf. of the European Physical Society*, Helsinki (1987).

76. Tsertos, H., *et al.*, 1987, *Z. Phys.* **A326**, 235.

77. Danzmann, K., *et al.*, 1988, *Phys. Rev. Lett.* **62**, 2353.

78. Danzmann, K., *et al.*, 1987, *Phys. Rev. Lett.* **59** , 1885.

79. El-Nady, M., & Badawy, O.E., 1988, *Phys. Rev. Lett.* **61**, 1271.

80. de Boer, F.W.N., & van Danzig, R., 1988, *Phys. Rev. Lett.* **61**, 1274.

81. Anand, B.M., 1953, *Proc. Roy. Soc. (London)* **A220**, 183.

82. Maeir, K., *et al.*, 1988, *Z. Phys.* **A 330**, 173.

83. Tsertos, H., *et al.*, 1987, *Z. Phys.* **A331**, 103.

84. von Wimmersperg, U., *et al.*, 1987, *Phys. Rev. Lett.* **59**, 266.

85. Koenig, W., *et al.* 1989, *Phys. Lett.* **B 218**, 12.

86. Diaz Alonso, J., 1989, in *Recent developments in gravitation*, Proceedings of *Relativity Meeting-89*, E. Verdaguer, J. Garriga, J. Céspedes, Eds., Barcelona (World Scientific), p. 275.

87. Caldi, D.G., Chodos, A., 1987, *Phys. Rev.* **D36**, 2876.

88. Peccei, R.D., Sola, J., Wetterich, C., 1988, *Phys. Rev.* **D37**, 2492.

89. Barut, A.O., 1990, Z. *Phys.* **A 336**, 317.

90. Geiger, K., *et al.*, 1989, in *XXIVth Rencontres de Moriond*, Tran, Ed. (Frontières).

91. Einstein, A., 1916, *Annalen der Physik* **49**, 769. English translation in "*The principle of relativity*", (Dover publications), p. 111.

92. Misner, C.W., Thorne, K.S., & Wheeler, J.A., *Gravitation* (Freeman, San Francisco, 1973).

93. Penrose, R., & Rindler, W., 1986, *Spinors and Space-Time*, Vol. II (Cambidge Univ. Press), p. 427.

94. Wigner, E. P., 1964, *Physics Today* **17**, 34.

95. Einstein, A., 1917, *Sitz. Preus. Akad. Wiss.* English translation in "*The principle of relativity*", (Dover publications), p. 175.

96. Nottale, L., 1990, in *The Early Universe and Cosmic Structures*, 10th Moriond Astrophys. meeting, A. Blanchard & J. Tran Thanh Van Eds., (Frontières), p. 13.

97. Paturel, G., private communication.

98. Nobili, A., & Milani, A., in 5th *force - neutrino physics*, Rencontres de Moriond (Frontière, 1988) p. 569.

99. Guerra, F., & Morato, L.M., 1983, *Phys. Rev.* **D27**, 1774.

Note added

It is only after the completion of this book that I became aware of the interesting work of G.N. Ord (1983, *J. Phys. A: Math. Gen.* **16**, 1869). In this paper, Ord introduces the concept of fractal space-time as a geometric analog of quantum mechanics. He obtains results quite similar to those described in Chapter 4 of the present book, including the introduction of a fractal time of dimension 2, and the fractal interpretation of particle-antiparticle virtual pairs that is developed in Chapter 5. His paper also contains the description of a random walk model which yields the free particle Klein-Gordon and Dirac equations. In this respect, note that relativistic stochastic quantum mechanics has been developed by B. Gaveau, T. Jacobson, M. Kac and L.S. Schulman (1984, *Phys. Rev. Lett.* **53**, 419), M. Serva (1988, *Ann. Inst. Henri Poincaré* **49**, 415) and T. Zastawniak (1990, *Europhys. Lett.* **13**, 13). The concepts of fractal quantum paths and space-time have also been recently considered by Sornette (1990, *Euro. J. Phys.* **11**, 334) and El Naschie (1992, *Chaos, Solitons and Fractals* **2**, 91; 211; 437).

Chapter 6

TOWARDS A SPECIAL THEORY OF SCALE RELATIVITY

6.1. Principle of Scale Relativity.

After having reached the conclusion that the fundamental dependence of microphysical laws on scale came under the principle of relativity, we were guided by Einstein's general theory of relativity to conjecture that a new geometry of space-time, namely its fractal structure, was an adequate physico-mathematical tool to develop a theory of scale relativity. However, one may wonder whether an essential step has not been jumped over in such a thought process. Consider, indeed, the evolution of the principle of relativity as applied to laws of motions. We now know that it was impossible to construct a self-consistent general relativity *before* the discovery of special relativity. The special theory of relativity yields a universal constraint, namely the limitation of any velocity in Nature, that is translated into the locally Minkowskian nature of space-time and must be taken into account before any attempt of generalization. We shall demonstrate in this Chapter that the same situation holds for scale relativity, so that we are led to proposing that there exists a universal, unpassable, lower scale in Nature.

Since the Galilean analysis of the nature of inertial motion, the theory of relativity has been developed by extending its application domain to coordinate systems involved in more and more general states of motion: this evolution culminated in Einstein's special and general theories of relativity. Hence the principle of general relativity states that "the laws of physics must be of such a nature that they apply to systems of reference in any kind of motion."[1]

However, as pointed out by Levy-Leblond,[2] the abstract *principle of relativity* should be distinguished from any of its possible realizations as concrete *theories of relativity*. This point of view may still be generalized into a framework in which relativity is considered as a general method of thinking in physical sciences[3]: it consists in analyzing how the results of measurements and (cor)relations between them are dependent on the particular conditions under which the measurements have been performed. Assuming that these results are measured in some "reference system" (e.g. coordinate systems in the case of position and time measurements), these conditions may be characterized as "states" of the reference systems.

Such states of reference systems play a special role in physics: they being defined as characteristics of the reference systems themselves, no absolute value can be attributed to them, but only *relative* ones, since they can be defined and described only with respect to *another* reference system. Two systems at least are needed to define them. As a consequence the transformation laws between reference frames will be of the greatest importance in a theory of relativity.

In such a perspective, the first relativity is the relativity of positions and instants, as recalled in Chapter 2. It is usually expressed in terms of homogeneity and isotropy of space and uniformity of time, and actually constitutes the basis of all physical theories. It states that there is no preferential origin for a coordinate system and is finally included in special relativity through the Poincaré group.

Then Einstein's relativity is, strictly speaking, a theory of "motion relativity", since the particular relative state of coordinate systems which the special and general theories of relativity have extensively analyzed (in the classical domain) is their state of motion.

However our main leading line in the present book is that the domain of application of Einstein's principle of relativity is even more general than we thought up to now. We claim that the principle of relativity applies not only to the laws of motion, but also to the laws of scale.

We therefore suggest that the resolution with which measurements have been performed *may be defined as a relative state of scale of reference systems*, and that *the laws of physics must be such that they apply to systems of reference whatever their state of scale*: this we call, in parallel with Einstein's statement of the principle of general relativity,[1] *the principle of scale relativity*. Its mathematical translation is the requirement of *scale covariance* of the *equations* of physics.

We shall show in this Chapter that the *quantum behaviour of microphysics* may to some extent be reinterpreted as a *manifestation of scale relativity*. But in its present form the quantum field theory corresponds to a *Galilean version* of such a theory of scale relativity, especially in the renormalization group approach.

This scale dependence of microphysics takes on reinforced importance in the case of the asymptotic behaviour of quantum field theories. The renormalization group methods yield one of the best description of such a behaviour.[4-8] This leads to some important results, like asymptotic freedom of QCD, the variation of coupling constants with scale and their convergence at the "Grand Unified Theory" scale $\approx 10^{14}$ GeV. These methods have for the first time explicitly introduced scale transformations into physical equations. Indeed it is remarkable that below the Compton length of the electron, there is a strong "degeneration" of space and time variables: the velocity becomes disqualified as a pertinent mechanical variable (all velocities are close to the velocity of light) and the classical laws of mechanics are actually replaced by dilatations in terms of Lorentz γ factors.Then, *at high energy, the laws of scale actually take the place of the laws of motion.*

There are, however, several additional elements in our proposals with respect to the standard renormalization group approach. The first is that it is argued that scale, like motion, may be considered as a state of coordinate systems which can never be defined in an absolute way, and thus comes under a relativity theory. In this respect we shall identify the theory of the renormalization group as a Galilean version of scale relativity. The second is that the renormalization group is, strictly, only a semigroup (one integrates the small scales to get the larger ones),[9] while one may hope to complete it in the future by an inverse transformation, at least for some elementary physical systems. This would enable us to deduce the small scale structure from the large scale: this is exactly what a fractal generator does.[3,10,11] The third new element is that the relativistic analysis of scale, once applied to space and time variables themselves, finally leads to a completely new structure of physical laws, as will be demonstrated in the following sections.

In this Chapter we shall first recall how the renormalization group may be applied to space-time itself, yielding an anomalous dimension for space and time variables, that comes from a reinterpretation of the fractal dimension. Then we shall demonstrate in a general way that the principle of relativity alone, in its Galilean form (i.e. without adding any extra postulate of invariance), is sufficient to derive the Lorentz transformation as a general solution to the (special) relativity problem. Once applied to scale, and owing to the fact that physical laws become explicitly scale-dependent only for resolutions below the de Broglie length and time (namely, that scale relativity is broken at the de Broglie transition), this reasoning leads to a demonstration of the existence of an absolute, universal scale which is invariant under dilatations and so cannot be exceeded. Then, after having identified this scale with the Planck scale, we attempt to develop a theory based on this new structure: the Einstein-de Broglie and Heisenberg relations are generalized, and first implications concerning the domain of high energy physics are considered.

But before going on with scale relativity, let us give a simplified reminder of the present state of high energy particle physics: we shall indeed use this domain as a framework for the possible applications of the new theory.

6.2. Masses, Charges and Renormalization Group.

The understanding of the physical origin of charges and masses (i.e. gravitational charges) is a long standing problem. While, concerning mass, this quest came to an end (at least partly) in the frame of Einstein's general theory of relativity, the question of the origin of the electric charge is still mainly open. Indeed general relativity naturally introduces energy-momentum (i.e., the conservative quantities which derive from the local uniformity of space-time) as the "charge" for the curvature of space-time, and explains gravitation as a manifestation of this curvature. Conversely the electric charge is still introduced in physics only from pure experimental grounds. The situation may be considered as improved in quantum physics, since charge results from the uniformity of the quantum phase and is related to the probability of emission/absorption of photons by electrons. But the situation is still very unsatisfactory: the present theory, when trying to find the theoretical expectation of the value of the electric charge and of the electromagnetic self-energy of the electron, finds infinite results. The phase symmetry is purely internal, and one may suspect that the inability of physics to explain the numerical value of the electric charge (i.e., equivalently, of the fine structure constant) and to properly solve its divergence problem is related in some way to the absence of space-time interpretation of its origin.

In classical electrodynamics the electrostatic energy of a system of point charges is given by

$$U = \frac{1}{2} \sum_i e_i \varphi_i \quad .$$

Once applied to the self-interaction of one electron, this gives one an electrodynamical self-energy

$$E_{em} = \frac{1}{2} \frac{e^2}{r} = \frac{1}{2} \frac{\alpha \hbar c}{r} \quad .$$

So in classical electrodynamics, when $r \to 0$, the electromagnetic contribution to mass becomes infinite while the charge e (or, in a similar way, the coupling constant α) remains constant. Lorentz, in his attempt at solving this problem, introduced an extension for the electron, the now so-called "classical radius of the electron"

$$r_0 = \frac{e^2}{m\,c^2} = \alpha\,\frac{\hbar}{m\,c} = \alpha\,\lambda_c\ ,$$

where λ_c is the Compton length. The quantum theory formulates the problem in a completely different way: the electron in the standard model is both totally point-like and structured by internal virtual particles from the Compton scale onwards: one may then think that the classical radius of the electron has lost any physical meaning in the quantum theory. One may remark, however, that it might not be only pure chance that this scale, $\approx$137 times smaller than the Compton length of the electron, is of the order of the Compton length of the muon. Indeed the muon/electron mass ratio is m_μ/m_e=206.768262(30).[12] This remark leads to the hope, reinforced by the $e/\mu/\tau$ universality (the muon and tau leptons have all the properties of the electron itself, except for their masses and lifetimes), that the muon mass is determined by electrodynamics alone, and then depends on α alone. In this connection the following empirical relation:

$$\frac{2}{\pi}\,\alpha + \pi\,\alpha^2 + \frac{1}{\pi}\,\alpha^3 \mp \alpha^4 = \{207.76816 \pm 0.00012\}^{-1} \approx \frac{m_e}{m_e + m_\mu}\ ,$$

may serve as a guide in such a search. This may be pure numerology, but its form is however suggestive of well-known α expansions deduced from radiative corrections: for example, the magnetic moment of the electron is expanded[13,14] as $a = (e\hbar/2mc)\,[1 + \frac{1}{2\pi}\,\alpha - \frac{0.328}{\pi^2}\,\alpha^2 + ...]$.

Quantum Electrodynamics.

Let us now recall the state of the question of masses and charges in the framework of quantum electrodynamics (QED). When relative distances become smaller than the Compton scale of the electron, the nature of the interaction between two nearby charges changes radically. Indeed, while the electromagnetic interaction was mediated only by photons at scales larger than λ_c, this is no longer the case when the distance between charges becomes smaller than λ_c. The new behaviour is due to the phenomenon of electron-positron pair creation and annihilation, which mainly occurs, as expected from the Heisenberg relations, for time intervals smaller than $\Delta t \approx \hbar/2m_e c^2$, i.e., distances smaller than $\approx \lambda_c/2$.

This phenomenon is often expressed in terms of vacuum polarisation by virtual particle-antiparticle pairs. One may also describe it by considering that the so-called "electromagnetic" field is no more purely electromagnetic at these scales, but becomes mediated by two spin-one "bosons", photons and e^+-e^- pairs: this view allows one to expect that the new pair contribution to the coupling will decrease at large scale following the Yukawa factor $exp(-2m_e r)$. This is confirmed by explicit calculations. When accounting for leading order radiative corrections (contribution of one electron-positron loop), the Coulomb potential becomes[15,13]

$$e\,\varphi(r) = \frac{e^2}{r}\left\{1+\frac{2\alpha}{3\pi}\int_1^\infty e^{-2mr\zeta}\,\left(1+\frac{1}{2\zeta^2}\right)\frac{(\zeta^2-1)^{1/2}}{\zeta^2}\,d\zeta\right\}.$$

This formula may be integrated to yield a low energy approximation, valid for $r>\lambda$ (in the following formulae, λ denotes the Compton length of the electron), that includes the expected Yukawa term (curve 1 in Fig. 6.1)[13]

$$e\,\varphi(r) = \frac{\hbar c}{r}\,\alpha\left\{1+\frac{\alpha}{4\pi^{1/2}}\frac{e^{-2r/\lambda}}{(r/\lambda)^{3/2}}\right\}.$$

A better approximation, which remains good up to $r \approx \lambda/4$, is (2 in Fig. 6.1)

$$e\,\varphi(r) = \frac{\hbar c}{r}\,\alpha\left\{1+\frac{\alpha}{4\pi^{1/2}}\left(\frac{\lambda}{r}\right)^{1/2}\frac{e^{-2r/\lambda}}{2+r/\lambda}\right\}.$$

The high energy asymptotic behaviour ($r\to 0$) is found to be, still to first order (curve 3 in Fig. 6.1)[13-15]

$$e\,\varphi(r) = \frac{\hbar c}{r}\,\alpha\left\{1+\frac{2\alpha}{3\pi}\left[\ln\frac{\lambda}{r}-\left(\gamma+\frac{5}{6}\right)\right]\right\},$$

where γ=0.577 215... is Euler's constant. This is often expressed in terms of a running coupling constant $\alpha = \alpha(r)$, which is related to the low energy coupling (i.e., the usual fine structure constant α_0) by a relation whose first order expansion is

$$\alpha = \alpha_0 \left\{ 1 + \frac{2\alpha_0}{3\pi} \; ln \frac{\lambda/Q}{r} \right\} , \tag{6.2.1}$$

where $Q = e^{\gamma+5/6} = 4.098\ 204...$.

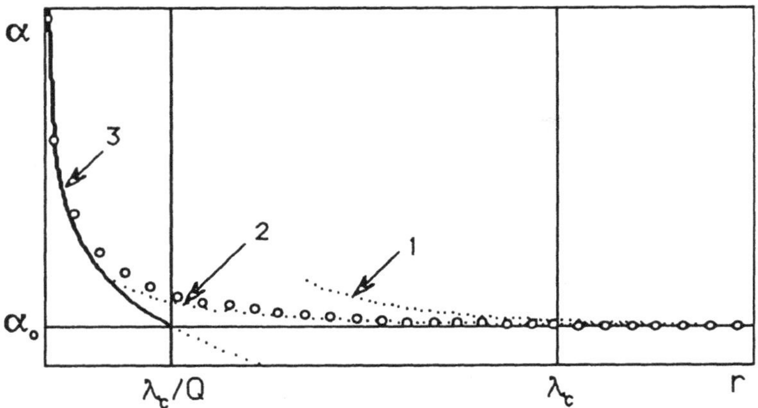

Figure 6.1. Variation with scale of the fine structure constant below the Compton length of the electron, $\lambda_c = \hbar/mc$. The open symbols show the result of a numerical calculation from the leading order formula. The numbered curved are various asymptotic formulae (see text).

Hence the electric charge, which was constant in the classical theory, becomes logarithmically divergent in the quantum theory. This loss is compensated by an improvement of the electromagnetic self-energy status: while linearly divergent classically, it becomes also logarithmically divergent in quantum electrodynamics[16]

$$m \;= m_0 \left\{ 1 + \frac{3\alpha_0}{2\pi} \left[ln \frac{\lambda}{r} + \frac{1}{4} \right] \right\} .$$

To second order, we find the electromagnetic running constant to be given, in terms of the *position* variable, by

$$\alpha^{-1} = \alpha_0^{-1} \left\{ 1 - \left(\frac{2\alpha_0}{3\pi} + \frac{\alpha_0^2}{2\pi^2} \right) ln \frac{\lambda/Q_1}{r} \right\}$$

with

$$lnQ_1 = \gamma + \frac{5}{6} - \frac{3\alpha_0}{2\pi} \left[\zeta(3) + \frac{5}{24} + \frac{2\pi^2}{27} \right]$$

where ζ is here the Riemann ζ function: $\zeta(3)=1.202\,056...$ (the value of Q_1 is tentative and needs an independent check). This form for the second order result will find its complete justification in the renormalization group approach, which we shall consider in the following. Note that in most textbooks the coupling constant and rest mass variation is given in terms of *energy scale* E ($E^2=p^2c^2+m^2c^4$) rather than length scale r. The passage from one representation to the other implies a Fourier transform, which reduces to a *Laplace* transform in the asymptotic case.

This leads to the same fundamental scale dependence (since $\ln E \equiv -\ln r$ from the Heisenberg relation, i.e., from the fact that the position and momentum representations are Fourier transforms one relative to the other), but also to *additional constant terms:* it is from the Laplace transform that constant terms like the Euler constant γ find their origin. These terms, which do not change the asymptotic behaviour, are most of the time considered unimportant.[14] However we have reasons to believe that they may actually play an important role in our understanding of the structure of the electron (this will be specified in a forthcoming section). Recall indeed our general point of view here that space-time should not be given up as a relevant tool for a fundamental description of the world, even in the microphysical domain. It is true that, with the methods presently used in quantum theory (particularly in high energy particle physics), the momentum representation is simpler to deal with and closer to experimental results than the position representation. As a consequence one gets the space-time results from (sometimes complicated) calculations additional to those which lead to the momentum solutions. This could seem to be unnecessary, but for the idea, which we support here, that the knowledge of detailed *space-time* structures is unavoidable in our understanding of elementary phenomena in Nature.

The hereabove results are correct when electron-positron pairs only are to be considered. But when the distance becomes smaller than the Compton length of the muon, muon-antimuon pairs begin to contribute to the interaction (at energies larger than ≈210 MeV). The *u* (5.1 MeV), *d* (8.9 MeV) and *s* (175 MeV) quarks, whose effective masses are of the order of ≈0.1 GeV,[17] intervene also at about the same energy. Then contributions by the *c* quark (1.25 GeV), the τ lepton (1.78 GeV), the *b* quark (4.5 GeV), the *t* quark (>80 GeV), the *W* boson (79.9 GeV) and *Z* boson (91.18 GeV) are to be taken into account. So an increasing slope is expected for the variation of α, with threshold effects at the energy of each new elementary particle pair (see Fig. 6.2). The masses of elementary particles then emerge as rest energies at which phase transitions occur. We shall come back to this point in the frame of the renormalization group approach, which we consider now.

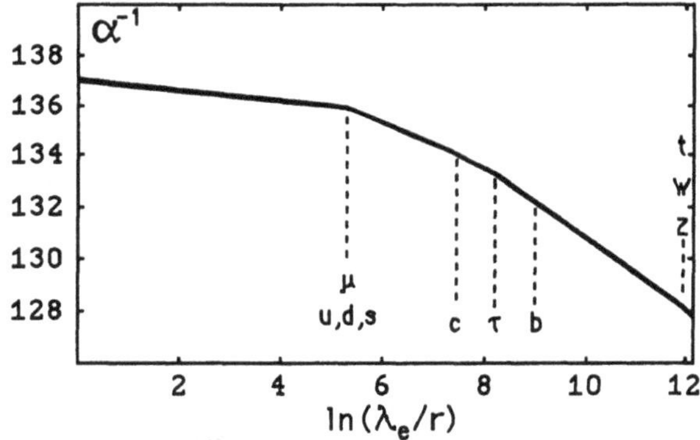

Figure 6.2. Variation of the inverse fine structure constant with scale, from the electron scale to the W/Z scale.

The renormalization group approach.

Let us briefly remind the reader of the historical road which finally led to the renormalization group concept. The problem of infinities which appeared in electrodynamics when one tried to compute the various physical quantities has been partly solved by a procedure called renormalization, introduced by Tomonaga,[18] Schwinger,[19] Feynman,[20] and Dyson.[21] It mainly consists in replacing the calculated infinite values of mass, charge and field by their observed finite values, and then compute again all the other physical quantities. The remarkable result is that, once these three fundamental quantities are renormalized, the predictions for other physical quantities (Lamb shift, anomalous magnetic moments, radiative corrections and so on) become finite and in extraordinary precise agreement with experiments. After renormalization, quantum electrodynamics becomes the second best experimentally verified theory among all physical theories: for example the predicted value for the magnetic moment of the electron to fourth order in α is[22] a = 1.001159652164(108), while the experimental value is[23] 1.001159652188(4). This is a 10^{-5} ppm precise result. Only general relativity is able to yield an even more precise agreement between theory and experiment: the motion of the *binary pulsar* is verified[24] to 10^{-14}. (Note that some *negative* assertions may be verified to a still better precision: hence the absence of anisotropy of inertia is verified to $\delta m/m<10^{-20}$).[25]

However, as impressive as these results may be, one remains dissatisfied with the present state of the theory. Renormalization may be experienced as some kind of principle of renounciation,[26] since in its frame one renounces to compute the "most interesting" quantities (from the view point of fundamental physics): the elementary masses and charges, which remain formally infinite at zero distance, i.e., in their "bare" version.

The question of the low energy numerical values of rest masses and charges is now asked in a new way: they could be predicted provided their value at some given very high energy-momentum scale was known from fundamental principles. But the remaining divergences seems to be a key point in our inability to understand the nature of the electric charge (and so to compute its value). These problems will be considered at lengths in the following, and possible solutions will be suggested in the framework of the theory of scale relativity.

Let us go on with renormalization. It is straightforward to understand that, since renormalization consists in subtracting infinite counterterms to the infinite values of mass and charge in order to get the finite observed values at low energy, there is not only one method to do that. There are several ways to renormalize, and clearly the final physical results should not depend on the particular method chosen. This was a new kind of invariance law, which was at first considered as a trivial statement. But only a few years after the renormalization procedure was understood and expressed in a practical way, it was realized by Gell-Mann and Low,[4] and independently by Stueckelberg and Peterman[5] (see also Landau *et al.*),[27] that the sole statement of renormalizability allowed one to deduce very important results about the asymptotic behaviour of QED and other renormalizable theories. Namely, the transformations among the various renormalization procedures follow a group law: from this comes the name "renormalization group".

The renormalization group approach indeed allows one to gain some knowledge about the global behaviour of the scale dependence of physical quantities like fields and couplings, while the perturbative approach in terms of coupling constant expansions, gives *a priori* no (or very incomplete) information about the asymptotic behaviour. This success is demonstrated by the fact that the lowest order solution to the renormalization group equations automatically includes infinite sums of terms of the form $\alpha^n ln^n(\lambda/r)$ from the "radiative correction" perturbation result.[14]

As discussed in Chapter 2 and throughout this book, the renormalization group method is already very close to the point of view adopted in scale relativity, since it describes a physics explicitly dependent on scale; we additionally require in our own approach that it is made scale-covariant. A renormalization group transformation, which allows one to go from a scale λ to another scale $\lambda'=s\,\lambda$ thanks to the application of a dilatation s, is decomposed into 3 steps:[28]

1) Integration on wave vector k: $1/s\,\lambda \le k \le 1/\lambda$.

2) Dilatation of length unit: $x \to x' = x/s$,
$k \to k' = s\,k$.

3) Field renormalization: $\varphi(x) \to \varphi'(x') = s^{\delta}\,\varphi(x)$,
$\tilde{\varphi}(k) \to \tilde{\varphi}'(k') = s^{\delta-E/2}\,\tilde{\varphi}(k)$,

where δ is the anomalous dimension of the field and E the space or space-time dimension. Note in this respect how δ=1 and E=4 are remarkable values. When E=4, which is the observed topological dimension of space-time, the requirement of symmetry between the position and momentum descriptions, $\delta_x = -\delta_\kappa$, yields precisely $\delta = 1$ (i.e., the value which corresponds to Brownian motion and Heisenberg relations, and, in the end, to fractal dimension 2).

The renormalization group equations has been put in a particularly simple form by Callan and Symanzik.[6,7] We shall only consider here a simplified version of these equations, in order to easily understand their physical meaning. Assume that some "field" φ varies with scale in such a way that the field itself is the only relevant parameter on which physical laws depend whatever the scale. This means that the variation of φ with scale will depend on φ alone through some function $\beta(\varphi)$, the so-called Callan-Symanzik β function:

$$\frac{d\varphi}{d(\ln \frac{r}{\lambda})} = \beta(\varphi) \quad .$$

Now let us place ourselves in the weak field approximation, $\varphi << 1$, which allows us to expand the β function in terms of powers of φ:

$$\frac{d\varphi}{d(\ln \frac{r}{\lambda})} = \beta(\varphi) = a - \delta\,\varphi + \beta_0\,\varphi^2 + \beta_1\,\varphi^3 + \ldots \quad .$$

Very different behaviours will be obtained depending on the lowest non-vanishing order in this expansion. One may first remark that it is not necessary to consider the case $\beta(\varphi) = a \neq 0$, since it is *formally* identical with the case $\beta(\varphi) = \beta_0\,\varphi^2$, for the substitution $\varphi \to 1/\varphi$. Two main situations can be encountered:

* That of a *relevant* field, which corresponds to $\delta \neq 0$. The renormalization group equation is integrated in this case as

$$\varphi = \varphi_0\,(\lambda/r)^{\delta} \ ,$$

i.e., the usual power law form in terms of anomalous dimension δ. As previously remarked (Sec. 3.8), keeping the constant term $[\beta(\varphi) = a - \delta\varphi]$ allows one to describe not only scaling, but also the breaking of scale invariance beyond the characteristic scale λ. Setting $a = \delta\varphi_0$ yields the solution $\varphi = \varphi_0[1 + (\lambda/r)^{\delta}]$.

* That of a *marginal* field: In this case, the expansion begins only at second order. This is precisely the situation for the electromagnetic coupling constant variation as demonstrated by the perturbative approach, which implies the vanishing of the zero and first order terms. The renormalization group equation for the coupling α writes

$$\frac{d\alpha}{d(\ln \frac{r}{\lambda})} = \beta(\alpha) = \beta_0\,\alpha^2 + \beta_1\,\alpha^3 + \dots \quad . \tag{6.2.2}$$

Now introducing the notation

$$\bar{\alpha} = \alpha^{-1} \; ,$$

we get the following differential equation for the *inverse coupling*:

$$\frac{d\bar{\alpha}}{d(\ln \frac{\lambda}{r})} = \beta_0 + \frac{\beta_1}{\bar{\alpha}} + \dots \quad .$$

The first order solution writes

$$\bar{\alpha} = \bar{\alpha}_0 + \beta_0 \ln \frac{\lambda_0}{r} \quad .$$

Coming back to direct coupling, one gets the remarkable scale-invariant result

$$\alpha(r') = \frac{\alpha(r)}{1 + \beta_0\,\alpha(r)\,\ln \frac{r}{r'}} \quad .$$

The usual perturbative result, $\alpha(r') = \alpha(r)\,[1 - \beta_0\,\alpha(r)\,\ln \frac{r}{r'}]$, now stands merely as a first order expansion of the renormalization group result.

But there is more to that: the success of the renormalization group approach is demonstrated by the fact that its lowest order solution also automatically includes infinite sums of terms of the form $\alpha^n \ln^n(\lambda/r)$, which correspond to arbitrarily high orders in the "radiative correction" perturbative method.[14]

To second order, the solution may be written as

$$\bar{\alpha} = \bar{\alpha}_0 + \beta_0 \, ln \frac{\lambda_0}{r} + \frac{\beta_1}{\beta_0} \, ln \, [1+\beta_0 \, \alpha_0 \, ln \frac{\lambda_0}{r} \,] + ... \, . \qquad (6.2.3)$$

These remarkable results, obtained from such a simple method, apply not only to the coupling of QED (from the electron to the *W/Z* scale), but also to the two couplings of the electroweak theory, α_1 [U(1) group] and α_2 [SU(2) group], and to the coupling α_3 of quantum chromodynamics (QCD) at high energies, for which the condition $\alpha_3 << 1$ remains fulfilled. Note that, to the second order, the actual renormalization group equations for α_1, α_2 and α_3 are coupled[29] : $d\bar{\alpha}_i / dln(\lambda/r) = \beta_i + \Sigma_j \beta_{ij} \, \alpha_j$, and so take a tensorial form, the detailed expression of which will be given in what follows. This non-independence of the 3 fields is an additional indication that they should, in the end, be treated together in the context of a unified theory.

Let us come back to the divergence problem. The renormalization group results, (connected to the success of the electroweak and QCD theories, see hereafter) allows one to make important progress in our understanding of its properties. The QED lowest order result

$$\alpha(r) = \frac{\alpha(\lambda)}{1 - \frac{2\alpha(\lambda)}{3\pi} \, ln \, \frac{\lambda}{r}} \qquad (6.2.4)$$

still leads to a small scale divergence of the electric charge, but now at the "Landau ghost," the scale which makes the denominator vanish. Note that the above factor $2/3\pi$ corresponds to considering the contribution of only one lepton. This holds between the electron and muon scales only. In a pure QED theory with 3 leptons (e, μ, τ) it would become $2/\pi$, yielding a Landau ghost at $r \approx \lambda \, e^{-2\alpha/\pi}$ of the order of 10^{-38} times the electron scale. This is far smaller than the Planck scale, at which physics is already expected to radically change: but one may also remain disatisfied if electrodynamics alone could not be formulated as a self-consistent theory. Unfortunately, the present standard quantum theory still fails to reach such self-consistency.

Consider now the question of self-energies. Neither in the perturbative approach nor in the renormalization group one, the origin of the mass is known: it could well be of pure electromagnetic origin, at least in a pure QED theory. Only its variation with scale (said to be due to "radiative corrections" in the current interpretation) can be described. The mass variation, being due to its electromagnetic self-energy, is expected to depend on the coupling and the scale, $m = m(lnr,\alpha)$, so that the renormalization group equation writes[30]

$$\frac{d}{dlnr}(m) = (\frac{\partial}{\partial lnr} + \beta(\alpha)\frac{\partial}{\partial\bar{\alpha}})(m) = \gamma(\alpha)\, m \quad .$$

To lowest order, the γ function is given by $\gamma(\alpha) = \gamma_0\, \alpha$, while $d\bar{\alpha} = \beta_0\, dlnr$. Then the renormalization equation takes the form

$$\frac{dm}{m} = \frac{\gamma_0}{\beta_0}\frac{d\bar{\alpha}}{\bar{\alpha}} \ ,$$

and is solved as

$$\frac{m}{m_0} = (\frac{\alpha_0}{\alpha})^{\gamma_0/\beta_0} \quad .$$

One thus finds in QED a rest mass varying with scale as $m/m_0 = (\alpha/\alpha_0)^{9/4}$, which diverges when $r \to 0$.

Anyway, the problem of these divergences was formulated in a completely renewed way in the framework of the electroweak theory and the standard model. Indeed one important result has led to the hope that it may be solved in a unified theory of electroweak and strong forces at high energy:[31-33] namely, the three running couplings have been found to converge at some high energy scale, the so-called Grand Unified Theory (GUT) scale (see Fig. 6.4). Let us first briefly recall the present status of the standard model, based on ElectroWeak Dynamics [of group structure U(1) x SU(2)] which unifies QED and the weak interaction, and Quantum Chromodynamics [group SU(3)], the theory of the color interaction between quarks, from which the strong interaction is derived.

Electroweak Theory and the Standard Model.

Let us first recall the lowest order QCD result[34]

$$\alpha_3 = \frac{\pi}{(\frac{11}{2} - \frac{n_f}{3})\, ln\frac{\lambda_s}{r}} ,$$

where n_f is the number of quark families ($n_f = 6$ in the present standard model) and λ_s is an integration constant whose experimental estimate presently lies around 150 MeV.[34] This expression for α_3, a consequence of the SU(3) group structure (which has $3^2-1=8$ generators, identified as the eight intermediate gluons), led to the important discovery of QCD asymptotic freedom,[35-37] $\alpha_3 \to 0$ when $r \to 0$. The integration constant may also be expressed in terms of the value of α_3 at some given scale, often taken as the W boson scale ($\approx$80 GeV). With this choice and adopting $n_f = 6$ (quarks u, d, s, c, t and b), one gets an inverse coupling given to the lowest order by

$$\bar{\alpha}_3 = \bar{\alpha}_3(\lambda_W) + \frac{7}{2\pi}\, ln\frac{\lambda_W}{r} , \tag{6.2.5}$$

where[44] $\bar{\alpha}_3(\lambda_W) = 9.35 \pm 0.80$.

These results do not *a priori* apply to the low energy case. Indeed, at large scale the strong coupling increases so much that the condition $\alpha \ll 1$ needed for expanding the β function in powers of α is no longer fulfilled. However, disregarding this fact, it is remarkable that Eq. (6.2.5) leads, in addition to asymptotic freedom, to "infrared slavery", i.e., $\alpha \to \infty$ when $r \to \lambda_s$. If this result is correct (i.e., if it is also a consequence of the still unknown form of the full β function) then it would explain the observed confinement of quarks, since an infinite energy would be needed to separate them.

Concerning the electromagnetic and weak interactions, they are now unified into the electroweak Weinberg-Salam theory.[38-40] The U(1) x SU(2) group structure of this theory implies 1 generator, B, and 3 generators (2^2-1), W^1, W^2, W^3, which, once mixed, yield the W^+ and W^- bosons on the one hand and the Z boson and the photon (i.e., the electromagnetic field A) on the other:

$$W^+ = (W^1 - i\, W^2) / \sqrt{2}\ , \qquad A = B \;\; cos\theta_W \; + \; W^3 \;\; sin\theta_W\ ,$$

$$W^- = (W^1 + i\, W^2) / \sqrt{2}\ , \qquad Z = -B \;\; cos\theta_W \; + \; W^3 \;\; sin\theta_W\ .$$

The "charges" are the SU(2) weak isospin (T_1,T_2,T_3) and the U(1) weak hypercharge Y, from which the U(1) QED electric charge may be recovered, as defined by

$$Q = T_3 + Y/2$$

The relation between the low energy and the high energy structures is described in Fig. 6.3. The symmetry breaking which occurs at an energy scale of about 250 GeV is ensured by the Higgs mechanism. The masses of the W and Z bosons are acquired by incorporation of some components of the Higgs doublet field (Fig. 6.3).

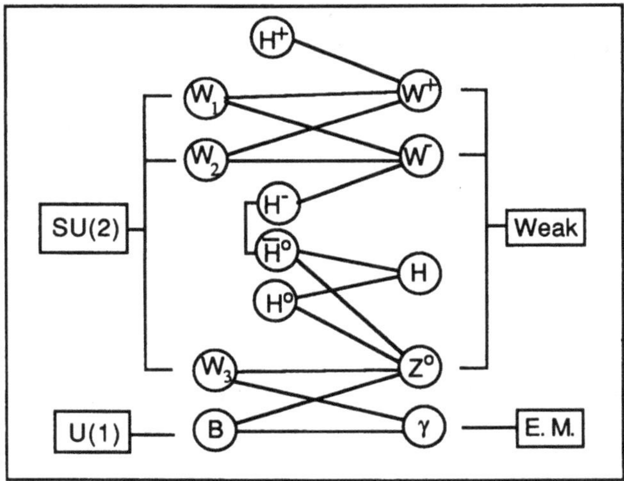

Figure 6.3. Structure of the electroweak transition, from high energy (left) to low energy (right). The photon, which includes no Higgs component, remains massless.

Thus QEWD still needs two coupling constants, α_1 and α_2, which are related to the electromagnetic coupling via the *weak mixing angle* θ_W. Several nearly equivalent definitions of the mixing angle exist, one of the most frequently used being related to the masses of the weak bosons:

$$cos\theta_w = M_W/M_Z \ .$$

The two U(1) and SU(2) coupling constants are thus given by

$$\bar{\alpha}_1 = \frac{3}{5}\ \bar{\alpha}\ cos^2\theta_w \quad , \quad \bar{\alpha}_2 = \bar{\alpha}\ sin^2\theta_w \ .$$

The inverse fine structure constant is therefore (formally) given by $\bar{\alpha} = \bar{\alpha}_2 + \frac{5}{3}\bar{\alpha}_1$, while the Fermi weak constant is itself related to the electroweak quantities by the relation

$$G_F = \frac{\sqrt{2}\ \alpha}{8\ M_W^2 sin^2\theta_w} \ .$$

Current (mid-1991) experimental determinations of the basic parameters of the model at LEP at CERN give[41] M_Z =91.177(31) GeV, M_W=79.9(4) GeV and $\sin^2\theta_w$=0.2302(21). Still unknown, in the model as well as experimentally, are the number N_H and the mass(es) of Higgs boson doublets.[34]

The two "running" (i.e., variable with scale) couplings are given to the lowest order for $r<\lambda_W$ by[42]

$$\bar{\alpha}_1 = \bar{\alpha}_1(\lambda_W) - (\frac{2}{\pi}+\frac{N_H}{20\pi})\ ln\frac{\lambda_W}{r} \ , \tag{6.2.6}$$

$$\bar{\alpha}_2 = \bar{\alpha}_2(\lambda_W) + (\frac{5}{3\pi}-\frac{N_H}{12\pi})\ ln\frac{\lambda_W}{r} \ , \tag{6.2.7}$$

while the variation of $\bar{\alpha}$ from the electron to the W scale is given by (see Fig. 6.2)[17]

$$\bar{\alpha}(M_W) = \bar{\alpha} - \frac{2}{3\pi}\sum_f Q_f^2\ ln\frac{M_W}{M_f} + \frac{1}{6\pi} \ , \tag{6.2.8}$$

where the sum is taken over all elementary particles of charges Q_f and masses M_f. With three leptons of charge 1, nine quarks of charge 1/3 (3 x 3 colours) and nine quarks of charge 2/3, one has ΣQ_f^2=8. This yields[43,44] $\bar{\alpha}(M_W) - \bar{\alpha} = -9.2\pm0.3$ (but see Sec. 6.11 for a slightly different determination). Combining this result with the measured value of Fermi's constant, the values of the couplings at the W scale are thus estimated to be[42,44] $\bar{\alpha}_1(\lambda_W)$=59.17±0.35 and $\bar{\alpha}_2(\lambda_W)$= 29.07±0.58.

To second order the renormalization group equations for the three fundamental couplings become mutually coupled, as recalled earlier. For completeness, one should also include in the equations the Yukawa couplings associated with the various Fermions. These couplings will have *a priori* a negligible influence, except when concerning a possible elementary particle of mass $\geq m_W$. Since it is now experimentally known that this indeed occurs for the t-quark, whose mass is now constrained[45] to be higher than about 80 GeV, one must, strictly speaking, also include its Yukawa coupling.

So the renormalization group equations for the three fundamental couplings write, assuming a number H of Higgs doublets and accounting for the Yukawa coupling associated with the t-quark[29,46]

$$\frac{d\bar{\alpha}_1}{dln(\lambda/r)} = -\frac{40+H}{20\pi} - \frac{1}{8\pi^2}\left\{ \frac{190+9H}{50}\,\alpha_1 + \frac{18+9H}{10}\,\alpha_2 + \frac{44}{5}\,\alpha_3 - \frac{17}{10}\,\kappa_t \right\},$$

$$\frac{d\bar{\alpha}_2}{dln(\lambda/r)} = \frac{20-H}{12\pi} - \frac{1}{8\pi^2}\left\{ \frac{6+3H}{10}\,\alpha_1 + \frac{22+13H}{6}\,\alpha_2 + 12\,\alpha_3 - \frac{3}{2}\,\kappa_t \right\},$$

$$\frac{d\bar{\alpha}_3}{dln(\lambda/r)} = \frac{7}{2\pi} - \frac{1}{8\pi^2}\left\{ \frac{11}{10}\,\alpha_1 + \frac{9}{2}\,\alpha_2 - 26\,\alpha_3 - 2\,\kappa_t \right\}. \qquad (6.2.9)$$

where, as above, $\bar{\alpha}_i = \alpha_i^{-1}$. One has defined here a running coupling $\kappa_t = \frac{1}{2}\,\alpha_2(m_W)\,(m_t/m_W)^2$, whose renormalization group equation writes to lowest order (neglecting the lighter fermion couplings)[42]

$$\frac{d\kappa_t}{dln(\lambda/r)} = \frac{9}{4\pi}\,\kappa_t^2 - \kappa_t\left\{ \frac{17}{40\pi}\,\alpha_1 + \frac{9}{8\pi}\,\alpha_2 + \frac{4}{\pi}\,\alpha_3 \right\}.$$

This equation assumes an explicit Higgs mechanism. In the case of dynamical symmetry breaking, the hereabove terms $(9/4\pi)\kappa_t^2$ and $-(9/8\pi)\kappa_t\alpha_2$ are replaced[42] respectively by $(3/2\pi)\kappa_t^2$ and $-(3/4\pi)\kappa_t\alpha_2$.

Assuming κ_t to be a function of α_i alone and neglecting in a first step α_1 and α_2 allow one to find the general family of solutions[42,47]

$$\kappa_t = a\,\alpha_3 \frac{1}{1+b\,\bar{\alpha}_3^{\,1/7}} \quad .$$

where a=2/9 or 1/3, depending on the symmetry breaking mechanism, and b is an arbitrary constant. Owing to the fact that κ_t occurs in (6.2.9) to second order and with coefficients <10% of other second order coefficients, $\bar{\alpha}_3^{\,1/7}$ may be approximated by a constant (it indeed varies only from 1.37 to 1.69 from the W to the GUT scale). So we may safely take $\kappa_t = k_3\alpha_3$ with $k_3 \approx 1/3$, and, more generally, reincluding the electroweak corrections[42]

$$\kappa_t = k_1\alpha_1 + k_2\alpha_2 + k_3\alpha_3 \;, \tag{6.2.10}$$

with $k_2 = -1/12$ and $k_3 = -17/540$ in the Higgs case. As a consequence the t-quark Yukawa coupling will finally contribute by less than 0.01 in the final estimation of the values of the inverse couplings. It is neglected in the following: however the form of (6.2.10) would allow one to easily account for it if necessary, since it yields only simple constant numerical corrections to the coefficients of (6.2.9).

The second order solutions to (6.2.9) are easily found by reinserting the first order solutions in the second order terms. This yields (i=1 to 3)

$$\frac{d\bar{\alpha}_i}{d\ln(r/\lambda)} = \beta_0^{\,i} + \sum_j \frac{\beta_1^{\,ij}}{\bar{\alpha}_{jW} + \beta_0^{\,j}\,\ln(r/\lambda_W)} \quad .$$

These equations are solved in terms of double logarithm as

$$\bar{\alpha}_i = \bar{\alpha}_{iW} + \beta_0^{\,i}\,\ln\frac{r}{\lambda_W} + \sum_j \frac{\beta_1^{\,ij}}{\beta_0^{\,j}}\,\ln\left(1 + \beta_0^{\,i}\,\alpha_{iW}\,\ln\frac{r}{\lambda_W}\right) .$$

We recall that the low energy electromagnetic coupling is related to the U(1) and SU(2) couplings of the electroweak theory. Strictly speaking, this coupling (the "running" fine structure constant) is defined only at energies smaller than the W boson energy. However let us introduce a *formal* running electromagnetic coupling *extrapolated* to high energies, defined by the same relation $\bar{\alpha} = \bar{\alpha}_2 + \frac{5}{3}\bar{\alpha}_1$ .Then we find explicitely, in matrix form, for the 4 couplings $\bar{\alpha}_1$, $\bar{\alpha}_2$, $\bar{\alpha}_3$ and $\bar{\alpha}$

$$\begin{bmatrix} \bar{\alpha}_1(\lambda) \\ \bar{\alpha}_2(\lambda) \\ \bar{\alpha}_3(\lambda) \\ \bar{\alpha}(\lambda) \end{bmatrix} = \begin{bmatrix} \bar{\alpha}_1(\lambda_o) \\ \bar{\alpha}_2(\lambda_o) \\ \bar{\alpha}_3(\lambda_o) \\ \bar{\alpha}(\lambda_o) \end{bmatrix} + \begin{bmatrix} -\frac{40+H}{20\pi} \\ \frac{20-H}{12\pi} \\ \frac{7}{2\pi} \\ -\frac{10+H}{6\pi} \end{bmatrix} ln(\frac{\lambda_o}{\lambda}) + \begin{bmatrix} \frac{190+9H}{20\pi(40+H)} & \frac{-54-27H}{20\pi(20-H)} & \frac{-11}{35\pi} \\ \frac{6+3H}{4\pi(40+H)} & \frac{-22-13H}{4\pi(20-H)} & \frac{-3}{7\pi} \\ \frac{11}{4\pi(40+H)} & \frac{-27}{4\pi(20-H)} & \frac{13}{14\pi} \\ \frac{104+9H}{6\pi(40+H)} & \frac{-20-11H}{2\pi(20-H)} & \frac{-20}{21\pi} \end{bmatrix} \begin{bmatrix} ln\{1-\frac{40+H}{20\pi}\alpha_1(\lambda_o)ln(\frac{\lambda_o}{\lambda})\} \\ ln\{1+\frac{20-H}{12\pi}\alpha_2(\lambda_o)ln(\frac{\lambda_o}{\lambda})\} \\ ln\{1+\frac{7}{2\pi}\alpha_3(\lambda_o)ln(\frac{\lambda_o}{\lambda})\} \end{bmatrix} \tag{6.2.11}$$

where H is the number of Higgs doublets.

Let us come back to the question of self-energies. We have recalled hereabove that the renormalization group approach leads to the conclusion that their variation was given in terms of a variation of couplings. But now each of the three interactions contribute to them: the variation with scale of rest masses is then expected to depend on the three couplings α_1, α_2 and α_3. Buras *et al.*[48] find the following rest mass variations beyond the W scale, respectively for quarks u, c and t (charge −2/3), quarks d, s and b (charges 1/3) and leptons e, μ and τ (charges −1, no strong interaction):

$$\begin{aligned} m_{uct} &\propto \alpha_3^{4/7} \ \alpha_2^{27/40} \ \alpha_1^{-1/20} \ , \\ m_{dsb} &\propto \alpha_3^{4/7} \ \alpha_2^{27/40} \ \alpha_1^{1/40} \ , \\ m_{e\mu\tau} &\propto \alpha_2^{27/40} \ \alpha_1^{-9/20} \ . \end{aligned} \tag{6.2.12}$$

As an illustration, let us point out that these variations amount to about 3.1 (u, c, t), 3.0 (d, s, b) and 1.35 (e, μ, τ) from the W scale (≈80 GeV) to the scale 10^{14} GeV, for which, as we shall see later, the three fundamental couplings are found to become equal.

Grand Unification.

Let us finally come to grand unified theories. From the above mentioned solutions to the renormalization group equations (Eqs. 6.2.5, 6.2.6, 6.2.7), one may plot the variation of the three couplings from the W scale to higher energies, i.e. smaller resolutions (see Fig. 6.4).

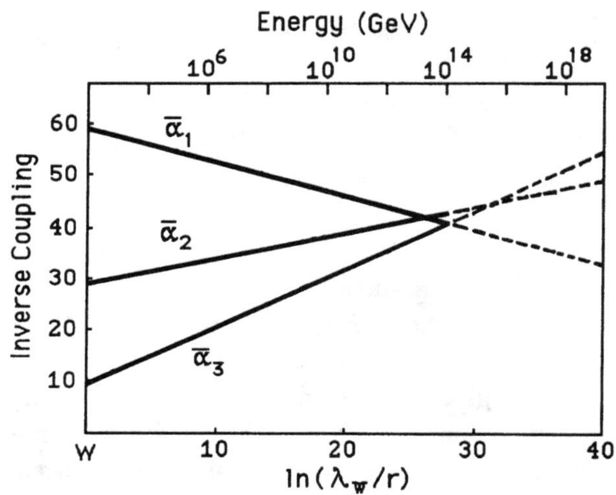

Figure 6.4. Convergence of the inverse coupling "constants" at the GUT scale in the standard model.

This yields the remarkable result that the three couplings converge at some high energy of the order of $m_X \approx 10^{14}$-10^{15} GeV/c^2. This is a very strong argument in favor of a complete unification of electromagnetic, weak and strong (colour) forces at this scale.[31-33] This convergence is ensured under the "great desert hypothesis" which assumes that there is no new particle (no new physics) between the electroweak scale $\approx m_W$ and the unification scale m_X. Second order terms in the solutions to the renormalization group coupled equations[29,46] do not change these conclusions, their contribution being presently smaller than the errors on the couplings at the W scale. (From Eqs. (6.2.11), we find these contributions to be respectively −0.19, −0.01 and 0.23 to the inverse couplings).

The simplest possible unification group is SU(5),[32] but more complicated structures are also possible. GUT's obtained at first a lot of successes:[49]

(1) In their framework, the quantization of the electric charge finds a natural explanation.[32]
(2) The value of the b quark mass may be predicted from its expected equality with the τ lepton at energy m_X, and from its evolution with scale deduced from the renormalization group equations (6.2.12);[48] one finds that M_b/M_τ(pred) = 2.75±0.37, to be compared with the observed ratio M_b/M_τ(obs)=2.38±0.06. This is an impressive result, since this would be the first prediction of an elementary particle mass (strictly speaking, of a mass ratio) in the present quantum theory.
(3) The number of generations of leptons and quarks is constrained to be n=3 (a larger number would have made the above prediction unacceptably high):[48] this has been confirmed for light neutrinos by primordial nucleosynthesis[50] and definitively by LEP.[41]
(4) The possible values of the low energy fine structure constant are constrained[51] to be $<1/25$, and even better[52], $1/120<\alpha<1/170$.
(5) The value of Weinberg's mixing angle may be predicted: at the unification scale, one has $\alpha_1 = \alpha_2$, and may introduce a running effective angle such that $sin^2\hat{\theta}_W(m_X) = \bar{\alpha}_2(m_X)/\bar{\alpha}(m_X) = 3/8$, while the renormalization group yields a scale variation given by[17]

$$sin^2\hat{\theta}_w(m_W) = \frac{3}{8}\left[1 - \frac{109}{9}\ \frac{\alpha(m_W)}{2\pi}\ ln\frac{m_X}{m_W}\right]. \qquad (6.2.13)$$

The effective and measured values are related by[44] $sin^2\hat{\theta}_w(m_W) = 0.9907\ sin^2\theta_w$ (assuming m_t=45 GeV; the ratio becomes 1.02 for m_t=125 GeV). This allowed physicists to predict[49,44] that $sin^2\theta_w(m_W) = 0.210$, which was in good agreement with the measured value at the time of the prediction (≈1980), 0.23±0.02.
(6) The decay of the proton was predicted[31] with a lifetime $\approx m_X^4/M_P^5 \gtrsim 10^{30}$ years for $m_X \approx 10^{15}$ GeV, also consistent with known experiments at that time.

Recall that most of these results hold in a large class of theories, not only in the simplest GUT, based on the SU(5) group.[32,49] Unfortunately these great hopes were soon dashed: an increase of the precision of experimental results led to unacceptable disagreements with the predictions. The present value of $sin^2\theta_w(m_W)$ as measured by LEP is[41] $sin^2\theta_w(m_W)=0.2302\pm0.0021$, more than 10σ off the theoretical prediction; the experimental proton lifetime is now known to be $>10^{32}$ years, thus requiring that $m_X>>10^{15}$ GeV, while agreement for the mixing angle would require $m_X\approx10^{13}$ GeV.

Physicists have tried to evade this dilemna by increasing the model complexity: introduction of supersymmetry, new symmetry breaking scale, additional particles with respect to the standard model... Most of these attempts can be criticized since they introduce *ad hoc* elements and break the simplicity of the initial scheme. It is rather astonishing to see so wonderful and simple ideas yield at first so remarkable predictions ($sin^2\theta_w$ could have fallen anywhere between 0 and 1) and finally disagree with experimental results.

It looks as if some essential point was missing in the present theory. Some authors have pointed out that the "great desert hypothesis", i.e. the absence of new particles or new physics between the *W/Z* scale and the GUT scale, over 12 decades, was not easy to accept. However this is a somewhat anthropocentric position, based on our own classical knowledges and feelings about scale transformations. The whole physical laws known in the high energy quantum domain ($\lambda<\hbar/mc$), in particular the renormalization group, proves that the relevant variables for these very small scales take a *logarithmic* form. So what seems to us to be very large scale ratios may well not be so large from the viewpoint of relevant physical laws (for example the whole scale domain in the Universe covers ≈60 decades, see Sec. 7.1).

Let us remark, in order to conclude this reminder, that what indeed seems to be lacking in the present state of physics is an authentic theory of scales. While the current theory is able to deal with scale invariance (e.g., the renormalization group), it is still unable to account for the emergence of particular scales in Nature. In the quantum domain, the electron scale (i.e., the value of the electron mass), the various scales corresponding to the elementary particle masses, the electroweak W/Z scale, the GUT scale are not predicted by any theory. (I mean here prediction of their very existence). For example the GUT scale itself has been discovered from a combination of experimental results (observed values of charges at the W scale) and theory (scale variation of charges as derived from the renormalization group equations, themselves based on the observed number and properties of known elementary particles).

Predicting the numerical values of fundamental scales in nature can be done only by referring them to a still more fundamental and universal scale. In the microphysical domain, the Planck scale plays this role. Being defined from the three constants G, $\hbar$, and c, (the numerical values of which are themselves arbitrary, since they depend on our choices of length, time and mass units), the Planck length, time and mass indeed define some absolute scale which constitutes a natural unit of scales. A prediction of scale would then consist in predicting the value of a pure number: the ratio of this scale to the Planck scale. This points towards a higher level of physical theory relative to its present state, dealing with dimensionless quantities. It is worth pointing out in this respect that dimensionless structures have already appeared in physics with the Poincaré-Einstein special theory of relativity. The v/c and $\sqrt{1-v^2/c^2}$ factors are indeed dimensionless and as such, impossible to obtain from pure dimensional analysis: only a theory based on *fundamental principles* has been able to derive them.

We shall argue in the remainder of this book that the *principle of relativity*, once applied, not only to motion, but also to *scale* transformations, allows one to attack the question of fundamental scales in Nature in a powerful and efficient manner. We shall demonstrate that many of the problems considered above, the divergence of mass and charge, the disagreement of GUT predictions with experimental results, and the nature of the GUT scale, find a natural solution in the framework of the new theory of scale relativity.

6.3. Galilean Scale Relativity.

One of the main characteristics of scale which points toward the need for a scale relativity theory is the nonexistence of an *a priori* absolute scale. Just as one may write for velocities in Galilean motion relativity

$$v = v_2 - v_1 = (v_2 - v_0) - (v_1 - v_0) \tag{6.3.1}$$

one may, in present physics, write for a scale ratio

$$\rho = \Delta x_2 / \Delta x_1 = (\Delta x_2 / \Delta x_0) / (\Delta x_1 / \Delta x_0) \quad . \tag{6.3.2}$$

It is indeed clear that one can never define the length of an object without comparing it to another object: only scale ratios, i.e. dilatations, have physical meaning. The expression (6.3.2) may be written in the same additive group form as Eq. (6.3.1) in a logarithmic representation:

$$ln\ \rho = ln(\Delta x_2/\Delta x_1) = ln(\Delta x_2/\Delta x_0) - ln(\Delta x_1/\Delta x_0). \tag{6.3.3}$$

So the "scale state" $V=ln(\Delta x_2/\Delta x_1)$ is formally equivalent to a "velocity", in accordance with our basic assumption which treats states of scale and states of motion on the same footing. Just as one can speak only of the velocity of a system *relative* to another, the scale of a system can be defined only by its ratio to the scale of another system. Equation (6.3.3) may now be written in exactly the same form as Eq. (6.3.1):

$$V = V_2 - V_1 = (V_2 - V_0) - (V_1 - V_0) \quad .$$

Concerning the problem of units, notice the difference of status between motion and scale laws. While velocity is expressed in terms of a physical unit (e.g., $m{\cdot}s^{-1}$), the state of scale is expressed in terms of a mathematical unit, i.e., the adopted base of logarithms. Indeed the same behaviour is obtained for any base *b* using the more general definition:

$$V = \frac{ln(\Delta x_2/\Delta x_1)}{ln(b)} = log_b \frac{\Delta x_2}{\Delta x_1} \quad .$$

It will be seen hereafter that this leads to a new kind of dimensional analysis.

Consider now a field φ which transforms under a dilatation $q=\Delta x/\Delta x'$ as a power law:

$$\varphi' = \varphi\, q^{\delta} \ . \tag{6.3.4}$$

In a renormalization group description, the power δ is identified with the anomalous dimension of the field[9] φ. In a fractal interpretation of the same phenomenon, we get $\delta=D-D_T$, where D is the fractal dimension and D_T the topological dimension.[3,11]

We are particularly interested here in the case where the "field" φ is space-time itself. We have seen in Chapters 4 and 5 that the length of a quantum particle path diverges as

$$\mathcal{L} = \mathcal{L}_0\, (\lambda/\Delta x)^{\delta} \ , \tag{6.3.5a}$$

where $\delta=D-D_T=1$, corresponding to a fractal dimension $D=2$ below the de Broglie length $\lambda=\hbar/p$.

The same behaviour is found for the total proper time of a particle (see Chapter 4).[3] When $\Delta t<\tau=\hbar/E$, one finds

$$\mathcal{T} = \mathcal{T}_0\, (\tau/\Delta t)^{\delta} \ , \tag{6.3.5b}$$

with the temporal anomalous dimension $\delta=1$ also in this case. Recall that this makes Lorentz covariant the reinterpretation of the de Broglie scale as a universal space-time transition from an anomalous dimension $\delta=0$ to $\delta=1$, since it applies to all the four space-time coordinates. In terms of the renormalization group, the de Broglie scale may be identified with the correlation length of space-time.

Keeping all these results in mind, let us write Eq. (6.3.4) in a linear form by taking once again the logarithm

$$ln(\varphi'/\varphi_0) = ln(\varphi\, /\varphi_0) + \delta\ ln(\Delta x/\Delta x') \ , \tag{6.3.6}$$

which is assumed to hold when $\Delta x<<\lambda$ (here Δx is the resolution of any coordinate, spatial or temporal, and $\varphi=\mathcal{L}$ or $\mathcal{T}$). We can go on with the analogy of scale relativity with motion relativity. The Galilean transform between two coordinate systems reads:

$$x' = x + vt \ ,$$

$$t' = t \ .$$

We may now get a consistent description in which, as conjectured, resolution acts as a "scale velocity", while the anomalous dimension (i.e. the fractal dimension minus the topological dimension) plays the role of a "scale time". Indeed, defining

$$X = ln(\varphi / \varphi_0)$$

we obtain the linear relation

$$X = X_0 + \delta \ ln(\lambda / \Delta x) \ .$$

Then we may define the state of scale V as

$$V = ln(\lambda / \Delta x) = \frac{d(ln\varphi)}{d\delta} = \frac{dX}{d\delta} \ ,$$

in the same way as the velocity of an object is defined as $u=dx/dt$ in motion relativity. Note the different approach with respect to the usual definition for the fractal dimension, $\delta = \partial ln(\varphi) / \partial ln(\lambda / \Delta x)$. The scale law (6.3.6) is structurally equivalent concerning scales to free motion at constant velocity, which is at the root of the definition of inertial motion. Likewise we suggest that a coordinate supersystem[3] (i.e. defined by its states of motion *and of scale*) in which Eq. (6.3.6) holds may be called "scale-inertial," and that we may set a principle of (special) scale relativity, which states that *the laws of nature are identical in all scale-inertial supersystems of coordinates.*

The anomalous dimension δ is assumed to be invariant (for space-time coordinates we find the universal value $\delta = 1$, itself coming from the universality of the Heisenberg relations), as time is invariant in Galilean relativity. This is translated by the equations of the Galilean scale-inertial transformation:

$$X' = X + V\,\delta \ , \tag{6.3.7a}$$

$$\delta' = \delta \ . \tag{6.3.7b}$$

In such a Galilean frame, the law of composition of scale states is the direct sum

$$W = U + V, \tag{6.3.7c}$$

which corresponds to the product $\Delta x''/\Delta x = (\Delta x''/\Delta x')\cdot(\Delta x'/\Delta x)$ for resolutions. Finally, with the three equations (6.3.7a)-(6.3.7c), we have put the scale relativity problem in exactly the same mathematical form (Galileo group) as that of motion relativity in classical mechanics.

But one should also keep in mind that the above "inertial scaling" holds only under some upper cutoff λ_0, contrary to motion relativity which is universal. This may be expressed by writing, instead of (6.3.6), a formula including a transition from scale-dependence to scale-independence, such as

$$X = X_0 + \delta \ ln\,[1+(\lambda_0/\Delta x)^2]^{1/2} \ .$$

More generally one may introduce a parameter k which characterizes the speed of transition and replace V in (6.3.7a) by Z, defined as

$$Z = \frac{1}{k}\, ln\, [1+(\frac{\lambda_0}{\Delta x})^k\,] = ln\,(1 + e^{kV})^{1/k} \ .$$

Then for $\Delta x >> \lambda_0$, Z=0, and for $\Delta x << \lambda_0$, $Z=V$. For k small, the transition between these two regions is slow, while it is sudden (singular point) for $k \rightarrow \infty$. This description of the transition is only a model, since the actual details of the transition depend on the physical system considered: for example in many situations the transition may imply exponentially decreasing "Yukawa-like" terms. Different methods may be considered to include the transition in the equations. One method consists in including it in δ itself by replacing the constant δ by a function of resolution $\delta(\Delta x)$ such that $\delta(\Delta x)$=0 for $\Delta x >> \lambda_0$ and $\delta(\Delta x)$=1 for $\Delta x << \lambda_0$. Another method consists in coming back to the renormalization group differential equation in which the constant lowest order term is kept. As seen in Sec 6.2, the solution then writes

$$\mathcal{L} = \mathcal{L}_0 [\, 1 + (\lambda/\Delta x)^{\delta}] \ .$$

6.4. A New Derivation of Lorentz Transformation.

As remarked by Levy-Leblond,[2] very little freedom is allowed for the choice of a relativity group, so that the Poincaré group is an almost unique solution to the problem.[53] In his original paper, Einstein derived the Lorentz transformation from the (sometimes implicit) successive assumptions of (i) linearity, (ii) invariance of c, the light velocity in vacuum (iii) existence of a composition law, (iv) existence of a neutral element, and (v) reflection invariance.

But one may demonstrate that the postulate of the invariance of some absolute velocity is not necessary for the construction of the special theory of relativity. Indeed it was shown by Levy-Leblond[2] that the Lorentz transformation may be obtained through six *successive* constraints: {1} homogeneity of space-time (translated as the linearity of the transformation of coordinates), {2} isotropy of space-time (translated as reflection invariance), {3} group structure (i.e. {3.1} existence of a neutral element, {3.2} of an inverse transformation, and {3.3} of a composition law yielding a new transformation which is a member of the group, viz, which is internal), and {4} the causality condition. The last group axiom, associativity, is in fact straightforward in this case and leads to no new constraint.

Actually this set of hypotheses is still overdetermined to derive the Lorentz transformation. We shall indeed demonstrate hereafter that the Lorentz transformation may be obtained from *only the assumptions of* {a} *linearity,* {b} *internal composition law, and* {c} *reflection invariance.* All the other assumptions, in particular the postulate of the existence of an inverse transformation which is a member of the group, may be *derived as consequences* of these purely mathematical constraints. The importance of this result, especially concerning scale relativity, is that we do not have to postulate a full group law in order to get the Lorentz behaviour: the hypothesis of a semigroup structure is sufficient.

Let us start from a linear transformation of coordinates:

$$x' = a(v)\, x - b(v)\, t \quad , \tag{6.4.1a}$$

$$t' = \alpha(v)\, t - \beta(v)\, x \quad . \tag{6.4.1b}$$

In these equations and in the whole section, the coordinates x and t do not denote *a priori* lengths and times, but may refer to any kind of variables having the mathematical properties considered. Equation (6.4.1) may be written as $x'=a(v)[x-(b/a)t]$. But we may *define* the "velocity" v as $v = b/a$ (in the case of motion laws, this is indeed the velocity in the usual meaning; in the case of scale laws, this is the state of scale or "scale velocity"). Then, without any loss of generality, linearity alone leads to the general form

$$x' = \gamma(v)\,[\, x - v\, t\,] \;, \tag{6.4.2a}$$

$$t' = \gamma(v)\,[\; A(v)\, t - B(v)\, x\,] \;, \tag{6.4.2b}$$

where $\gamma(v)$ now stands for $a(v)$, and where A and B are new functions of v. Let us now perform two successive transformations of the form (6.4.2):

$$x' = \gamma(u)\,[\, x - u\, t\,] \;, \tag{6.4.3a}$$

$$t' = \gamma(u)\,[\; A(u)\, t - B(u)\, x\,] \;, \tag{6.4.3b}$$

$$x'' = \gamma(v)\,[\, x' - v\, t'\,] \;, \tag{6.4.3c}$$

$$t'' = \gamma(v)\,[\; A(v)\, t' - B(v)\, x'\,] \;. \tag{6.4.3d}$$

This results in the transformation

$$x'' = \gamma(u)\,\gamma(v)\,[\;1 + B(u)\, v\,]\,[\, x - \frac{u + A(u)\, v}{1 + B(u)\, v}\, t\,] \;, \tag{6.4.4a}$$

$$t'' = \gamma(u)\,\gamma(v)\,[\,A(u)\,A(v) + B(v)\,u\,]\,[\,t - \frac{A(v)\,B(u) + B(v)}{A(u)\,A(v) + B(v)\,u}\,x\,]\ . \qquad (6.4.4b)$$

Then the principle of relativity tells us that the composed transformation (6.4.4) keeps the same form as the initial one (6.4.2), in terms of a composed velocity w given by the factor of t in (6.4.4a). We get four conditions:

$$w = \frac{u + A(u)\,v}{1 + B(u)\,v}\ , \qquad (6.4.5a)$$

$$\gamma(w) = \gamma(u)\,\gamma(v)\,[\,1 + B(u)\,v\,]\ , \qquad (6.4.5b)$$

$$\gamma(w)\,A(w) = \gamma(u)\,\gamma(v)\,[\,A(u)\,A(v) + B(v)\,u\,]\ , \qquad (6.4.5c)$$

$$\frac{B(w)}{A(w)} = \frac{A(v)\,B(u) + B(v)}{A(u)\,A(v) + B(v)\,u}\ . \qquad (6.4.5d)$$

Our third postulate is reflection invariance. It reflects the fact that the choice of the orientation of the x (and x') axis is completely arbitrary, and should be indistinguishable from the alternative choice $(-x, -x')$. With this new choice, the transformation (6.4.3) becomes $\{-x'=\gamma(u')(-x-u't)$, $t'=\gamma(u')\,[A(u')t + B(u')x]\}$ in terms of the value u' taken by the relative velocity in the new orientation. The requirement that the two orientations be indistinguishable yields $u'=-u$. This leads to parity relations[2] for the three unknown functions γ, A and B:

$$\gamma(-v) = \gamma(v), \qquad A(-v) = A(v), \qquad B(-v) = -B(v)\ . \qquad (6.4.6)$$

Combining Eqs. (6.4.5a), (6.4.5b) and (6.4.5c) yields the relation

$$A\left[\frac{u + A(u)\,v}{1 + B(u)\,v}\right] = \frac{A(u)\,A(v) + B(v)\,u}{1 + B(u)\,v}\ . \qquad (6.4.7)$$

Taking $v = 0$ in this equation gives

$$A(u)\,[\,1 - A(0)\,] = u\,B(0)\ . \qquad (6.4.8)$$

Taking u=0 yields only two solutions, $A(0)$=0 or 1. The first case gives $A(u)=uB(0)$. $B(0)\neq0$ is excluded by reflection invariance (6.4.6); then $A(u)$=0. But (6.4.5d) becomes $A(w)=B(w)u$, so that $B(w)$=0: this is a case of complete degeneration to only one efficient variable since t'=0 for any u, which can thus be excluded, since we are looking for two-variable transformations. We are left with $A(0)$=1, which implies $B(0)$=0, and the existence of a neutral element is demonstrated. Let us now take $v=-u$ in (6.4.7) after accounting for (6.4.6), and introduce a new even function $F(u)=A(u)-1$, which verifies $F(0)$=0. We obtain

$$2\ F(u)\ \frac{1+F(u)/2}{1-u\ B(u)}=F\left[\ \frac{u\ F(u)}{1-u\ B(u)}\ \right]\ . \tag{6.4.9}$$

We shall now use the fact that B and F are continuous functions and $B(0)$=0. This implies that there exists $\eta_0>0$ such that in the interval $-\eta_0<u<\eta_0$, $1-uB$ and $1+F/2$ become bounded to $k_1<1-uB(u)<k_2$ and $k_3<1+F(u)/2<k_4$ with k_1, k_2, k_3 and $k_4>0$. The bounds on $1+F/2$ and $1-uB$ allow us to bring the problem back to the equivalent equation, $2F(u)=F[uF(u)]$. The continuity of F at u=0 reads, owing to the fact that $F(u)$=0,

$$\forall\varepsilon,\ \exists\eta\ /\ |u|<\eta\ \Rightarrow|F(u)|<\varepsilon\ .$$

Start with some $u_0<\eta$ and define n such that $F(u_0)=F_0=2^{-n}<\varepsilon$. Then $F(u_0F_0)=2F_0$. Set $u_1=u_0F_0$ and iterate. After p iterations, one gets $F(u_p)=F[2^{p[(p-1)/2\ -n]}u_0]=2^{p-n}$. In particular one gets, after n iterations, $F[2^{-n(n+1)/2}u_0]=1$ if n is an integer. (In the general case where n is not an integer, one gets after Int[n] iterations a value of F larger than 1/2.) This is in contradiction with the continuity of F, since $u_n<u_0<\eta$ while $F(u_n)>\varepsilon$. Then the only solution is F=0 in a finite non-null interval around the origin, and by extension whatever the value of u, so that

$$A(u)=1\ . \tag{6.4.10}$$

As a consequence (6.4.7) becomes $B(u)v=B(v)u$, a relation which finally constrains the B function to be

$$B(v)=\kappa\ v\ , \tag{6.4.11}$$

where κ is a constant. At this stage of our demonstration, the law of transformation of velocities is already fixed to the Einstein-Lorentz form:

$$w = \frac{u + v}{1 + \kappa\, u v} \; , \qquad (6.4.12)$$

and it is easy to verify that a full group law is obtained, i.e. the existence of an identity transformation and an inverse transformation is demonstrated rather than postulated. Consider now the γ factor. It satisfies the condition

$$\gamma\left(\frac{u + v}{1 + \kappa\, u v}\right) = \gamma(u)\, \gamma(v)\, (1 + \kappa\, u\, v) \; . \qquad (6.4.13)$$

Let us consider the case $u=-v$. Equation (6.4.13) reads $\gamma(0) = \gamma(v)\gamma(-v)(1-\kappa v^2)$. For $v = 0$ it becomes $\gamma(0) = [\gamma(0)]^2$ implying $\gamma(0)=1$, and we get

$$\gamma(v)\, \gamma(-v) = \frac{1}{1 - \kappa\, v^2} \; , \qquad (6.4.14)$$

The final step to the Lorentz transformation is straighforward from reflection invariance, which implies that $\gamma(v)=\gamma(-v)$ (see Eq. 6.4.6) and fixes the γ factor in its Lorentz-Einstein form:

$$\gamma(v) = \frac{1}{\sqrt{1 - \kappa\, v^2}} \; . \qquad (6.4.15)$$

The case $\kappa<0$ yields a non-ordered group (applying two successive positive velocities may yield a negative one), and we are left with only two physical solutions, the Galileo ($\kappa = 0$) and Lorentz ($\kappa = c^{-2} > 0$) groups. Three of their properties (existence of a neutral element and of an inverse element, commutativity in case of one space dimension) have not been postulated, but deduced from our initial axioms.

Let us end this section with a brief but important comment. We have shown that, once linearity is assumed, the Lorentz transformation may be obtained through only the postulates of internal composition law and reflection invariance. Linearity is not a constraint by itself: indeed it corresponds to the simplest-possible choice (i.e. when searching for a transformation which would satisfy a given law, one may first look for a linear one, and then for non-linearity only in case of failure, or later as a generalization). With regard to the other two postulates, they may be seen as a *direct translation of the Galilean principle of relativity.* Indeed the hypothesis that the composed coordinate transformation ($K \rightarrow K''$) and the transformation in the reversed frame ($-K \rightarrow -K'$) must keep the same form as the initial one ($K \rightarrow K'$) is nothing but an application of the Galilean principle of relativity ("the laws of nature must keep the same form in different inertial reference systems") *to the laws of coordinate transformation themselves,* which are clearly part of the laws to which the principle should apply. So the general solution to the problem of inertial motion, without adding any postulate to the way it might have been stated in the Galileo and Descartes epoch, is actually Einstein's special relativity, of which Galilean relativity is a special case ($c=\infty$).

6.5. Lorentzian Scale Transformation.

In the preceding section, we have recalled that the general solution to the (special) relativity problem is the Lorentz group. In the case of motion relativity the Lorentz transformation for systems in inertial motion is now one of the most solid bases of physics. What about scale?

We have argued in Chapter 2 and Secs. 6.1–6.3 that scale (resolution) also came under a relativity theory. Formulated in a general way, the problem of scale transformation now consists in looking for a two-variable transformation, $ln\varphi' = f_1(ln\varphi,\delta)$, $\delta' = f_2(ln\varphi,\delta)$, depending on one parameter, the state of scale $V=ln(\lambda_o/\Delta x)$.

Let us analyze how the mathematical axioms on which was founded the above derivation of the Lorentz transformation are physically translated into the case of scale. In the theory of motion relativity, *linearity* may be derived from the homogeneity of space and time (which is itself an application of the principle of relativity to positions and instants). In scale relativity, the logarithm of some field, $ln\varphi$ (specifically, $\varphi = \mathcal{L}$ and $\mathcal{T}$) and the anomalous dimension δ play the roles of lengths and times respectively. The uniformity of these variables is not *a priori* straighforward, even though it is already assured in the scale laws of present physics. But *linearity*, as already mentioned, is the simplest choice to make, and thus comes as a provisional specialization of the theory. A generalization to non-linear transformations will be considered in the future: in such a generalization, we shall be obliged to deal in more detail with the fractal structure of space-time,[3] but this departs from the scope of the present book.

The second axiom, existence of an *internal composition law*, is a direct application of the principle of relativity: there is no difference of status here between motion and scale. *Reflection invariance* means that one may work equally with either $ln(\varphi/\varphi_0)$ or $ln(\varphi_0/\varphi)$, to which would respectively correspond scale states $ln(\rho)$ and $ln(\rho^{-1})$: this is indeed straightforward. Finally, the case $\kappa<0$ is clearly also excluded for scales, since when applying two successive dilatations, we indeed expect the final product not to be a contraction.

So from our result that the general solution to the linear relativity problem is the Lorentz transform, *we conclude that the laws of scale transformation must also take a Lorentzian form*, instead of the Galilean form which was up to now assumed to be self-evident.

Let us now explicitly compare Lorentzian scale transformations with motion transformations. While the composition of velocities follows an additive group law, the composition of scales follows a multiplicative group law. It is easy to come back to a multiplicative group by taking the logarithm of scale ratios, as shown in Sec. 6.3 (Galilean case, which is also the case of the standard renormalization group).

Start with the Einstein-Lorentz law of composition of "velocities"

$$w = \frac{u + v}{1 + \frac{u\ v}{c^2}}\,, \tag{6.5.1}$$

where u, v, and w are *dimensioned* quantities and c is a universal *constant*. This may be written in a dimensionless way by setting $U=u/c$, $V=v/c$, and $W=w/c$:

$$W = \frac{U + V}{1 + U\ V}\ . \tag{6.5.2}$$

Let us now write U, V and W, which are pure numbers, as logarithms of other dimensionless quantities. This may be done *in any base* for the logarithms, say K, by setting $U=log_K \nu$, $V=log_K \rho$, and $W=log_K \mu$, i.e.,

$$u = \frac{c}{lnK} \; ln\nu , \tag{6.5.3}$$

with similar formulas for v and w. So (6.5.2) now becomes

$$log_K \mu = \frac{log_K \rho + log_K \nu}{1 + log_K \rho \;\; log_K \nu} . \tag{6.5.4}$$

We may now divide both members of this equation by $\ln K$ and get

$$ln\mu = \frac{ln\rho + ln\nu}{1 + \dfrac{ln\rho \;\, ln\nu}{ln^2 K}} . \tag{6.5.5}$$

This formula is formally identical with the initial one, Eq. (6.5.1), and with the general structure (6.4.12), with the difference that lnK is itself a pure number while c was a dimensioned quantity. Now identifying μ, ν, ρ and K with scale ratios, we see that Eq. (6.5.5) becomes the scale-relativistic generalization of the usual law of dilatation: this means that the successive application of two dilatations ρ and ν now yields the dilatation μ instead of the usual product $\rho\nu$.

We get a new law for the transformation of the field φ, which generalizes (6.3.6):

$$log_K(\frac{\varphi'}{\varphi_0}) = \frac{log_K(\dfrac{\varphi}{\varphi_0}) + \delta \; log_K \rho}{[1 - log_K^2 \rho]^{1/2}} . \tag{6.5.6}$$

The anomalous dimension, which was previously invariant, now becomes a function of the resolution and the field:

$$\delta' = \frac{\delta + log_K \rho \;\, log_K(\dfrac{\varphi}{\varphi_0})}{[1 - log_K^2 \rho]^{1/2}} . \tag{6.5.7}$$

However, these laws cannot yet be considered as the definitive laws of scale relativity, since they do not incorporate the classical / quantum transition. Its inclusion is performed in the following section.

6.6. Breaking of Scale Relativity.

As already pointed out, scale relativity, contrary to motion relativity, is not a universal principle of nature. The fact that scales (or resolutions) can be defined only by their ratios is indeed universal, but this is of no consequence in the classical domain ($\Delta x >> \lambda_{dB}$, $\Delta t >> \tau_{dB}$). There, resolution reduces to precision, and improving the precision of measurements improves the precision of results, but does not change the physics. The situation changes in the quantum and quantum-relativistic domains, i.e. below the de Broglie scale of transition.

Hence scale relativity must be a broken symmetry at resolutions larger than the de Broglie scales $\lambda_{dB} = (\hbar/mv)(1-v^2/c^2)^{1/2}$ and $\tau_{dB} = (\hbar/mc^2)(1-v^2/c^2)^{1/2}$. In order to simplify the argument, let us look at the high energy degenerate case, where only one space-time variable may be considered, say $\Delta x \approx c\Delta t$. Then $c\tau_{dB}$ ($= \hbar c/E$) becomes equal to the Compton length $\lambda_o = \hbar/mc$ in the rest frame of a system of mass m. Let us explicitly introduce this particular scale in the composition law (6.5.5).

Scale laws must be Galilean for scales $\lambda \geq \lambda_o$ and Lorentzian for scales $\lambda < \lambda_o$. The explicit writing of a dilatation as a scale ratio implicitly assumes a Galilean structure: indeed, if $ln\rho = ln(\lambda_2/\lambda_1)$, then $ln\rho = ln(\lambda_2/\lambda_0) + ln(\lambda_0/\lambda_1)$. So we shall express the fact that the de Broglie scale makes the transition between Galilean and Lorentzian laws by assuming that it is the smallest scale (for a given system) which may still be used as reference to construct a Galilean scale ratio.

We start from the scale λ_o and apply a contraction which leads to a new scale λ; then we start again from λ_o and apply another contraction leading to another scale λ'. What is the dilatation ρ between λ and λ'? The Galilean character of λ_o allows us to take it as reference for scale ratios (with the exception of ρ which relates λ to λ'). We thus set in (6.5.5)

$\mu=\lambda'/\lambda_o$, $v=\lambda/\lambda_o$ and $K=\mathbb{A}/\lambda_o$: as noticed above, this explicit writing of dilatations as scale ratios introduces a Galilean structure, and the composition law now takes the form

$$ln\frac{\lambda'}{\lambda_o} = \frac{ln\dfrac{\lambda}{\lambda_o} + ln\rho}{1 + \dfrac{ln\rho \;\; ln\dfrac{\lambda}{\lambda_o}}{ln^2\dfrac{\mathbb{A}}{\lambda_o}}} \quad , \tag{6.6.1}$$

plus a similar relation for time scales.

We verify that the dilatation which relates λ_o to any scale λ remains equal to their ratio, as in the classical case, while it is no longer true of two scales that are both different from λ_o. In fact (6.6.1) may be inverted and understood as the function $\rho(\lambda,\lambda';\lambda_o)$ which yields the dilatation allowing one to go from one scale λ to another scale λ' (see Eq. 6.8.1). This dilatation factor now depends on the initial de Broglie scale λ_o.

Before going on, let us attempt to clarify the physical meaning of the scales λ and λ'. One must take care not to interpret them classically. These are *virtual* length and time scales which cannot in general be measured directly: what is actually measured in the quantum domain is energy and momentum, while length and time scales are *deduced* from the Heisenberg and de Broglie relations. So in Eq. (6.6.1) they are meant to describe in a virtual way the internal structure of a quantum system. Moreover, λ and λ' are purely *relative* scales, in particular relative to the system considered and to its de Broglie length and time. It would then be of no meaning to compare these scales between two different systems. In the case of actual measurement, one of these two scales may be achieved, but not both. Indeed, starting from a system prepared at $\lambda_{dB}=\lambda_o$, we may perform a measurement with a resolution $\delta x=\lambda<\lambda_o$ (this resolution is, e.g., the de Broglie length of the radiation or particle beam used as microscope). Throughout this work, we conjecture that, in the case of a *single* measurement, the result of this measurement gives us information about the state of the system just before the measurement (see Sec. 5.5). But it is clear that, after the measurement, the system jumps to a new state of de Broglie length $\lambda_{dB}=\lambda_o'$, which is of the order of λ from Heisenberg's relation. So a subsequent measurement with a resolution $\delta x'=\lambda'$ would be made on a *different* system and cannot be compared with λ using Eq. (6.6.1). The scale λ' which appears in (6.6.1) corresponds to preparing the system again at $\lambda_{dB}=\lambda_o$ and making a single measurement with $\delta x=\lambda'$. Equation (6.6.1) tells us that the internal dilatation ρ that transforms the scale λ into λ' is different from the ratio λ'/λ. Though this cannot be checked directly from length and time measurements, the fundamental change that it implies for the structure of space-time is expected to have observable consequences for energy and momentum measurements, as we shall see later.

Consider now the behaviour of the particular length $\mathbb{A}$. It has been introduced for the moment in a purely formal way: when referred to λ_o, the dilatation K defines this new length $\mathbb{A}$, which could *a priori* depend on λ_o itself. However, assume that we start with this length, i.e. $\lambda=\mathbb{A}$, and that we apply to it a dilatation or contraction ρ. From (6.6.1), we find that this results in a length given by $ln(\lambda'/\lambda_o)=ln(\mathbb{A}/\lambda_o)$, i.e. $\lambda'=\mathbb{A}$, *whatever the value of the initial scale* λ_o. Starting from any scale larger than $\mathbb{A}$, and applying any finite contraction, we get a scale larger than $\mathbb{A}$. The scale $\mathbb{A}$ can be the result only of an infinite contraction or of an infinite product of contractions, i.e. it plays the same role as the zero point of the old theory. In terms of the renormalization group theory, it is a fixed point for space-time itself.

Hence the principle of relativity, once applied to scales and combined with the existence of the Einstein-de Broglie transition, leads to the existence of an absolute length in nature, which is invariant under dilatations and contractions. Motion relativity immediately ensures that this will also be true for time, and that an invariant time interval $\mathbb{T}=\mathbb{A}/c$ exists in Nature. The Lorentz contraction of length and dilatation of time is themselves particular cases of scale transformations: as a consequence it is straighforward that $\mathbb{A}$ and $\mathbb{A}/c$ will also be invariant under a Lorentz transformation, i.e. independent of the relative velocity of the reference system in which they are observed.

One might be disturbed by the fact that K is not an absolute constant, contrary to the structure expected from a pure special relativity theory. However, once λ_o is fixed (and it is fixed by the state of motion of the system considered, since the de Broglie length and time are Lorentz-covariant by construction), λ_0/Λ is a constant: *scale relativity relies on motion relativity*. Conversely it is rather satisfactory that, in the same way as motion relativity led to the existence of an absolute and unexceedable velocity, scale relativity leads to the existence of an absolute, invariant limit for all lengths and times. The final point to be elucidated is the nature of Λ. We suggest in the following section that it is nothing but the Planck scale.

6.7. On the Nature of Planck Scale.

The Planck length already plays a very special role in physics: it is the characteristic scale for which all forces of nature are expected to become equivalent, while the concept of a space-time continuum seems to lose its physical meaning for smaller resolutions. It has been proposed[54,55] that the topology of space-time may become extremely complicated (foamlike) at that scale, the continuum itself being broken.

Even though a bundle of physical arguments makes clear that the Planck scale must play a central role in microphysics, all the previous approaches to the problem were worked out in a framework where the scaling laws themselves were not questioned, i.e. in which it was considered self-evident that applying a dilatation q to a scale Δx yields a new scale, $\Delta x' = q\, \Delta x$. This is reminiscent of classical Galilean physics, in which it seemed also self-evident that throwing an object with a velocity v with respect to a body moving with velocity u relative to the ground finally yields a velocity $w = u+v$.

Let us now consider the question from the point of view, adopted here, of scale relativity. The Planck scale $(\hbar G/c^3)^{1/2}$ is particular in that *its expression depends on no particular physical object*, but only on the three fundamental constants of physics, G, $\hbar$ and c. While we have insisted at the beginning of this book on the relativity of all scales, the Planck scale is the only one which is in fact absolute in its definition, i.e. independent of particular physical bodies or systems.

If we admit that the three constants G, $\hbar$ and c are indeed universal and unvarying, even at the time and length scales of the Universe ($\hbar$ is known to vary by less than $4 \cdot 10^{-13}$ per year and G by less than 10^{-11} per year, i.e. respectively less than ≈4‰ and 10% over the age of the Universe),[56] one may be upset by the fact that, in present physics, a "Planck rod" $(\hbar G/c^3)^{1/2}$ when submitted to a dilatation q becomes $q(\hbar G/c^3)^{1/2}$ in spite of its universality, and when observed from a reference system in which it moves with velocity v, is submitted to Lorentz contraction and becomes $\sqrt{(\hbar G/c^3)(1-v^2/c^2)}$. Even if it is admitted that physics may drastically change when the Planck scale is crossed, it is still admitted in present physics that scales smaller than the Planck scale do exist in Nature.

We take here a radically different position: based on the absolute character of the definition of the Planck scale, we suggest identifying it with the invariant scale $\mathbb{A}$, which was derived above from the application to scale of the principle of relativity. *The Planck length becomes a universal scale which remains invariant under dilatations.* It now plays for scale the same role as the velocity of light already plays for motion. The concept of a resolution smaller than the Planck length also loses physical meaning, since the Planck length is now a lower limit which cannot be exceeded.

Strictly, $\mathbb{A}$ could be identified with the Planck length times a pure and constant number, but this would destroy the formal simplicity of the construction: only the exchanges $G \rightarrow 2G$ (the Schwarzschild radius is $r_s=2Gm/c^2$) and $\hbar \rightarrow h$ remain uncertain, but $\hbar$ is preferable to h since the actual transition lengths around which physics changes are actually the reduced Compton length $\hbar/mc$ rather than h/mc); so we set, while waiting for possible future experimental verification,

$$\mathbb{A} = \sqrt{\frac{\hbar G}{c^3}} . \tag{6.7.1}$$

It should be noted that, taken together, the three fundamental constants G, $\hbar$ and c do nothing but determine the arbitrary part of our units of length, time and mass. Motion relativity has supplied us with a conceptual framework in which lengths and times are logically related: one now deduces length units from time units with c fixed. (It would be more consistent in fact to take $c = 1$ and definitively measure lengths, for example in nanoseconds.) Scale relativity, if confirmed by experiments, will achieve the same for times themselves (at least in principle, since the present poor precision on G prevents one from doing this explicitly for the moment; a precise determination of the constant of gravitation now becomes an urgent task). Setting the Planck time $\Lambda/c = 1$, *all lengths and time intervals in nature become dimensionless real numbers larger than* 1. In such a system, one would get $G = 1/\hbar$. The final step, setting also the Planck mass to 1, demands the determination of the ratios of the (low energy) masses of all elementary particles over the Planck mass, and, if one wants to be completely consistent, the understanding of the values of the three remaining coupling constants at a given scale. It will be shown hereafter that scale relativity allows one to take some steps towards the achievement of this grand program.

6.8. The New Scale-Relativistic Transformation.

Before going further, let us write the complete new transformation, in the case where all the considered scales are smaller than the de Broglie scale λ_o (assumed fixed) of the system. Let ρ be the dilatation which allows one to go from Δx to $\Delta x'$ ($\Delta x \leq \lambda_o$ and $\Delta x' \leq \lambda_o$); let $\varphi = \varphi(x, \Delta x)$ be some scale-dependent field [in the first place, $\varphi = \mathcal{L}(s, \Delta x)$ or $\mathcal{T}(s, \Delta t)$] and $\delta = \delta(\Delta x)$ its anomalous dimension; we set $\varphi' = \varphi(\Delta x')$, $\delta' = \delta(\Delta x')$ and define an arbitrary reference value for the field, φ_0; Λ is the Planck length scale $(\hbar G/c^3)^{1/2}$; the new transformations for the dilatation, field and anomalous dimension read

$$ln\, \rho = \frac{ln \dfrac{\Delta x}{\Delta x'}}{1 - \dfrac{ln \dfrac{\lambda_o}{\Delta x}\; ln \dfrac{\lambda_o}{\Delta x'}}{ln^2 \dfrac{\lambda_o}{\Lambda}}} \,, \tag{6.8.1a}$$

$$ln\frac{\varphi'}{\varphi_0} = \frac{ln\frac{\varphi}{\varphi_0} + \delta\, ln\,\rho}{\sqrt{1 - \frac{ln^2\rho}{ln^2\frac{\lambda_0}{\Lambda}}}}, \quad \delta' = \frac{\delta + \frac{ln\,\rho\; ln\frac{\varphi}{\varphi_0}}{ln^2\frac{\lambda_0}{\Lambda}}}{\sqrt{1 - \frac{ln^2\rho}{ln^2\frac{\lambda_0}{\Lambda}}}} \; . \qquad (6.8.1b)$$

Letting $\Lambda \to 0$ gives back the standard ("Galilean") scaling transformation $\varphi' = \varphi\, \rho^{\delta}$ and $\delta' = \delta$. The standard transformation is also obtained as an approximation in the limit $ln\rho << ln(\lambda_0/\Lambda)$.

Equations (6.8.1) hold only for the quantum case ($\Delta x \leq \lambda_0$ and $\Delta x' \leq \lambda_0$). Going to a scale larger than the de Broglie scale leads to scale independence: δ vanishes, φ becomes independent of scale and a "Galilean" dilatation law $\rho = \Delta x/\Delta x'$ is recovered. As was already noted in Sec. 6.3, a precise description of this transition from scale dependence to scale independence depends on the particular physical system considered. A useful model consists in replacing in Eqs. (6.8.1) $ln(\lambda_0/\Delta x)$ by $\frac{1}{k}\, ln[1+(\lambda_0/\Delta x)^k]$ (and similar changes for all scales referred to λ_0), where k is a parameter which allows one to describe the steepness of the transition. Indeed for *fixed* λ_0 and $\Delta x << \lambda_0$, one gets $\frac{1}{k}\, ln[1+(\lambda_0/\Delta x)^k] \approx ln(\lambda_0/\Delta x)$, while for $\Delta x >> \lambda_0$, $\frac{1}{k}\, ln[1+(\lambda_0/\Delta x))^k] \approx 0$. To be complete, one should also replace δ by some function $\Delta(\Delta x)$, with $\Delta = \delta$ for $\Delta x << \lambda_0$ and $\Delta = 0$ for $\Delta x >> \lambda_0$.

We recall once more that Eqs. (6.8.1) apply in the first place to the case where φ represents either the length $\mathcal{L}(x, \Delta x)$ measured along a quantum particle trajectory (nonrelativistic case) or its integrated proper time $\mathcal{T}(t,\Delta t)$ (relativistic case + reinterpretation of particle-antiparticle pairs as part of a fractal trajectory running backward in time), and more generally any of the four space-time coordinates (once integrated along the fractal path). Then *Δx may represent the resolution of any of the four coordinates,* Δx^i, i=0 to 3, and more generally of some combination of them, in particular the resolution of the classical invariant, Δs.

To conclude this section, let us finally specialize (6.8.1) in order to get the scale dependence of the field φ ($\mathcal{L}$ and $\mathcal{T}$) itself. What is the "Lorentzian" generalization of our initial "Galilean" relation, $\varphi=\varphi_0\ (\lambda_0/\Delta x)$ when $\Delta x<\lambda_0$? It is obtained by taking in (6.8.1) the de Broglie length as reference, i.e., $\varphi=\varphi_0=\varphi(\lambda_0)$, $\rho=\Delta x/\lambda_0$, $\delta(\lambda_0)=1$. The anomalous dimension becomes a function of resolution:

$$\delta_x(\Delta x) = \frac{1}{\sqrt{1-\dfrac{ln^2\,(\lambda_0/\Delta x)}{ln^2(\lambda_0/\Lambda)}}}\ , \tag{6.8.2a}$$

$$\delta_t(\Delta t) = \frac{1}{\sqrt{1-\dfrac{ln^2\,(\tau_0/\Delta t)}{ln^2(c\tau_0/\Lambda)}}}\ , \tag{6.8.2b}$$

i.e. *the Lorentz-like "scale γ factor" is directly given by the anomalous dimension at the new scale*, so that we get the new laws

$$\mathcal{L} = \mathcal{L}_0\ (\lambda_0/\Delta x)^{\delta(\Delta x)}\ , \tag{6.8.3a}$$

$$\mathcal{T} = \mathcal{T}_0\ (\tau_0/\Delta t)^{\delta(\Delta t)}\ , \tag{6.8.3b}$$

to be compared with the standard one, $\varphi = \varphi_0(\lambda_0/\Delta x)^{\delta(\lambda_0)}$.

The new structures found in scale relativity imply profound changes of many other fundamental basic relations. Indeed the requirement that any space and time resolution in Nature be larger than the Planck length and time implies that this should be required also of any length and time interval. This is a radical change of the nature of space-time itself, which is expected to have consequences for physics as a whole. In particular, the scale transformation (6.8.1) relies on the concept of de Broglie length and time. But it is immediately clear that the theory cannot be self-consistent if their usual definition is retained. Indeed for masses larger than the Planck mass, the Compton length (i.e. c times the rest frame de Broglie time) would become smaller than Λ, which is a now-forbidden behaviour. The next section is devoted to this crucial problem of the mass (more generally, energy-momentum) scale in the new theory of scale relativity.

6.9. Scale-Relativistic Mechanics.

A new invariant.

Let us attempt to clear up the question which is now asked to us. In the classical and relativistic theories of motion, the laws of mechanics relate energy-momentum to the essential variable in the inertial case, the velocity. Our claim here is that, in the quantum domain, the classical concept of motion becomes inoperative, to the advantage of the concept of scale, and velocity is disqualified as an essential variable, its place being taken by resolution. So it becomes logical to expect the energy-momentum / velocity relations to be replaced in the quantum theory by energy-momentum / scale relations: and this is precisely what the de Broglie ($<p> \lambda = \hbar$) and Heisenberg ($\sigma_p \cdot \sigma_x \geq \hbar/2$) relations tell us. The way by which one may obtain these relations as *consequences* of the principle of scale relativity and then generalize them in the Lorentzian case is clearly to construct a scale-relativistic mechanics.

In the framework of standard quantum mechanics, we have recalled that the Heisenberg relations $\Delta p \cdot \Delta x \approx \hbar$ and $\Delta E \cdot \Delta t \approx \hbar$ can be reinterpreted in terms of some internal length $\mathcal{L}$ which becomes scale-dependent (fractal) as $\mathcal{L} \approx \mathcal{L}_0(\lambda/\Delta x)^\delta$ for $\Delta x<\lambda$ (λ being the de Broglie length $\hbar/<p>$), and of some internal time $\mathcal{T}$ such that $\mathcal{T} \approx \mathcal{T}_0(\tau/\Delta t)^\delta$ for $\Delta t<\tau$ (τ being the de Broglie time $\hbar/<E>$), with $\delta=1$ in both cases. Let us consider the one-dimensional case, with φ always denoting either $\mathcal{L}$ or $\mathcal{T}$ in what follows. Assuming that the classical coordinate system is well determined (origin, axes orientation and state of motion), $d\ln\varphi$ and $d\delta$ are independent of each other and invariant in this "Galilean" framework.

Consider now the framework of scale relativity. The variables $\ln\varphi$ and δ respectively play in scale relativity the roles played by position x and time t in motion relativity. It is well known that, in the formulation of special (motion) relativity, the requirement of Lorentz covariance is equivalent to the requirement of invariance of the Minkowski metric element $ds^2=c^2dt^2-dx^2$ (we remain in the one-dimensional case in order to simplify the argument). In the same way, neither $d\delta$ nor $d\ln\varphi$ remains invariant in scale relativity. The new scale-invariant is (for λ_o fixed and resolution $\Delta x<\lambda_o$)

$$d\sigma^2 = \ln^2\frac{\lambda_0}{\Lambda}\; d\delta^2 - \frac{d\varphi^2}{\varphi^2} \,. \tag{6.9.1}$$

In this form, an intuitive physical interpretation of the new invariant is difficult, since δ is not a directly measurable quantity. However, the Minkowski invariant may also be expressed in terms of velocity as $ds^2=c^2dt^2(1-v^2/c^2)=(c^2/v^2-1)dx^2$. In scale relativity, the state of motion v is replaced by the state of scale $ln(\lambda_0/\Delta x)$, so that the new invariant may be expressed in terms of quantities which are measurable (at least in principle), namely the integrated length (or time) φ and the measurement resolution Δx (or Δt):

$$d\sigma^2 = \left[\, \frac{\ln^2\frac{\lambda_0}{\Lambda}}{\ln^2\frac{\lambda_0}{\Delta x}} - 1 \,\right] \frac{d\varphi^2}{\varphi^2} \,. \tag{6.9.2}$$

This result confirms our initial conjecture[3] that the space-time of microphysics is of a radically new nature as compared with the classical space-time: a proper description of its properties involves equations in which the resolution appears explicitly.

Let us proceed further in our construction of a relativistic "scale mechanics." The experience of special (motion) relativity may still be followed advantageously.[57] We first assume that scale physical laws emerge from a least action principle. Once we fix the state of motion, we expect the action to be the integral over $d\delta$ of some Lagrange function $L = L(ln\varphi, ln(\lambda_0/\lambda), \delta)$, [to be compared with $L = L(x, v, t)$ in motion-relativity] and its differential $Ld\delta$ to be given[57], up to some multiplicative constant μ, by the invariant $d\sigma$: $Ld\delta = \mu d\sigma$.

If we denote the scale-state by $V= ln(\lambda_0/\lambda)$, "conserved" quantities (prime integrals) $\partial L/\partial V$ and $V\partial L/\partial V-L$ will emerge from the uniformity of $ln\varphi$ and δ respectively. But note that these quantities are not "conserved" in terms of time independence: here "conserved" means that these quantities do not depend explicitly on the anomalous dimension δ, which plays for scale laws the structural role played by time for motion laws.

Generalized de Broglie and Compton relations.

Let us consider first the uniformity of $ln\varphi$. It implies the existence of a prime integral, a "scale momentum" $\mathcal{P}$, which is a function of the scale state $ln\frac{\lambda_o}{\lambda}$:

$$\mathcal{P}(\lambda) = \mu \frac{ln\frac{\lambda_o}{\lambda}}{\sqrt{1 - \frac{ln^2\frac{\lambda_o}{\lambda}}{ln^2\frac{\lambda_o}{\Lambda}}}}, \tag{6.9.3a}$$

where μ is a constant, to be determined later, which comes from the fact that the action and the metric invariant are equal only to some proportionality factor (this factor is equal to $-mc$ in motion relativity[57]). A similar relation is obtained for the time variable in terms of de Broglie (τ_o) and Planck (Λ/c) times:

$$\mathcal{E}(\tau) = \mu \frac{ln\frac{\tau_o}{\tau}}{\sqrt{1 - \frac{ln^2\frac{\tau_o}{\tau}}{ln^2\frac{c\tau_o}{\Lambda}}}}. \tag{6.9.3b}$$

These two relations are the scale-relativistic equivalent of the motion-relativistic equation for momentum, $p=mv/\sqrt{1-v^2/c^2}$.

In order to know the meaning of this result, one first notes that physics must be invariant under the choice of the logarithm base. Then the form of (6.9.3) implies that $\mathcal{P}$ is itself a logarithm of some dimensionless quantity. Now (6.9.3a) has been obtained from the uniformity of a space variable, from which the usual (motion) momentum also derives as a conserved quantity in classical mechanics, and (6.9.3b) from the uniformity of time, from which the concept of conserved energy derives in classical mechanics. We then suggest that $\mathcal{P}$ is related to the classical momentum (case of a space variable), leading us to write $\mathcal{P} = ln(p/p_o)$, and

$\mathcal{E}$ to the classical energy (case of a time variable), so that $\mathcal{E} = ln(E/E_0)$.

Consider now the limit $\mathbb{A} \rightarrow \infty$: this limit should give us back the standard quantum mechanics, i.e., (6.9.3) must be identifiable with already known equations of quantum mechanics. Indeed (6.9.3) becomes $p/p_0=(\lambda_0/\lambda)^\mu$ for the space variable and $E/E_0=(\tau_0/\tau)^\mu$ for the time variable. We recognize in these equations the two Einstein-de Broglie relations, $p\lambda = p_0\lambda_0 = constant$ and $E\tau = E_0\tau_0 = constant$, *provided the constant μ is definitely fixed to the value μ=1*. Since λ and τ are themselves defined up to some multiplicative factor, we may choose them in such a way that the universal constants $p_0\lambda_0$ and $E_0\tau_0$ are the same. This defines the reduced Planck constant $\hbar$ (or h with a different choice for the remaining arbitrary scale factor) and we get

$$p\,\lambda = p_0\,\lambda_0 = \hbar \ , \tag{6.9.4a}$$

$$E\,\tau = E_0\,\tau_0 = \hbar \ . \tag{6.9.4b}$$

Hence the de Broglie relations are nothing but the Galilean scale equivalent of Descartes' equation relating momentum and velocity (which characterizes the state of motion of the system) $p = mv$: This is clearly apparent in their logarithmic form $ln(p/p_0) = \mu\, ln(\lambda_0/\lambda)$, with μ=1. This completes the already noticed similitude from which we started, concerning the composition of dilatation and the scale transformation of internal space-time coordinates, compared with the composition of velocities and the transformation of inertial systems of coordinates.

Consider now the Lorentzian case where $\mathbb{A}\neq 0$: this leads us to infer that the full equations (6.9.3 a, b), in which $\mathcal{P} = ln(p/p_0)$, $\mathcal{E} = ln(E/E_0)$ and μ=1 must be assumed, are the scale-relativistic generalization of the de Broglie relations which we were seeking. They indeed possess the expected property that momentum and energy now tend to infinity when the generalized de Broglie length and time tend to the Planck length and time. We may sum up these results by a comparison between the four structures of Galilean/Lorentzian, motion/scale relativity: Galilean motion relativity yields the momentum/velocity Descartes relation $p = mv$, whose scale equivalent is the momentum/wavelength de Broglie relation $\mathcal{P} = \mu V$, (with $\mathcal{P} = ln(p/p_0)$, $V = ln(\lambda_0/\lambda)$ and μ=1 it writes $p/p_0=\lambda_0/\lambda$); Einstein motion relativity yields $p=mv/\sqrt{1-v^2/c^2}$ while scale relativity generalizes the de Broglie relations as

$$p = p_0 \left(\frac{\lambda_0}{\lambda}\right)^{\delta_x} , \tag{6.9.5a}$$

$$E = E_0 \left(\frac{\tau_0}{\tau}\right)^{\delta_t} , \tag{6.9.5b}$$

where $\delta_x = [1-ln^2(\lambda_0/\lambda)/ln^2(\lambda_0/\mathbb{A})]^{-1/2}$ and $\delta_t = [1-ln^2(\tau_0/\tau)/ln^2(c\tau_0/\mathbb{A})]^{-1/2}$.

How can we interpret such a result? The above energy is, *a priori*, a *potential* energy contained into the scale structure of a particle: it can be eventually transformed into kinetic energy by effective interaction with a measurement apparatus. But it is already present even in the rest frame of the particle, so that it seems logical to identify it with the mass of the particle: this is in particular consistent with our interpretation of the rest mass as a geometrical fractal structure of the trajectory of the particle (Sec. 5.3). Then Eq. (6.9.5b) applies only in the quantum relativistic domain, $\Delta t<\hbar/m_e c^2$, where m_e is the electron mass, i.e. the smallest known mass for an elementary particle. In order to define the domain of application of (6.9.5a), we recall that a non-zero momentum is deduced from a vanishing momentum by a Lorentz transformation of the energy-momentum quadrivector in rest frame, $(mc^2,0,0,0)\rightarrow(E,p,0,0)$. A Lorentz transformation of (6.9.5b) yields

$$p = mc\ [\left(\frac{\lambda_0}{\lambda}\right)^{2\delta} - 1]^{1/2} . \tag{6.9.5c}$$

We shall assume this formula to be exact. As a consequence, (6.9.5a) is only an asymptotic formula, since it coincides with (6.9.5c) only when $v\approx c$, that is, when λ is smaller than the Compton length $\lambda_0=\hbar/mc$. Then we expect that the strict relation $p\lambda_{dB}=\hbar$ remains true in the quantum nonrelativistic domain, $\lambda_{dB} > \lambda_0$.

Equations (6.9.5) actually provides us with a new relation between the momentum-energy scale and the length scale. In standard high-energy quantum mechanics, the length and mass scales are directly inverse: there is an inverse correspondence between any mass scale m (equivalently an energy mc^2) and a length r through the relation $mr \approx \hbar/c$. So the asymptotic behaviour of the various quantum theories, which is so crucial for their ultimate validity, corresponds to both $r \to 0$ and $p \to \infty$. In scale relativity it now corresponds to $r \to \Lambda$ and $p \to \infty$: the experimental consequences of this new length-momentum relation will be considered in Sec. 6.10.

Let us present another reasoning that leads to the same conclusion, namely that (6.9.5) applies only below the Compton length of the electron. The fact that (6.9.5) is supposed to apply to the de Broglie lengths themselves, while the de Broglie scale is used as the starting point of the scale transformations, implies some difficulty of interpretation. We are now comparing the de Broglie lengths with one another, rather than assuming λ_o fixed and then measuring the system at some resolution Δx. So we need one new universal scale to serve as reference for all other scales. The Compton length of the electron clearly plays this role in microphysics. It corresponds to the less massive of all elementary particles and thus to the transition from non-motion-relativistic to relativistic quantum behaviour: at this scale, all velocities become relativistic, and the concept of well-defined position loses its physical meaning, since the first occurrence of particle-antiparticle pair creation-annihilation starts the domain of elementary particle physics: *it is from the electron Compton length* λ_e *onwards that the fundamental coupling constants and the particle rest masses begin to vary.* The electron Compton length clearly plays the role of a zero point for the whole domain of relativistic quantum fields, so that we can write (6.9.5) in terms of a new relation between mass $m > m_e$ and Compton length $\lambda_c < \lambda_e$:

$$\ln \frac{m}{m_e} = \frac{\ln \frac{\lambda_e}{\lambda_c}}{\sqrt{1 - \frac{\ln^2 \frac{\lambda_e}{\lambda_c}}{\ln^2 \frac{\lambda_e}{\Lambda}}}} \ . \tag{6.9.6}$$

This introduces the fundamental number

$$\mathbb{C}_e = ln \frac{\lambda_e}{\mathbb{A}} = ln \frac{m_P}{m_e} .$$

Its value is $\mathbb{C}_e$=51.527967(65) from the presently known values[12] of $\hbar$, c and G, which are used to define the Planck mass $m_P=(\hbar c/G)^{1/2}$. (Note that the number in parentheses following the experimental numerical results here and after is by convention the error on the last digits.) It is straightforward to verify on (6.9.6) that now the Compton length is limited by $\mathbb{A}$ when the mass tends to infinity.

Generalized Heisenberg relations.

The problem of generalization of the Heisenberg relations is somewhat different from the de Broglie and Compton problems since we now deal with inequalities rather than with strict equalities: however a similar behaviour is expected for them, i.e., we expect σ_p to tend to infinity when σ_x now tends to the Planck scale. Actually a full treatment of the problem would imply a proper generalization of the whole structure of quantum mechanics; this huge technical problem lies outside the scope of this book. Our hope is that a demonstration of the self-consistency of our generalization of the two basic quantum relations, de Broglie's and Heisenberg's, will ensure the possibility of a self-consistent generalization of the whole formalism of quantum mechanics. Let us briefly consider a possible way in this direction.

We have already shown in Ref. 3 and Sec. 5.2 that, from the hypothesis that the quantum space-time lines which define the possible particle trajectories have a fractal dimension $D=1+\delta$, one gets the generalized Heisenberg inequality: $(\sigma_p/p_o)\,(\sigma_x/\lambda_o)^{\delta} \gtrsim 1$, which holds for all four space-time coordinates. The usual Heisenberg relation is recovered, as expected, for the particular value δ=1. This generalization, which was purely formal in Sec. 5.2, is endowed with physical meaning now that we have introduced a fractal dimension (equivalently, an anomalous dimension in a renormalization group approach) which is allowed to vary. Note that such a relation is not incompatible with Heisenberg's: the usual Heisenberg inequality remains true even for δ>1, though the inequality becomes stronger as δ increases, i.e., $\sigma_p\, \sigma_x >> \hbar$. Since we have demonstrated above that, after a dilatation (or contraction), the "scale δ factor" is precisely equal to the anomalous dimension at the new scale, the generalized Heisenberg relations finally keep, as expected, a form similar to de Broglie's in the new theory ($\sigma_x \leq \lambda_o$):

$$ln \frac{\sigma_p}{p_o} \geq \frac{ln \frac{\lambda_o}{\sigma_x}}{\sqrt{1 - \frac{ln^2 \frac{\lambda_o}{\sigma_x}}{ln^2 \frac{\lambda_o}{\mathbb{A}}}}} , \tag{6.9.7}$$

with an equivalent expression holding for time and energy.

Transformation of probabilities.

Let us now make a first attempt at understanding how probabilities (more precisely probability densities) transform in the new theory. Consider the following expression for the one-dimensional normalized probability of presence of a particle:

$$dP = p_k(x)\, dx = \frac{k}{\sigma\, \Gamma(1/2k)}\, exp\{-(\frac{x^2}{\sigma^2})^k\}\, dx .$$

This expression has the advantage that it allows us to describe a large range of shapes which generalize the Gaussian function (k=1) with only one parameter, going from a loophole ($k\to\infty$) to sharp functions (k<<1). Forgetting the unessential normalizing term $k/\sigma\,\Gamma(1/2k)$, we consider the function

$$P_k(x) = exp\{-(\frac{x^2}{\sigma^2})^k\} .$$

Taking its logarithm yields

$$lnP_k^{-1}(x) = (\frac{x^2}{\sigma^2})^k ,$$

and taking once again the logarithm:

$$ln ln P_k^{-1}(x) = 2k \; ln \frac{x}{\sigma} \; .$$

Following the standard interpretation of quantum mechanics, we may identify σ with a resolution, i.e., the standard error of the mean of position measurement results obtained for a system described by a probability amplitude $P_k^{1/2} e^{i\varphi}$. Then, if the parameter σ is changed from σ to σ', $ln(\sigma/\sigma')$ measures the transformation of relative "states of scale". The above expression transforms in standard quantum mechanics as

$$[ln ln P_k^{-1}(x)]' = ln ln P_k^{-1}(x) + 2k \; ln \frac{\sigma}{\sigma'} \; ,$$

$$2\,k' = 2\,k \; .$$

We are once again faced with a couple of quantities, namely $ln ln P_k^{-1}(x)$ and $2k$, whose scale transformation law is nothing but the Galilean transformation law for the position and time variables with the relative resolution in logarithm form, $ln \frac{\sigma}{\sigma'}$, still playing the part of a "scale velocity". In scale relativity, a dilatation ρ applied to a resolution σ yields a new resolution σ' which is no longer given by $\rho\sigma$, but rather by the new relation (6.6.1). The transformation now takes a Lorentzian form.:

$$[ln ln P_k^{-1}(x)]' = \frac{ln ln P_k^{-1}(x) + 2k \; ln \, \rho}{\sqrt{1 - ln^2 \rho \, / \, \mathbb{C}^2}} \; ,$$

$$2\,k' = \frac{2k + ln\rho \;\; ln ln P_k^{-1}(x) \, / \, \mathbb{C}^2}{\sqrt{1 - ln^2 \rho \, / \, \mathbb{C}^2}} \; .$$

Thus the transformed probability may be written as

$$P'_k(x) = exp -\left(\frac{x^2}{(\rho\sigma)^2} \right)^{k\,(1 - ln^2 \rho / \mathbb{C}^2)^{-1/2}} \; .$$

The effect of the scale δ–factor is equivalent to increasing the parameter k when resolution decreases. At the limit when resolution tends to the Planck scale, the new power tends to infinity, and this corresponds to a loophole shape for the probability. This result is consistent with what could be expected in such a theory: the Planck length scale being an absolute limit for any resolution in Nature, one expects that, even statistically, positions cannot be localized more precisely than Λ. Indeed, when contracting indefinitely the dispersion σ, one finds from the above formula that we do not go to a δ function as in standard mathematics, but that the probability distribution cannot become narrower than the Planck length.

A new prime integral of equations of scale.

Let us come back to our construction of a scale-relativistic mechanics and build a prime integral independent of δ which comes from the homogeneity of the anomalous dimension itself, i.e. $V\partial L/\partial V-L$. *This is a completely new quantity* which had no theoretical existence in Galilean scale relativity, since δ was either undefined (classical case: scale independence) or constant (standard quantum case: scale invariance). This quantity is the equivalent for scales (from the viewpoint of the mathematical structure) of the relativistic expression for energy, $E=mc^2/\sqrt{1-v^2/c^2}$. It reads:

$$\mathbb{E} = \frac{ln^2 \dfrac{\lambda_o}{\Lambda}}{\sqrt{1 - \dfrac{ln^2 \dfrac{\lambda_o}{\lambda}}{ln^2 \dfrac{\lambda_o}{\Lambda}}}} \quad . \tag{6.9.8}$$

This expression should *a priori* involve an arbitrary multiplicative factor μ, but this factor was already set to 1 by identification of the "scale-momentum" with the logarithm of the ratio of motion momenta p_o/p. Once again the requirement of invariance under the logarithmic form of this equation leads us to set $\mathbb{E} = ln^2 \dfrac{\Phi}{\Phi_o}$. Then one gets

$$\frac{\Phi}{\Phi_o} = \left(\frac{\lambda_o}{\Lambda}\right)^{\delta^{1/2}} , \tag{6.9.9}$$

where $\delta= 1/\sqrt{1-ln^2(\lambda_o/\lambda)/ln^2(\lambda_o/\Lambda)}$. The remarkable result, which is reminiscent of what happened in motion special relativity, is the *emergence of a nonzero value at large scale for this new physical quantity*, i.e., of a physical quantity which must still exist in the classical scale-independent domain:

$$\Phi_\infty = \frac{\lambda_0 \Phi_0}{\Lambda} . \tag{6.9.10}$$

We do not know, *a priori*, what the dimensional equation of Φ is. Let us tentatively express it in units of energy. It then seems logical to identify Φ_0 with the rest energy mc^2 of the system and λ_0 with the Compton length $\hbar/mc$, so that $\lambda_0 \Phi_0 = \hbar c$ and $\Phi_\infty = \hbar c/\Lambda$. We arrive at the conclusion that any massive quantum system, in particular any massive particle, owns a physical property which is in some way equivalent to an internal energy given by the Planck energy:

$$\Phi_\infty = m_P c^2 = (\hbar c^5/G)^{1/2} . \tag{6.9.11}$$

How can this be interpreted ? Let us consider the Newtonian gravitational force between two bodies of masses m_1 and m_2. It reads $F_g = Gm_1m_2/r^2$. The Coulomb force between two electric charges e_1 and e_2 reads $F_{em} = \alpha\hbar c Z_1 Z_2/r^2$ in units where Z_1 and Z_2 are dimensionless integers. But let us now write the Coulomb force in a form such that the charges are expressed in units of *masses*. We get: $F_{em} = G\,(Z_1\sqrt{\alpha}\, m_P)(Z_2\sqrt{\alpha}\, m_P)\,/\,r^2$. A similar transformation may be (at least formally) made for the electroweak and strong forces, with the fine structure constant α replaced respectively by the U(1), SU(2) and SU(3) coupling constants α_1, α_2 and α_3. On this form we see that, indeed, the various charges in Nature other than the gravitational one correspond to an internal energy of order $m_P c^2$: they would even be strictly equal to the Planck energy provided the high energy common bare coupling constant was equal to 1 [see Sec. 6.10 and 11 about the convergence of U(1), SU(2) and SU(3) couplings at high energy].

This fact, often expressed in terms of the high value of the ratio of electric over gravitational forces ($\approx 4.10^{42}$), is a well-known structure of present physics. The new point here is the following: the mere existence of the electromagnetic field presently relies mainly on experimental grounds[57] and does not seem to have been made necessary from fundamental principles. (It can be deduced from gauge invariance: but gauge invariance itself relies on the existence of the quantum phase, which has been imposed to us by experiments). This is to be compared to the present status of the gravitational field: the principle of relativity, once applied to accelerated motion, leads to Einstein's general relativity whose equations are the simplest and most general ones which are invariant under *continuous* and *differentiable* transformations of coordinates; they introduce space-time curvature as a universal property of nature whose manifestations are what we call gravitation. In this sense, one may say that the principle of (motion) relativity leads to a demonstration of the unavoidable existence of gravitation in Nature. We suggest that the above result is a first step towards a similar demonstration concerning electromagnetism (and the other fundamental interactions). Indeed one may interpret it by saying that, once applied to scale dilatations, the principle of relativity implies the existence of some force of nature additional to gravitation, of strength $F_{em} \approx Gm_{\mathbb{P}}^2/r^2 = \hbar c/r^2$. However, the full understanding of these structures must clearly await a proper generalization of scale relativity to fields, i.e., to nonlinear scale transformations.

Eqs. (6.9.8-11) also add realism to the idea that there is indeed a huge amount of (potential) internal energy tied to the electric charge (2.3895(2) 10^{22} times the mass energy of the electron). The quantization of charge implies that charge cannot be divided; the prospective transformation of this "scale-energy" into inertial (motion) energy would imply a breaking of the law of charge conservation (in terms of some process of, e.g., the form "$e \rightarrow photons$") which is forbidden in the framework of present physics, but which should be seriously considered at the Planck energy scale.

This is not a new result in itself. Indeed the convergence of couplings at high energy in the standard quantum theory already led to such a conclusion: in the standard model, there is a first convergence of the electroweak and color (strong) couplings at a "Grand Unified" scale of about 10^{14} GeV. This means that at higher energies, what are known as weak, electromagnetic and strong interactions become different aspects of the same phenomenon: the three U(1), SU(2) (from which the weak and electromagnetic charges are deduced by mixing) and SU(3) charges are replaced by a unique charge. But this unique charge itself varies with scale and finally becomes equal to the gravitational charge (i.e. mass-energy-momentum) at the Planck scale. In other words, the standard theory already told us that at the Planck scale the electric charge and the energy become two aspects of a same phenomenon: this leads to contemplating the possibility that they may be exchanged. The main difference in scale relativity is that, as will be demonstrated in the following sections, the GUT and Planck scales become identical in terms of mass-energy; so the convergence of the three couplings occurs in conjunction with their convergence with the gravitational charge (mass-energy-momentum). This common convergence of the 4 existing couplings in Nature clearly strengthens our argument (see Figs. 6.5 and 6.6).

Let us finally remark, before jumping to the analysis of some possible experimental implications of the new theory, that we also predict a slow variation with scale of our new "scale integral" (6.9.10) with a factor $(1 + ln^2 \frac{\lambda_0}{r} / 4\mathbb{C}_e)$ to leading order.

6.10. Implications for High Energy Physics.

Introduction.

It is well known that Galilean motion relativity is recovered in the limit $c \to \infty$ of Einstein's special relativity. Is it strictly true? Starting from special relativity, one gets an expanded formula for energy given by $E = mc^2 + (1/2)mv^2 + (1/c^2)...$ .Taking the limit $c \to \infty$ indeed makes all the last terms vanish and yields the classical kinetic energy, but *it also yields a term of infinite internal energy*. Thus, if one admits our argument of Sec. 6.4, according to which special relativity could have been derived from the Galilean principle of relativity alone, classical mechanics was already faced with the problem of energy divergence (even if this was not explicitly realized) which the Einstein-Poincaré-Lorentz theory of relativity has solved. Does scale relativity, which aims at understanding from first principles the quantum behaviour of microphysics, and which clearly has something to do with electromagnetism (see previous Section), solve the old problem of the divergences of electric charge and electromagnetic self-energy ?

The present section is aimed at analyzing this important issue and at first considering a no less important question, that of possible experimental verifications of the theory. At this point of our argument, the reader may indeed legitimately ask himself whether scale relativity is a pure theoretical construction whose consequences are to be looked for only at the presently unobservable Planck scale, or whether experimental consequences are to be expected in the energy range reached by existing particle accelerators. We shall show that scale relativity is able to yield new predictions in the observable energy range ($E \lesssim 100$ GeV), and that these theoretical predictions may be used to test the theory.

Solution to the divergence problem of charges and masses.

Because of the previous quasi-identity in standard quantum theory between length scale and mass(-energy) scale [$ln(m/m_0) \approx ln(r_0/r)$ for high energy in the rest frame], the renormalization group equations are currently written indifferently in terms of $ln\, m$ or $ln\, r$. Actually the momentum representation, easier to work out and closer to experimental data (lengths and times are no more directly measured in quantum physics, but only deduced from energy and momentum measurements), is systematically used in quantum field theory rather than the position one (which may be obtained from the momentum representation through a Fourier transform). In today's quantum mechanics, the momentum and position solutions to (6.2.2) differ only by some constants. This is no longer true in scale relativity, and one should now specify the changes to be brought to the renormalization group equations.

One must in this respect distinguish between the cases of *relevant* fields and of *marginal* fields. Fields which vary with scale as power laws, for which we have been able to establish a parallel with motion relativity laws, are "relevant" fields. The lowest order term in their renormalization group equation is linear in the asymptotic behaviour:

$$\frac{d\varphi}{d\ ln\frac{\lambda_o}{r}} = \delta\ \varphi\ , \tag{6.10.1}$$

yielding the "Galilean" solution $\varphi = \varphi_o(\lambda_o/r)^{\delta}$. Note that the scale-relativistic solution (6.8.1) would correspond to the equation

$$\frac{d\varphi}{d\ ln\frac{\lambda_o}{r}} = \delta_o\ \varphi\ [\ 1 + \frac{ln^2(\varphi/\varphi_o)}{\delta_o{}^2\ ln^2(\lambda_o/\Lambda)}\ ]^{3/2} \tag{6.10.2}$$

with $\delta_o = cst$. The first nonlinear term in the right hand side of this equation is in $(\varphi\ ln^2\varphi)$, so that such a form of the β function could not have been guessed from the usual pure power expansion.

Conversely the lowest order term in the β function for marginal fields (which is the case of coupling constants and masses) is to the square:

$$\frac{d\alpha}{d\ ln\frac{\lambda}{r}} = \beta_o\ \alpha^2\ . \tag{6.10.3}$$

This means that this lowest order term is beyond any order of the expansion of (6.10.2), so that we conclude that the renormalization group equation for the coupling constants is unchanged in special scale relativity (case of linear scale transformations), at least *to leading order approximation* (see Sec. 6.11 for additional comments on this point). However, this conclusion holds only for the renormalization group equation expressed in terms of *length* scale. Indeed the renormalization group method, in its general definition,[4-9] as well as the scale-relativistic approach, essentially aims at describing the way physical laws change when going from one *spatio-temporal* scale to a larger one. But the relation from length-time scale to mass-energy-momentum scale is generalized in scale relativity (Eq. 6.9.6). So, while we obtain the usual solution (6.2.3) in terms of length scale, we get in terms of mass scale the new relation (to lowest order)

$$\bar{\alpha}(m) = \bar{\alpha}(m_o) + \beta_o \frac{ln(m/m_o)}{\sqrt{1 + ln^2(m/m_o)/\mathbb{C}_o{}^2}}\ , \tag{6.10.4}$$

where $\mathbb{C}_o = ln(\lambda_o/\mathbb{A})$. This relation applies in particular in the possible "great desert" beyond the W boson scale. In this case, we must take $\mathbb{C}_o = \mathbb{C}_W = ln(\lambda_W/\mathbb{A}) = 39.876(6)$. *At the absolute limit* $r = \mathbb{A}$, i.e., $m \rightarrow \infty$, *the charge is now finite*. So pure QED becomes a self-consistent theory in the framework of scale relativity. In the same way, masses which were previously divergent even to the first order as

$$m = m_0 \left[1 + \kappa \alpha_0 \, ln \frac{\lambda_0}{r} \right]$$

now remain finite in the new theory. We shall hereafter let the question of mass determination be open for future works, and shall focus mainly on the coupling constant problem.

New predictions.

In order to fix the ideas about the way scale relativity is expected to yield new testable predictions, let us consider some numerical values. The amplitude of scale-relativistic corrections will be given by Lorentz-like "scale δ-factors", which we have identified with the anomalous dimension now varying with scale, depending on "$V/\mathbb{C}$" ratios, i.e., $ln(\lambda_0/\lambda) / ln(\lambda_0/\mathbb{A})$. For example, from the electron scale (0.511 MeV) to the W scale (79.9 GeV), one already gets $V/\mathbb{C}_e = 0.232$, i.e. $\delta = 1.028$, which is small but not negligible. From the W scale to the GUT scale ($\approx 10^{14}$ GeV), one gets $V/\mathbb{C}_W = 0.7$, leading to a large value of the anomalous dimension, $\delta = 1.4$.

This last result allows us to introduce our first new prediction. In the standard model, the reason for the emergence of a scale, namely the GUT scale, in addition to the electroweak one, is not understood, and so no purely theoretical prediction of its value is available. On the other hand, scale relativity naturally introduces a new fundamental scale in Nature. Indeed the new relation between the mass scale and the length scale (case $r \approx ct$) is such that the Planck mass m_P does not correspond any more to the Planck length $\mathbb{A}$. We know that an infinite mass now corresponds to the Planck length and thus expect a new fundamental length λ_P to emerge (see Fig. 6.5). Let us compute it starting from the W scale (we recall that in the new theory lengths are no more absolute; now only scale dilatations from one scale to another have physical meaning).

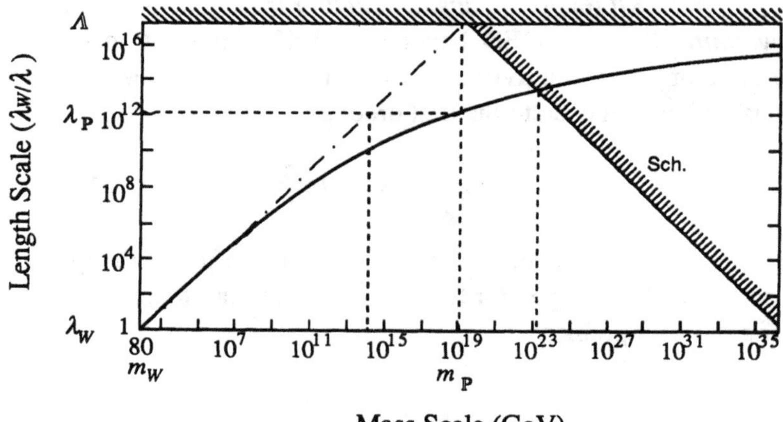

Figure 6.5. The mass-energy / length relation in scale relativity. In the new theory the Planck length, $\mathbb{A} = (\hbar G/c^3)^{1/2}$, becomes an absolute and unpassable limit for all lengths in nature. The new relation is such that now the Planck space-time scale corresponds to infinite energy-momentum. The Planck mass (m_P= 1.22105(8) x 10^{19} GeV) then allows one to define a new universal length scale, λ_P, which is found to be 10^{-12} times the W scale. This length scale corresponds, in the previous standard theory (dot-dashed line), to energy of 10^{14} GeV, i.e. precisely the value of the GUT scale deduced from the convergence of coupling constants. To the right of the diagram the standard relation which corresponds to the Schwarzschild horizon $r_s = 2Gm/c^2$ is shown. It is now reached at an energy of $\approx 10^{23}$ GeV (however, see Sec. 7.1 and Fig. 5.17).

We obtain from (6.9.6)

$$ln \frac{\lambda_W}{\lambda_P} = \frac{ln(m_P/m_w)}{\sqrt{1 + ln^2(m_P/m_w)/C_w{}^2}} \, . \qquad (6.10.5)$$

A first estimate of this scale may be obtained by neglecting the fact that the W length scale should itself be subject to a scale-relativistic correction. To this approximation we have $C_w = ln(\lambda_W/\mathbb{A}) \approx ln(m_P/m_w)$, so that the denominator of (6.10.5), i.e. the δ-factor of the new length λ_P is found to be $\delta = \sqrt{2}$. So, since $m_P/m_w = 10^{17.2}$, λ_P is expected to be $10^{17.2/\sqrt{2}}$ = 1.4 x 10^{12} times smaller than the W scale, which would correspond to 10^{14} GeV in the current standard theory; in other words, the new fundamental scale introduced by scale relativity is precisely the GUT scale where the U(1), SU(2) and SU(3) couplings are known to converge.

A more precise computation, accounting for the fact that the ratio λ_e/λ_W is itself no more equal to m_W/m_e, yields essentially the same result. Indeed at such scales the correction is still small: one gets $log(\lambda_e/\lambda_W)$ = 5.059(3), $log(m_W/m_e)$ = 5.194(3), so that $\delta = 1.409$ instead of $\sqrt{2}$.

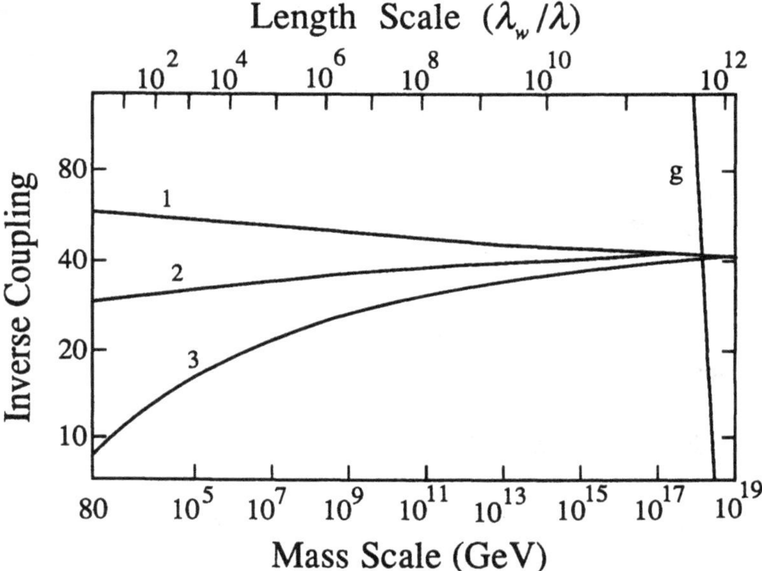

Figure 6.6. Variation with length and mass scales of the inverse of the four coupling constants, U(1), SU(2), SU(3) and gravitational (g). The mass scale goes from the W boson mass (79.9±0.4 GeV) to the Planck mass, m_P = 1.22105(8) x 10^{19} GeV. The convergence point of couplings (i.e. the GUT scale in the standard model), at a length scale about 10^{-12} times the W length scale, now corresponds in scale relativity to the Planck energy (see Fig. 6.5 and text). In the new theory the *four* fundamental couplings converge at the same scale.

This result has an important consequence. Indeed one may express it in another way, by computing the energies corresponding to the unification scale. As expected from the fact that the mass corresponding to λ_P is the Planck mass, we find that the energy at which the α_1 and α_3 couplings cross (Eqs. 6.2.5, 6.2.6 and 6.10.4) is $m_{13} = 1.1 \times 10^{19}$ GeV/c^2, in excellent agreement with the value of the Planck energy $m_P =$

$1.22105(8) \times 10^{19}$ GeV/c^2. This means that now, not only the three electromagnetic, weak and color couplings converge at about the same energy, but also the gravitational one (see Fig. 6.6). Indeed the gravitational coupling α_g varies with mass scale as $\alpha_g = (m/m_P)^2$ when $m = (m_o{}^2+p^2/c^2)^{1/2} >> m_o$, so that $\bar{\alpha}_g$ reaches the common value $\bar{\alpha}_1 \approx \bar{\alpha}_2 \approx \bar{\alpha}_3 \approx 40$ at energy $\approx m_P/\sqrt{40} \approx 1.9 \times 10^{18}$ GeV. One may also directly study the crossing of the gravitational coupling with the three other couplings. We find that they cross at a scale $\lambda = 1.8 \times 10^{-12}\ \lambda_W$ (which corresponds in the standard theory to 4.4×10^{13} GeV), at $\bar{\alpha} = 42.4$, 42.8 and 39.7 respectively.

These results resolve the discrepancy concerning the GUT prediction of the mixing angle and proton lifetime: the solution comes from the fact that the weak angle theoretical prediction is made in terms of length scale, while the proton lifetime prediction depends on mass scale. The length scale where $\alpha_1 = \alpha_2$ (Eqs. 6.2.6 and 6.2.7) is found to be 10^{11} times smaller than that of the W (previously 10^{13} GeV), so that one gets a prediction $sin^2\theta_W(W) = 0.232 \pm 0.004$ (see Eq. 6.2.13), which compares well with the experimental value[41,44] 0.230 ± 0.002. (Note that this may imply that there is no strictly common unification point, as recently remarked by some authors[58]). But this unification range, 10^{-11}-10^{-12} times the W scale, now corresponds in mass to 10^{17}-10^{19} GeV/c^2. Hence the proton lifetime theoretical expectation, which varies as $m_X{}^4$, becomes larger than 10^{37} years, far greater than the present experimental lower limit.

Another comment about the new structure is that it points toward a common origin of all forces of nature. If Figure 6.6 is taken at face value, one starts at a very high energy from a purely gravitational regime, while when going to lower energies, one finds a decoupling first with the colour field, and finally with the electroweak field which separates in U(1) and SU(2).

Let us note another remarkable fact. In the standard theory, the Planck length is equal to the Compton length of a Planck mass, but also to the Schwarzschild radius of a Planck mass (to a factor of 2): $\Lambda = \hbar/m_P c = Gm_P/c^2$. This implies that a point mass larger than the Planck mass is a black hole of radius larger than the Planck length (see Sec. 5.11). Hence the Planck scale is not only a domain where the quantum and gravitational phenomena are expected to become of the same order, but also a domain where the gravitational field itself is very strong: this explains partly the difficulty of elaborating a theory of quantum gravity. This is radically changed in the new theory, so that the quantum gravity problem is now set in completely different terms. Indeed the Planck mass m_P now corresponds to the new scale λ_P, so that the gravitational potential at the Planck energy scale becomes in scale relativity $Gm_P/c^2\lambda_P \approx 10^{-5}$. This is a weak field situation, i.e. typically of the order of the potential at the solar limb. The mass-length relation (Eq. 6.9.6) eventually crosses the standard black hole horizon-mass relation ($r_S = 2Gm/c^2$), but at a far larger energy of 1.3×10^{23} GeV (see Fig. 6.5). However these estimates are tentative, since the Schwarzschild relation also needs to be generalized in scale relativity, owing to the fact that $r_S < \mathbb{A}$ is now excluded, while there is no limitation on mass-energy (see Fig. 5.17 and Sec. 7.1).

Let us now consider the question of theoretical determination of the values of the fundamental charges. The standard quantum theory, thanks to the renormalization group approach, already arrived at a magnificent clarification of the problem: the low energy charges (equivalently, the low energy couplings) result from some high energy unifying value at the GUT scale and from a variation from high to low energy which is completely determined by the renormalization group equations; these equations depend on the gauge group and the number of families of fermions, and their solutions on initial conditions set at the thresholds defined by the elementary particles masses (see Sec. 6.2 and Figs. 6.2 and 6.4). Hence one of the couplings may be estimated from the other two: for example, starting from $\alpha_1(m_W)$ and $\alpha_3(m_W)$, one may find, thanks to Eqs. (6.2.5) and (6.2.6), λ_X and α_X at the GUT scale and then predict $\alpha_2(m_W)$ from Eq. (6.2.7).

Scale relativity brings several improvements to the solution of this problem. The first comes from the fact that we may now use the intersections of the electroweak and color couplings with the gravitational coupling, which is completely known, in addition to their own intersections. Starting, for example, from the intersection point of α_g and α_3, and admitting as a first step that it also yields the zero point of α_1 and α_2, allows us to deduce the low energy fine structure constant to 6% of its measured value and the Fermi constant to 10%. We may also use the fact that one of the free parameters of the standard theory, the GUT scale, is theoretically known in scale relativity.

But the best improvement would come from the expectation of the value of the "bare" common coupling at high energy. The observation of new structures in the ($\bar{\alpha}$, *log r*) plane allows us to make a conjecture about this value.[59] Attempts at a theoretical understanding of the origin of this value will be presented in Sec. 6.11 and in forthcoming works. Our conjecture is that there exists a universal high energy charge whose value is equal to

$$\sqrt{\alpha_{\mathbb{P}}} = \frac{1}{2\pi} .$$

This would yield a common inverse coupling, $\bar{\alpha}_{\mathbb{P}} = 4\pi^2 = 39.478418...$. Let us consider the arguments in favour of this suggestion.

One argument is given by the (α_g, α_3) intersection for which we have obtained $\bar{\alpha}_{3g} = 39.7 \pm 0.8$. Another argument is given by the electroweak couplings. At first sight they do not support our conjecture, since they meet at an inverse coupling value of ≈ 42. Consider, however, the formal electromagnetic inverse coupling, $\bar{\alpha} = \bar{\alpha}_2 + \frac{5}{3}\bar{\alpha}_1$ (see Sec. 6.2). Below the unification scale, it becomes equal to 8/3 of the common inverse coupling $\bar{\alpha}_1 = \bar{\alpha}_2$, and the solution to its renormalization group equation is (to lowest order)

$$\bar{\alpha}(r) = \bar{\alpha}(\lambda_W) - \left(\frac{5}{3\pi} + \frac{N_H}{6\pi}\right) \ln \frac{\lambda_W}{r} . \tag{6.10.6}$$

where N_H stands for the (unknown) number of Higgs doublets. The value of the fine structure constant at the W scale is presently estimated to be $\bar{\alpha}(\lambda_W) = 127.8 \pm 0.3$, (see Eq. 6.2.8) so that we may compute the value of $\bar{\alpha}(\mathbb{A})$, the inverse formal fine structure constant at the absolute limiting Planck scale, and then of $\bar{\alpha}_1(\mathbb{A}) = \bar{\alpha}_2(\mathbb{A}) = 3\bar{\alpha}(\mathbb{A})/8$. We find, in this framework of a pure electroweak theory (one Higgs doublet assumed), that

$$\bar{\alpha}_1(\mathbb{A}) = \bar{\alpha}_2(\mathbb{A}) \approx \frac{3}{8}\,\bar{\alpha}(\lambda_W) - \frac{5}{8\pi}\,\mathbb{C}_W = 40.0 \pm 0.5 , \tag{6.10.7}$$

which also supports the $4\pi^2$ conjecture. Conversely the conjecture becomes testable, since it allows us to predict the low energy value of the SU(3) coupling with an improved precision over the presently known value. Starting from the hypothesis that $\bar{\alpha}_3(m_P) = 4\pi^2$ and going back to the W scale from the renormalization group equation, we find from a first order calculation $\bar{\alpha}_3(\lambda_W) = 0.113$. However, second order terms[29,46] are not negligible to this precision (see Sec. 6.2). Including them yields $\bar{\alpha}_3(\lambda_W) = 0.1165 \pm 0.0005$, where the error comes from a rough estimate of the contribution of third order terms. This prediction is compatible with present experimental results[41], 0.120 ± 0.012, and[44] $0.107^{+0.013}_{-0.008}$, but being far more precise, will allow one to test the theory when the experimental error decreases. (However, see Sec. 6.11 for a report on recent improvements of these experimental values and additional comments about the significance of the second order prediction).

Equation (6.10.7) does not lead to such a prediction, owing to the fact that the low energy fine structure constant is currently known to a high precision[22] ($\bar{\alpha} = 137.0359914(11)$). However it allows us to get its value to 5‰, namely $\bar{\alpha} = 137.7 \pm 0.7$ (to first order with one Higgs doublet assumed), 138.1 ± 0.7 (to second order with one Higgs doublet), 136.0 ± 0.7 (to second order with zero Higgs doublet), and to predict that there can be no more than 1 Higgs doublet, since each additional doublet would contribute + 2.1 to the final result.

Let us finally consider an additional argument in favor of the $1/2\pi$ conjecture. It is remarkable that well-defined structures seem to emerge in the new ($\bar{\alpha}$, *log r*) plane. Hence the fundamental ratios of the Planck to *W/Z* mass scales, which are now the two fundamental symmetry-breaking scales, yield from the current values of weak boson masses (see Sec. 6.2)

$$ln\frac{m_P}{m_W} = 39.567 \quad , \quad ln\frac{m_P}{m_Z} = 39.436 \quad ,$$

which enclose the $\bar{\alpha}_P = 4\pi^2$ value. Similar structures seem also to relate the *W/Z length* scale to the charge at the *electron* scale, which is the third fundamental scale in the theory. Indeed we find that

$$ln\frac{\lambda_e}{\lambda_W} = 11.650 \quad , \quad ln\frac{\lambda_e}{\lambda_Z} = 11.772 \quad ,$$

which enclose the inverse electric charge value, $\bar{e} = \sqrt{\bar{\alpha}} = 11.706$. More tentatively, note that the λ_P to Λ scale ratio may also be related to the low energy electric charge, since it is given to the first approximation by $\mathbb{C}_W(1-1/\sqrt{2}) = 11.679$. We shall show in the next section that these remarks may be turned into precise (though still empirical) formulas which allow one to predict with high precision the W and Z boson masses, and may bring new insight into the nature of the electric charge.

6.11. On the Nature of Charges and Masses.

We shall in this section attempt to push further the question of the nature and determination of elementary masses and charges. This is motivated in particular by the fact that, in the past months (end of 1991), our experimental knowledge of the numerical values of the three fundamental couplings, α_1[U(1)], α_2[SU(2)], α_3[SU(3)] at the electroweak boson scale has been improved. Recent determinations have given[60-63]

$$\begin{aligned} \alpha_1^{-1}(Z) &= 59.22 \pm 0.14 \, , \\ \alpha_2^{-1}(Z) &= 30.10 \pm 0.23 \, , \\ \alpha_3^{-1}(Z) &= 8.93 \pm 0.23 \; . \end{aligned} \tag{6.11.1}$$

They correspond to a value[58] of $\alpha^{-1}(Z) = 128.8 \pm 0.3$ significantly different from the previously accepted one[44] (127.8 ± 0.3).

We shall see that, with these improved values, the conjecture that the bare charge at infinite energy is $1/2\pi$ is still reinforced. Possible ways towards an understanding of the origin of this value are considered, and new empirical formulae for the Z and W boson masses are given hereafter.

Emergence of GUT and Electroweak scales.

The main new result from scale relativity that we want to stress here is the *natural emergence of two fundamental scales*, identified as the Grand Unification scale (as already pointed out in Sec. 6.9) and the electroweak symmetry breaking scale. It is remarkable that, while pure scale invariance seems to be unable to provide us with the observed fundamental scales, scale relativity in its simplest form (i.e, linear "special" case) seems to fill this serious gap of present physics. It must however be noticed that, though the *existence* of particular scales is provided by scale relativity, the theory says nothing about the way these structures are explicitely achieved by Nature. This is reminiscent of Einstein's special motion relativity in which the relation $E = mc^2$ is demonstrated as a universal result, without yielding any information about the way it may be achieved in practice. It is in the frame of a completely different theory, nuclear quantum physics, that explicit transformation between mass and energy has been achieved. Relativity, as already remarked, provides us with universal constraints about the world, which supersede the particular objects which achieve them.

Let us first recall how the electroweak and grand unified scales can be simply found from the Planck scale. Define, to lowest order approximation, a mass scale given by

$$m_{WP} = m_P \; e^{-4\pi^2} \quad . \tag{6.11.2}$$

This scale corresponds to a mass of 87 GeV, in good agreement with the *W-Z* scale of 80.0 - 91.2 GeV (especially owing to the fact that we start from a scale as large as the Planck mass scale, 1.22 x 10^{19} GeV). From this (semi-empirical) determination of the *WZ* scale, we may now derive the GUT length scale, still to lowest order approximation. Since we have identified it with the Planck mass scale, we may write $ln(\lambda_W/\lambda_P) \approx ln(\lambda_W/\mathbb{A})/\sqrt{2}$, so that

$$\lambda_P \approx \mathbb{A} \; e^{4\pi^2(1-\sqrt{2}/2)} \approx 10^5 \; \mathbb{A} \; . \tag{6.11.3}$$

In this framework, the GUT scale is then nothing but the equivalent in scale relativity (from the view-point of the mathematical structure) of the particular momentum $p = mc$ (i.e., of the Compton wavelength) in motion relativity. Indeed, this momentum is obtained for $mv/\sqrt{1-v^2/c^2} = mc$, i.e., $v/c = 1/\sqrt{2}$. In the framework of the standard model, (6.11.3) corresponds to 10^{-5} x 10^{19} GeV = 10^{14} GeV, the right order for the GUT scale. The fact that the GUT and EW scales are related is not unexpected in grand unified theories, since the Higgs bosons responsible for EW symmetry breaking are assumed to be residuals of the Higgs responsible for GUT symmetry breaking.

Though these results are already very encouraging as a rough estimate, we think that scale relativity is able to do even better, and to actually yield the precise scales themselves, as could be expected from a theory based on fundamental principles. In the new theory, there will indeed be a particular scale λ_v for which our new "constant" $\mathbb{C}$ takes the highly particular value $4\pi^2$ (i.e., the value of the formal U(1) and SU(2) coupling at infinite energy) :

$$\mathbb{C}_v = ln(\lambda_v/\Lambda) = 4\pi^2 \; . \tag{6.1.4}$$

We find from the current value of the Planck length that λ_v is 171022(12) times the electron length scale, i.e. 241147(17) times the electron mass scale from Eq. (6.9.6). This corresponds to a mass

$$m_v = 123.228(8) \text{ GeV} \; .$$

Assume that some fundamental particle of this mass exists in Nature, a particle-antiparticle pair of such an object would have a rest mass 246.46(2) GeV. This is a remarkable result, owing to the fact that the electroweak symmetry breaking mass scale is given by[42]

$$\mathrm{v} = \frac{m_W}{\sqrt{\pi \; \alpha_2(m_W)}} \; ,$$

that is, v = 246.9 ± 0.9 GeV from the recently measured values of m_W and α_2, in very good agreement with the above value. Candidates for a particle of mass 123 GeV in the framework of the standard model are the Higgs particle itself and the t-quark. Reversing the argument, assuming that $2m_v$ = 246.46(2) GeV is indeed the electroweak spontaneous symmetry breaking scale allows us to predict that

$$\alpha_2(m_W) = 0.03341(10) \; ,$$

a value in agreement with (and more precise than) the current experimental value[61] (0.03322(25)). This prediction is expected to be still improved in the near future when direct measurements of the W mass are available. Remember that behind this question of the determination of the electroweak scale lies that of the precise relation of the Fermi weak constant with the parameters of the electroweak theory. It depends on electroweak radiative corrections Δr, following Sirlin's formula:[64,65,58]

$$\frac{G_F\ (1-\Delta r)\ m_Z{}^2}{8\sqrt{2}\ \pi\ \alpha} = \frac{1}{16\ sin^2\theta_w\ cos^2\theta_w}$$

In this form the electroweak correction is dominated by the variation of the fine structure constant from the electron to the W scale. The value of Δr depends on the masses of the Higgs boson and the t-quark (see Halzen and Morris[45]). Note especially that a value 123 GeV for the mass of the t-quark seems to be consistent with the most recent theoretical expectations.[66]

Prediction of low energy couplings.

Let us now come back to the prediction of low energy charges from the $1/2\pi$ conjecture. We shall in the following use the Z boson scale as reference scale. We know that, in scale relativity, the length dilatation from the electron scale to the Z scale now slightly differs from m_Z/m_e. From[68,34] m_Z= 91.174(21) GeV/c^2, m_e = 0.51099906(15) keV/c^2, and $\mathbb{C}_e = ln(\lambda_e/\Lambda) = ln(m_P/m_e)$ = 51.527967(65), one gets $ln(\lambda_e/\lambda_Z)$ = 11.77212(23). Then we may set $\mathbb{C}_Z = ln(\lambda_Z/\Lambda)$ = 39.75585(24). The new scale λ_P is given, relatively to the Z scale, by

$$ln(\lambda_Z/\lambda_P) = \mathbb{C}_Z / \sqrt{2} = 28.11163(17)\ .$$

This length scale is thus 1.6×10^{12} times smaller than the Z length scale, and would correspond in the standard (non scale-relativistic) quantum theory to 1.07×10^{14} GeV.

Consider the first order renormalization group equations. The inverse couplings $\bar{\alpha}_1$ and $\bar{\alpha}_2$, and subsequently the combined "electromagnetic" inverse coupling $3\bar{\alpha}/8 = (3\bar{\alpha}_2 + 5\bar{\alpha}_1)/8$ (as defined above) all converge towards some GUT scale λ_{12}. Then for higher energy scales the coupling is expected to become unique. We assume that in a pure electroweak theory this common coupling is $\frac{3}{8}\bar{\alpha}$. Then our basic conjecture is that, at infinite energy, i.e. at length scale $\mathbb{\Lambda}$, we have $\bar{\alpha}_1 = \bar{\alpha}_2 = \frac{3}{8}\bar{\alpha} = 4\pi^2$, so that $\bar{\alpha}(\mathbb{\Lambda}) = 32\pi^2/3 = 105.27578...$.

Concerning the inverse colour coupling $\bar{\alpha}_3$, our conjecture is that its value is $4\pi^2$ at its intersection with the gravitational coupling $\bar{\alpha}_g = (m_P/m)^2$, i.e., for a mass $m_{g3} = m_P/2\pi$ which from (6.9.6) corresponds to a length scale $ln(\lambda_Z/\lambda_{g3}) \approx (\,\mathbb{C}_Z - ln\sqrt{2\pi}\,)/\sqrt{2}$. A more precise calculation yields numerically $ln(\lambda_Z/\lambda_{g3}) = 27.43879(17)$.

Recall the lowest order solutions to the renormalization group equations (see Sec. 6.2; N_H is the number of Higgs doublets):

$$\bar{\alpha}(\lambda) = \bar{\alpha}(\lambda_Z) - \frac{10+N_H}{6\pi}\, ln(\lambda_Z/\lambda) \quad ,$$

$$\bar{\alpha}_3(\lambda) = \bar{\alpha}_3(\lambda_Z) - \frac{7}{2\pi}\, ln(\lambda_Z/\lambda) \quad .$$

Then our lowest order predictions for the electromagnetic and color couplings at Z boson scale are

$$\bar{\alpha}(m_Z) = 126.36 \text{ (0 Higgs doublet)}, \quad \bar{\alpha}(m_Z)=128.47 \text{ (1 Higgs doublet)},$$

$$\bar{\alpha}_3(m_Z) = 8.91 , \qquad \text{or} \qquad \alpha_3(m_Z) = 0.112 \quad .$$

A comparison with (6.11.1) is very encouraging. In particular, if we accept the recent value[58] for $\bar{\alpha}(m_Z)$, 128.8 ± 0.3, we obtain a first interesting result: already to first order, *only the* 1 *Higgs boson case is acceptable*. The Higgs boson contribution being $+2.11\ N_H$ in the final result, we may reject as well the 0 Higgs case (e.g. dynamical symmetry breaking) as the case $N_H > 1$.

Consider now the second order solution. We may first predict the value of $\bar{\alpha}$ at the Z scale from (6.2.11), in which we insert $ln(\lambda_o/\lambda)=\mathbb{C}_Z$ and use in the second order terms the approximation $\alpha_1(\lambda_o) \approx \alpha_2(\lambda_o) \approx \alpha_3(\lambda_o) \approx 1/4\pi^2$. This yields $\bar{\alpha}(m_Z) = 126.85$ (0 Higgs doublet assumed) and 128.97 (1 Higgs doublet). However it would be more correct to use the Z values of the couplings in the second order terms. In order to base the calculation only on the high energy charge conjecture, this may be done by reinserting in the second order terms the first order solutions, and similar estimations for

$\bar{\alpha}_1(m_Z)$ and $\bar{\alpha}_2(m_Z)$ based on the hypothesis that the electroweak U(1)x SU(2) separation occurs at scale λ_{g3} (but see hereafter). This yields a second order correction larger by 0.26, the difference being due essentially to the large variation of $\bar{\alpha}_3$ from the Z to the GUT scale, i.e., $\bar{\alpha}(m_Z) = 129.22$ (one Higgs doublet), still compatible with the experimental value. Concerning $\bar{\alpha}_3$, the first method yields a second order solution $\bar{\alpha}_3(m_Z) = 8.79$, i.e., $\alpha_3(m_Z) = 0.1138$, and the second method $\alpha_3(m_Z) = 0.1174$. Both values are compatible with the recently determined experimental values[62,63] $\alpha_3(m_Z) = 0.112 \pm 0.03$, and 0.113 ± 0.04.

However these second order predictions should be considered with caution. Indeed, we have demonstrated that the renormalization group equations for running coupling constants are unchanged when written in terms of length scale *to lowest order*; but this is *a priori* no more the case to the following order in the α expansion (more precisely, one expects scale-relativistic corrections of order $V^2/\mathbb{C}^2$ while $V = ln(\lambda/r) \propto \alpha - \alpha_o$). A proper account of such scale relativistic corrections for fields implies a non-linear theory: this comes no longer under special scale relativity, but under a generalized theory that remains to be constructed.

Concerning possible predictions of $\bar{\alpha}_1(m_Z)$ and $\bar{\alpha}_2(m_Z)$ individually from their convergence at the GUT scale, the task is more difficult. Indeed the hypothesis that the three couplings should cross at exactly the same point, which is often used as a constraint on possible GUTs, is difficult to keep in scale relativity because we work in position representation and because of the intervention of gravitation. Concerning the first point, the example of the electron transition, which is known in details, is instructive. Recall that in terms of *length scale* the electromagneric coupling varies, for $r << \lambda_e = \hbar/m_e c$, as

$$\bar{\alpha}(r) = \bar{\alpha} - \frac{2}{3\pi} \left[ln\frac{\lambda_e}{r} - (\gamma + \tfrac{5}{6}) \right] , \tag{6.11.5}$$

where $\gamma = 0.577...$ is Euler's constant, while at larger scale it tends to the "zero energy value" α through a Yukawa term with mass $2m_e$ (at scale λ_e the inverse fine structure constant is found to be smaller by only 0.0076 from its zero energy value). But the point of convergence of (6.11.5) with the low energy value $\bar{\alpha}$ is at a scale λ_e / Q, with $Q = exp(\gamma + \frac{5}{6}) = 4.098$, rather than at the scale λ_e which the unbroken renormalization group equations would yield (see Fig. 6.1). If a threshold effect of comparable importance (which remains to be calculated) exists for the transition from 2 to 1 couplings, then λ_{12} is expected to be larger than λ_{g3} and $ln(\lambda_Z/\lambda_{12}) \approx 10^{11}$, which would yield results at the W scale in agreement with experiments.

Such threshold effects should also be accounted for if one wants to derive the fine structure constant from our prediction for $\bar{\alpha}(m_Z)$ and its variation from the Z to the electron scale. This variation was estimated[43,44] to be $\Delta\bar{\alpha}_{eZ} \approx 9.2 \pm 0.3$ while the full threshold effects encountered for every elementary particles which contribute to it may be estimated to be ≈ -1.2. Combined with our above estimate $\bar{\alpha}(m_Z) \approx 129.0$, we get a satisfying result: $\bar{\alpha} = 137.2 \pm 0.3$. Though we are still very far from effectively having a theoretical prediction of the fine structure constant which could compete with the experimental determination, this result may be used in another way: from the predicted $\bar{\alpha}(m_Z)$ and observed $\bar{\alpha}$ one may constrain the number and mass distributions of elementary particles in the electron to Z boson domain.

Another promising method is to use the value of $\bar{\alpha}_2(m_Z) = 0.03333(10)$ hereabove derived from our expectation for the symmetry breaking scale (the slight difference with the value quoted above comes from the passage from the W scale to the Z scale) and combine it with our prediction for $\alpha(Z)$ in order to deduce an expectation value for $\bar{\alpha}_1(m_Z)$ and constraints on $sin^2\theta_w$.

Masses of weak bosons.

Let us now turn to new predictions concerning the Z and W masses. We propose hereafter two empirical equations which relate the W/Z scale to the two other fundamental scales in the new theory, the electron scale and the Planck mass scale. We shall not attempt to justify them here in detail: let us only remark that they may have something to do with the equilibrium of quarks in the proton and, perhaps, with the equilibrium of the vacuum.

Before presenting these relations, let us make a remark on the charges of leptons and quark. Consider the two following equations:

$$e_1 + e_2 + e_3 = 1 \;, \tag{6.11.6a}$$

$$e_1^2 + e_2^2 + e_3^2 = 1 \ . \qquad (6.11.6b)$$

The first of these equations has a simple interpretation. This is an equation of conservation of charge in a process involving four particles, one of them having a unit charge. The second equation would then correspond to conservation of the square of charges; such conservation is not fulfilled by most systems. Consider however the particular case where two particles have the same charge, $e_1 = e_2$. The system of equations (6.11.6) then has only two solutions, $\{e_1 , e_2 , e_3 \}= \{0, 0, 1\}$ and $\{2/3, 2/3, -1/3\}$. These solutions correspond to the only possible values of electric charges (in terms of the electron charge) as observed in Nature, the charges of leptons in the first case and quarks in the second. The two equations (6.11.6) apply to processes as different as the decay of the muon into electron and neutrinos (plus every equivalent leptonic processes) and the constitution of the proton by 3 quarks (*u*, *u*, *d*) (plus every charge-one hadrons). However, if they are to stand in the future as equations for elementary processes, it will first be necessary to have an understanding of the conservation of coupling constants, (6.11.6b).

The empirical equations we propose hereafter for the weak boson masses have a similar structure. They also involve charges and square of charges, now as a scalar product with the masses of weak bosons, m_Z, m_{W-}, and m_{W+}.

The first equation involves the *u*, *u*, *d* quark/ electron charge ratios and the "bare" inverse coupling $\alpha_P^{-1} = 4\pi^2$:

$$ln\{ \ \tfrac{2}{3}\frac{m_Z}{m_P} + \tfrac{2}{3}\frac{m_{W-}}{m_P} + \tfrac{-1}{3}\frac{m_{W+}}{m_P} \ \} = \ -4\pi^2 \ . \qquad (6.11.7)$$

This identity is verified to a remarkable precision by the values $m_Z =$ 91.176(23) GeV and $m_W =$ 80.05(19) GeV that have recently been obtained at LEP: one gets −39.478(1) for the left-hand side of (6.11.7), to be compared with $-4\pi^2 = -39.478418...$. This agreement may also be expressed in another way: from the Planck mass, a new mass scale is defined (6.11.2):

$$m_{WP} = m_P \ e^{-4\pi^2} \ = \ 87.393(7) \ \text{GeV} \ ,$$

where the main source of uncertainty is the badly determined value of the constant of gravitation G (we recall that $m_P = (\hbar c/G)^{1/2}$). This is to be compared with $(2\,m_Z + m_W)/3 = 87.467(65)$ GeV, so that we get the remarkable result

$$e^{4\pi^2}\,\frac{2m_Z + m_W}{3m_P} = 1.0008 \pm 0.0009 \ .$$

The second equation relates the W/Z scale to the electron scale:

$$ln\left\{\left(\tfrac{2}{3}\right)^2 \frac{m_Z}{m_e} + \left(\tfrac{2}{3}\right)^2 \frac{m_{W-}}{m_e} + \left(-\tfrac{1}{3}\right)^2 \frac{m_{W+}}{m_e}\right\} = \frac{e^{-1}}{\sqrt{1 - e^{-2}/\mathbb{C}_e^{\,2}}} \cdot \qquad (6.11.8)$$

It involves the *square* of quark charges and the low energy dimensionless electric charge $e = \alpha^{1/2}$, which takes here the new meaning of a fundamental *length* dilatation $e^{-1} = ln(\lambda_e/\lambda_{we}) \approx 11.7$. Such an interpretation for the electric charge will be considered again at the end of this section.

The right-hand side of (6.11.8) defines a scale δ-factor $\delta = 1.02684979(8)$, from $\mathbb{C}_e = \ln(m_P/m_e) = 51.527967(65)$ and $e^{-2} = \alpha^{-1} = 137.0359914(11)^{22}$; the left-hand side times e yields a result in good agreement with this prediction, 1.02686(23). The agreement with experimental data may be also expressed by introducing a scale:

$$m_{we} = m_e \ exp\sqrt{\frac{\bar{\alpha}}{1 - \bar{\alpha}\,/\,\mathbb{C}_e^{\,2}}} = 84.89406(9)\,,$$

while $(4\,m_Z + 5\,m_W)/9 = 84.99 \pm 0.11$, so that we get the relation

$$\frac{4\,m_Z + 5\,m_W}{9\,m_e}\ exp\!-\!\sqrt{\frac{\bar{\alpha}}{1 - \bar{\alpha}\,/\,\mathbb{C}_e^{\,2}}} = 1.0011 \pm 0.0013\,.$$

In terms of the two new scales, the weak boson masses write

$$m_W = 3\,m_{we} - 2\,m_{WP}\ ,$$

$$m_Z = \frac{5}{2} m_{WP} - \frac{3}{2} m_{we}\ .$$

These relations may be used in another way. One may also start from the measured value[41] of $sin^2\theta_w(W) = 1-m_W{}^2/m_Z{}^2 =$ 0.2302(21) and derive an estimate for the fine structure constant: one finds $\alpha^{-1} =$ 137.040 ± 0.008. Conversely these equations may be used to predict precise values of m_Z, m_W and $sin^2\theta_w$:

$$m_Z = 91.139(8)\ \text{GeV},\ m_W = 79.898(8)\ \text{GeV}\ ,\ sin^2\theta_w\ (W) = 0.2314(3)\,,$$

which correspond to an improvement by the respective factors of 3, 25 and 10 over their present best experimental estimates.

One must however note that this prediction is based on the choice of the "zero energy" value (actually Bohr radius scale) of the fine structure constant as input in Eq. (6.11.8). Choosing its value at the Compton scale, (137.0284 from a lowest order calculation, see Sec. 6.2), would have given slighly different results:

$$m_Z = 91.184(8)\ \text{GeV},\ m_W = 79.809(8)\ \text{GeV}\ ,\ sin^2\theta_w\ (W) = 0.2339(3)\,.$$

Note also that, at this level of precision, the precise masses of *W and Z* bosons depend on their definition.[67] So a more profound understanding of the physical origin of the above equations will be necessary for a comparison with future high precision experimental determinations of the weak boson masses.

On origin of the bare charge $1/2\pi$.

The question of the nature of charges is asked in a new way if one admits the conjecture (now confirmed by experimental data to their current precision) that the equivalent "charge" $2\sqrt{2\alpha/3}$ is equal to $1/2\pi$. In standard quantum field gauge theory, the electric charge is a conservative quantity which results from the uniformity of the quantum phase, and thus from gauge invariance, and as such is directly related to the elementary probability of emission-absorption of a photon by an electron. Why are we still unable to fix its strengh, i.e., to compute the numerical value of the fine structure constant?

We suggest that the main obstacle to our understanding of the origin of the charge is the lack of spatio-temporal interpretation of its nature. In standard quantum theory, the quantum phase has no classical counterpart.

More generally, the three electric, weak and color charges are presently considered to be purely internal quantum numbers. As a consequence the dimensional equation of the charge remains completely arbitrary; this leads in practice to giving it a fourth independent unit. Fixing its dimensional equation, i.e., reducing it to M, L, and T units, could be a decisive step towards its understanding.

There is one quantity which, in this respect, lies exactly at the frontier between the classical and quantum domains: this is the quantum spin. Indeed, spin is *known* to be an angular momentum and is thus a conservative quantity derived from the isotropy of space, so that its dimensional expression is ML^2T^{-1}. But it is also assumed in standard quantum theory to be a purely internal quantum number, since its interpretation in terms of classical motion of an extended electron is forbidden, both from experiments, which set strong constraints on the extension of the electron, and from theory in which a rotating extended electron with angular momentum $\hbar/2$ is in contradiction with motion-relativity. We have seen in Sec. 5.4 however, that the fractal dimension 2 of quantum trajectories leads to geometrical emergence of quantum spin, and therefore to the evidence of its space-time interpretation, *but in the frame of a non-differentiable space-time.*

The case of spin is also quite instructive concerning quantization. Where does quantization come from? It comes in a general way from the existence of boundary conditions for a given quantum system in conjunction with the requirement of stationarity. Then one may distinguish two kinds of such conditions. Those which are universal and those which occur only in some particular situations. The quantization of the energy corresponds to the second case: the energy of a bound electron in an atom is quantized, but once the electrons free it comes back to a continuum. On the contrary, the quantization of spin and angular momenta, and more generally of action, is *universal*: this comes from the fact that its conjugate variable is an *angle*, and that it is an universal (and apparently trivial) law of Nature that angles are bounded. The fact that any action difference in Nature can only be a multiple of $\hbar$ is a direct consequence of the fact that angles cannot vary by more than 2π in locally Euclidean spaces.

Basing ourselves on this property, we shall make the conjecture here that all "internal" quantum numbers are conjugate variables of angles (and indeed the charges are conjugate of phase angles), with the additional postulate that these angles do have a space-time interpretation (steps towards such a space-time interpretation have been made in Sec. 5.6). This leads to attributing to charge the dimensions of action, and then to write the Coulomb force in the form

$$F = \frac{c}{\hbar}\,(Z_1\sqrt{\alpha}\,\hbar)\,(Z_2\sqrt{\alpha}\,\hbar)\,/\,r^2 \tag{6.11.9}$$

with Z_1 and Z_2 being positive or negative integers

Let us now attempt to reexpress our result on the emergence of a finite internal angular momentum from fractal dimension $D=2$ (Sec. 5.4), in terms of a renormalization group equation. Consider an angular momentum-like quantity of the form $S = mr^2 d\varphi/dt$ around some axis; assuming self-similarity of the fractal curve, in a change of scale by a factor q, *the product* $r^2\dot{\varphi}$ will change by a factor $q^{\delta-1}$. Then the average angular momentum of F_n with respect to some earlier approximation, F_m, is given by $S_n = S_m\, q^{(n-m)(\delta-1)}$. We set $\lambda/r = q^n$, so that $n = \log_q(\lambda/r)$. We shall now pass to a continuous description by writing that, from one scale to the following, $S\{1+\log_q(\lambda/r)\} = S\{\log_q(\lambda/r)\}\, q^{(\delta-1)}$. Let us now make $q \to 1$ and replace $\ln q$ by the differential $d\ln(\lambda/r)$. One gets $S\{\ln(\lambda/r)+d\ln(\lambda/r)\} = S\{\ln(\lambda/r)\}\, e^{(\delta-1)d\ln(\lambda/r)}$, and finally obtains the equation

$$\frac{dS}{d\ln(\lambda/r)} = (\delta-1)\,S\ .$$

But this result assumes perfect self-similarity, and under the assumption that S is << 1 (or S >> 1, which would give an identical result for its inverse), one may consider it as the first term of a power series expansion in terms of S:

$$\frac{dS}{d\ln(\lambda/r)} = (\delta-1)\,S + \beta_0 S^2 + \beta_1 S^3 + \ldots\ .$$

For $\delta = 1$, i.e. fractal dimension $D=2$, one recognizes precisely the Callan-Symanzik renormalization group equation for a marginal field, which holds for the fine structure constant, the electroweak couplings α_1 and α_2 and the colour coupling α_3 at high energy thanks to asymptotic freedom (see Sec. 6.2). (It would hold neither for α_3 at "low" energy, nor for spin itself which does not fulfill the condition $S << 1$). This result encourages us to go on with our working hypothesis that charge has the dimension of an action, i.e., corresponds, in some way to be specified, to a generalized angular momentum.

It is also remarkable that charge and spin are spatio-temporally complementary in the coupling of a source to the electromagnetic field. In the antisymmetric tensor F_{ik} of the electromagnetic field, the electric field corresponds to temporal coordinates tx, ty and tz, i.e. to the boost components of the momentum relativistic 4-tensor M_{ik},while the magnetic field is identified with components which corresponds to the non-relativistic angular momentum. Finally one may also remark, after several authors, that the Maxwell equation, which are completely symmetrical from the viewpoint of fields, are not symmetrical in the currents: this is usually interpreted as due to the absence of magnetic charges, and the usual solution consists in introducing magnetic monopoles.[49] One may instead search for a possible solution in which the electric charges would be removed as a parameter independent of mechanics.

We thus propose that the electric, weak and colour charges are conservative quantities which result in some way from the global *isotropy of space-time*. The unification at large energy of the (weak isospin) U(1), (weak hypercharge) SU(2) and (color) SU(3) structures may help as a guide in such a quest. This indeed allows one to demonstrate that charge must be quantized.[32] Consider the SU(3) group, which contains an SU(2) group as a subgroup. SU(3) is known to be *the dynamical symmetry group of the isotropic three-dimensional harmonic oscillator*.[68] (We recall that *dynamical* symmetry originates from particular structures of a given field).

The group SU(3) may be represented in terms of coordinates and momenta.[68] Since SU(2) is a subgroup, three of its eight generators will take the form of the usual angular momenta. The five remaining components may be associated with the components of a quadrupole tensor and may themselves be expressed as second order combinations of coordinates and momenta:

$$L_x = y\,p_z - z\,p_y \;,\quad L_y = z\,p_x - x\,p_z \;,\quad L_z = x\,p_y - y\,p_x \;,$$

$$Q_{xy} = a\,x\,y + b\,p_x\,p_y \;,\; Q_{yz} = a\,y\,z + b\,p_y\,p_z \;,\; Q_{zx} = a\,z\,x + b\,p_z\,p_x \;,$$

$$Q_0 = \frac{a}{2\sqrt{3}}\,(x^2 + y^2 - 2\,z^2) + \frac{b}{2\sqrt{3}}\,(p_x{}^2 + p_y{}^2 - 2\,p_z{}^2) \;,$$

$$Q_1 = \frac{a}{2}\,(x^2 - y^2) + \frac{b}{2}\,(p_x{}^2 - p_y{}^2) \;,$$

where a and b are real numbers, respectively of dimensional equations $M\,T^{-1}$ and $M^{-1}T$.

The 8 generators of SU(3) may also be represented by 3x3 matrices, three of which (corresponding to an SU(2) subgroup) are Pauli matrices (completed by a null row and a null column), say λ_1, λ_2 and λ_3, plus five remaining matrices[68] λ_4 to λ_8. It can be shown that the two SU(2) subgroups in the two representations considered do not coincide. The relation between the two representations is given by

$$L_x = \hbar\,\lambda_7 \;,\; L_y = -\hbar\,\lambda_5 \;,\; L_z = \hbar\,\lambda_2 \;,$$

$$Q_{xy} = \hbar\,\sqrt{a\,b}\;\lambda_1 \;,\; Q_{yz} = \hbar\,\sqrt{a\,b}\;\lambda_6 \;,\; Q_{zx} = \hbar\,\sqrt{a\,b}\;\lambda_4 \;,$$

$$Q_0 = \hbar\,\sqrt{a\,b}\;\lambda_8 \quad,\quad Q_1 = \hbar\,\sqrt{a\,b}\;\lambda_3 \;.$$

From their dimensional equations, one may set for a and b

$$a = \frac{\hbar}{\lambda_o{}^2} \;,\; b = \frac{\hbar}{p_o{}^2} \;.$$

Assume now that these equations characterize an elementary particle. The hereabove eight invariants can be derived as fractal structures of its trajectory, as was already achieved for spin. The natural periodical unit of the trajectory is the de Broglie *full* period, $\lambda_o = h/p_o$, so that one gets

$$a\,b = \frac{1}{4\pi^2} \quad .$$

We obtain finally

$$Q_{xy} = \frac{\hbar}{2\pi} \; \lambda_1$$

and similar formulae for the other components. Inserting this generalized angular momentum in (6.11.9) yields a coupling $(c/\hbar)(\hbar/2\pi)^2 = \hbar c/4\pi^2$, or, in other words a charge $1/2\pi$. Additional work is needed to decide whether such a remark indeed contains the seed of the solution to the problem of the nature of charge, or whether one must search along completely different roads.

Charge as fundamental dilatation.

In the hereabove approach, the low energy value of the electric charge is derived from its conjectured high energy value and from its variation with scale, as given by the renormalization group equations. This method leads to interesting successes, but the final value obtained for the electric charge is a function of several parameters which remain arbitrary in the standard model, in particular the masses of all elementary particles (i.e., they are not theoretically predicted, but only known by direct measurements). Is it the only possible method ? Cannot we hope to get one day a direct equation for the low energy fine structure constant ? Such a hope is motivated by the fact that, as stated above in this book, the fractal approach is not necessarily reductionistic, in the sense that it may also allow one to *set constraints at the high length scale in order to derive properties of the small scale*, contrary to the standard approach to physics which assumes that structures at a given scale may be reduced to structures at a smaller scale. We actually believe that, in the end, the fundamental structures observed in the world (masses and charges of elementary particles) derive from constraints which are *set at the smallest and highest scales.* This leads us to the cosmological question, which we shall briefly consider in Sec. 7.1. Concerning the elementary electric charge, we suggest hereafter a possible road towards such a goal of a non-reductionist physics, based on self-similar properties of electromagnetism and on a first attempt at connecting the spatial and temporal structures of scale relativity.

Let us first remark that the fine structure constant plays the part of a *fundamental scale ratio* in present physics. It is the ratio of the Rydberg over Bohr radii, of the Bohr radius over Compton length, and also of the Compton length over the classical radius of the electron.

Consider the Coulomb potential created by one electric charge. For scales larger than the Compton length of the electron, one has $\Phi = \alpha\hbar c/r$, so that by introducing the mass of the electron one obtains

$$ln(mc^2/\Phi) = ln\alpha_0^{-1} - ln(\lambda/r) \; . \tag{6.11.10}$$

Consider also the first order solution from the renormalisation group equation for the inverse coupling, at scales now smaller than λ. We may write it in exactly the same form (neglecting some small additive constants):

$$(3\pi/2)\, \alpha^{-1}(r) = (3\pi/2)\, \alpha_0^{-1} - ln(\lambda/r) \; . \tag{6.11.11}$$

This demonstrates the self-similarity of electromagnetism, from the nonrelativistic case ($r > \lambda$) to the relativistic case ($r < \lambda$), and thus shows some symmetry of physical laws on both sides of λ. This self-similarity may be extended approximatively to the next order (but this time in the relativistic domain for both Φ and α). Indeed the field Φ may be (at least formally) corrected below the Compton scale to account for vacuum polarization by introducing a correction precisely given by Eq. (6.11.11), and this may be compared with the second order solution of the renormalization group equation[29]

$$ln(mc^2/\Phi) = ln(\alpha_0^{-1}) - ln(\lambda/r) + ln\left[1 - \frac{2\alpha_0}{3\pi}\, ln(\lambda/r)\right] ,$$

$$(3\pi/2)\, \alpha^{-1}(r) = (3\pi/2)\, \alpha_0^{-1} - ln(\lambda/r) + \frac{9}{8}\, ln\left[1 - \frac{2\alpha_0}{3\pi}\, ln(\lambda/r)\right] .$$

This approximate self-similarity may be a key towards the solution of one of the most important open problems of microphysics, namely, the knowledge of the full β function.[69]

Let us now apply these results to an analysis of the nature of the low energy fine structure constant (denoted α hereafter). Consider the above variable $\xi = \Phi/mc^2$. Since $\lambda = \hbar/mc$, it may be written as

$$\xi = \alpha(\lambda/r) \ ,$$

while α itself may be *defined* as the fundamental scale ratio from the Compton to the Bohr scale:

$$\alpha = \frac{\lambda}{r_B} \ .$$

Thus ξ is the product of two scale ratios, which are made in the quantum nonrelativistic domain $r_B > r > \lambda$. But the Bohr radius plays for electromagnetism (e.g., in an hydrogen atom) precisely the role played by the de Broglie scale for a free particle. So we expect the hereabove relation for ξ in scale relativity to become:

$$ln\xi = \frac{ln\alpha + ln(\lambda/r)}{1 - \dfrac{ln(\alpha^{-1})\ ln(\lambda/r)}{\mathbb{C}_B{}^2}} \ , \tag{6.11.12}$$

where we have set $\mathbb{C}_B = ln(r_B/\mathbb{A})$. Let us express this result in terms of a "scale-relativistic correction". The numerator of (6.11.12) is the standard result, say $ln\xi_{st}$, so the correction is given by its denominator:

$$\frac{ln\xi}{ln\xi_{st}} = \frac{1}{1 - \dfrac{ln(\alpha^{-1})}{\mathbb{C}_B{}^2}\ ln\ \dfrac{\lambda}{r}} \ . \tag{6.11.13}$$

Consider now the QED variation of the fine structure constant beyond the Compton scale (see Sec. 6.2). It corresponds to a motion-relativistic (and quantum) correction given by

$$\frac{\alpha(r)}{\alpha} = \frac{1}{1 - \dfrac{2\alpha}{3\pi}\ ln\ \dfrac{\lambda}{r}} \ . \tag{6.11.14}$$

The similarity between the two formulas, Eqs. (6.11.13) and (6.11.14), is striking, especially when one remembers the already noticed self-similarity shown by $ln\xi$ and $\alpha(r)$ in the standard theory. It leads us to propose a new (lowest order) equation:

$$\frac{ln(\alpha^{-1})}{\mathbb{C}_{\mathrm{B}}^{\,2}} = \frac{2\alpha}{3\pi} \;,$$

or, in another form

$$\frac{\bar{\alpha}\; ln\; \bar{\alpha}}{(\mathbb{C}_e + ln\; \bar{\alpha})^2} = \frac{2}{3\pi} \;. \tag{6.11.15}$$

From the presently known values of the Bohr radius and the Planck length, one computes $\mathbb{C}_{\mathrm{B}} = 56.44821(7)$; solving Eq. (6.11.15) for α yields $\alpha^{-1} = 137.36$, a result to 2×10^{-3} of the measured value[22], $\alpha^{-1} = 137.0359914(11)$. This is an encouraging result, in agreement with what could be expected from a lowest order solution (since $\alpha/\pi \approx 2.3 \times 10^{-3}$). But one must clearly await a generalization of this equation to higher orders and the verification that it yields a value for α which converges towards the observed value (or better, to get an exact equation), before definitively claiming that this is an equation for the fine structure constant, rather than simply numerology.

If this is confirmed as a genuine equation for the electric charge, it is remarkable that it implies only low energy quantities. However we have seen that one has a still more founded hope to find the electric charge value from its infinite energy value (possibly $1/2\pi$), then integrate the renormalization group equations from high to low energy. The final value of the charge depends on the full structure of elementary particles, in particular their number, masses, charges (actually the ratio of their charges over the electron charge), and the group structure of their associated fields. Hence the combination of these two methods would finally yield constraints on the mass structure of elementary particles.

6 References

1. Einstein, A., 1916, *Ann. Phys.* **35**, 769.

2. Levy-Leblond, J.M., 1976, *Am. J. Phys.* **44**, 271.

3. Nottale, L., 1989, *Int. J. Mod. Phys.* **A4**, 5047.

4. Gell-Mann, M., & Low, F.E., 1954, *Phys. Rev.* **95**, 1300.

5. Stueckelberg, E.C.G., & Peterman, A., 1953, *Helv. Phys. Acta*, **26**, 499.

6. Symanzik, K., 1970, *Comm. Math. Phys.*, **18**, 227.

7. Callan, C.G., 1970, *Phys. Rev.* **D2**, 1541.

8. Wilson, K.G., 1969, *Phys. Rev.* **179**, 1499.

9. Wilson, K.G., 1983, *Rev. Mod. Phys.* **55**, 583.

10. Mandelbrot, B., 1982, *The fractal Geometry of Nature* (Freeman).

11. Nottale, L., & Schneider, J., 1984, *J. Math. Phys.* **25**, 1296.

12. Cohen, E. R. & Taylor, B.N., 1987, *Rev. Mod. Phys.* **59**, 1121.

13. Landau, L., & Lifchitz, E., *Relativistic Quantum Theory* (Mir, Moscow, 1972).

14. Itzykson, C., & Zuber, J.B., *Quantum Field Theory* (McGraw-Hill, 1980).

15. Uehling, E.A., 1935, *Phys. Rev.* **48**, 55.

16. Weisskopf, V.F., 1939, *Phys Rev.* **56**, 72.

17. Marciano, W.J. & Sirlin, A., 1981, *Phys. Rev. Lett.* **46**, 163.

18. Tomonaga, S., 1946, *Prog. Theor. Phys.* **1**, 27.

19. Schwinger, J., 1948, *Phys. Rev.* **74**, 1439.

20. Feynman, R.P., 1949, *Phys. Rev.* **76**, 769.

21. Dyson, F.,1949, *Phys. Rev.* **75**, 486, 1736.

22. Kinoshita, T., 1989, *IEEE Trans. Instr. Meas.* **38**, 172.

23. Van Dyck, R.S.Jr., Schwinberg, P.B., & Dehmelt, H.G., 1987, *Phys. Rev. Lett.* **59**, 26.

24. Will, C.M., 1989, in *Tests of Fundamental Laws in Physics*, Proceedings of the XXIVth Rencontre de Moriond, O. Fackler & J. Tran Thanh Van Eds. (Frontière, 1989), p. 3.

25. Weinberg, S., 1972, *Gravitation and Cosmology* (John Wiley and Sons, New York), p. 88.

26. Nambu, Y., *Quarks*, (World Scientific, 1984).

27. Landau, L.D., Abrikosov, A.A., & Khalatnikov, I.M., 1954, *Dokl. Akad. Nauk. SSSR*, **95**, 773.

28. Le Bellac, M., *Des Phénomènes Critiques aux Champs de Jauge*, (Savoirs Actuels, InterEditions, CNRS, Paris, 1988).
29. Babu, K.S. & Ma, E., 1986, *Z. Phys.* **C 31**, 451.
30. Nanopoulos, D.V., Ross, D.A., 1979, *Nucl. Phys.* **B157**, 273.
31. Pati, J.C., & Salam, A., 1973, *Phys. Rev. Lett.* **31**, 661.
32. Georgi, H. & Glashow, S.L., 1974, *Phys. Rev. Lett.* **32**, 438.
33. Georgi, H., Quinn, H.R., Weinberg, S., 1974, *Phys. Rev. Lett.* **33**, 451.
34. Hernandez, J.J., et al. (Particle Data Group), 1990, *Phys. Lett.* **B239**, 1.
35. Gross, D.J. & Wilczek, F., 1973, *Phys. Rev. Lett.* **30**, 1343.
36. Politzer, H.D., 1973, *Phys. Rev. Lett.* **30**, 1346.
37. Coleman, S. & Gross, D.J., 1973, *Phys. Rev. Lett.* **31**, 851.
38. Salam, A., in *Elementary Particle Physics* (Almqvist & Wiksell, Stockholm, 1968).
39. Weinberg, S., 1967, *Phys. Rev. Lett.* **19**, 1264.
40. Glashow, S., 1961, *Nucl. Phys.* **22**, 579.
41. Miller, D.J., 1991, *Nature* **349**, 379.
42. Marciano, W.J., 1989, *Phys. Rev. Lett.* **62**, 2793.
43. Marciano, W.J. & Sirlin, A., 1988, *Phys. Rev. Lett.* **61**, 1815.
44. Amaldi, U. *et al.*, 1987, *Phys. Rev.* **D36**, 1385.
45. Halzen, F., & Morris, D.A., 1991, *Particle World*, **2**, 10.
46. Jones, D.R.T., 1982, *Phys. Rev.* **D25**, 581.
47. Kubo, J., Sibold, K., & Zimmermann, W., 1985, *Nucl. Phys.* **B259**, 331.
48. Buras, A.J., Ellis, J., Gaillard, M.K., Nanopoulos, D.V., 1978, *Nucl. Phys.* **B135**, 66.
49. Nanopoulos, D.V., 1980, in *Nuclear Astrophysics*, International School of Nuclear Physics, Erice, Sicily.
50. Schramm, D.N., 1991, in *The Second International Symposium on Particles, Strings and Cosmology* (Northeastern Univ., Boston).
51. Glashow, S.L., & Nanopoulos, D.V., 1979, *Nature* **281**, 464.
52. Ellis, J., & Nanopoulos, D.V., 1981, *Nature* **292**, 436.
53. Bacry, H., & Levy-Leblond, 1968, *J. Math. Phys.* **9**, 1605.
54. Fuller, F.W., Wheeler, J.A., 1962, *Phys. Rev.* **128**, 919.
55. Hawking, S., 1978, *Nucl. Phys.* **B144**, 349.
56. Norman, E.B., 1986, *Am. J. Phys.* **54**, 317.
57. Landau,L., & Lifshitz, E., *Field Theory*, Ed. Mir, Moscow.
58. Amaldi, U., de Boer, W., & Fürstenau, H., 1991, *Phys. Lett.* **B 260**, 447.

59. Nottale, L., 1992, *Int. J. Mod. Phys.* **A7**, 4899.
60. Ellis, J., 1991, *CERN Preprint*-TH.5997/91.
61. Arason, H., *et al.*, 1991, *Phys. Rev. Lett.* **67**, 2933.
62. Martin, A.D., Roberts, R.G., & Stirling, W.J., 1991, *Phys. Rev.* **D43**, 3648.
63. Ellis, J., Nanopoulos, D.V., & Ross, D.A., 1991, *CERN Preprint*-TH.6130/91.
64. Sirlin, A., 1980, *Phys. Rev.* **D22**, 971.
65. Marciano, W.J., & Sirlin, A., 1984, *Phys. Rev.* **D29**, 945.
66. Marciano, W.J., 1990, *Phys. Rev.* **D41**, 219.
67. Sirlin, A., 1991, *Phys. Rev. Lett.* **67**, 2127.
68. Schiff, L.I., *Quantum Mechanics* (McGraw-Hill, 1968).
69. Yukalov, V.I., 1991, *J. Math. Phys.* **32**, 1235.

Chapter 7

PROSPECTS

7.1. Scale Relativity and Cosmology.

Introduction.

We have shown in the preceding chapter how the principle of scale relativity, even in its "special" version (i.e. *linear* in logarithm form), was able to shed new light on questions such as the charges and masses of elementary particles and the origin and values of some of the fundamental scales of microphysics. Scale invariance alone (as expressed in the standard renormalization group approach or in empirical models based on "scaling") does not permit one to derive the existence of universal characteristic scales in Nature: on the contrary such scales break the naive scale invariance, and they have to be postulated on the ground of observations. We have demonstrated with scale relativity that going from scale invariance to *scale covariance* provides us with fundamental scales: the GUT scale emerges from the splitting of the Planck scale (since the Planck *length scale* and the Planck *mass scale* become different in the new theory) and the electroweak scale emerges from the postulated value $1/2\pi$ of the charge at infinite energy.

There is however an important question, which we already considered briefly in the previous section and which we want to address here more thoroughly: is it possible that every fundamental scale in Nature emerges from constraints which are set at the smallest scales ? We shall adopt a more general view: namely that the fundamental scales in Nature *are determined by constraints which are set at both the small and the large scales*. This will lead us to considering the largest scales in Nature, i.e., to study the relations of scale relativity with cosmology.

What are the arguments in favour of our conjecture ? One may first recall that, from a methodological point of view, the fractal approach allows one to do that naturally. As remarked above, the standard renormalization group is only a semi-group since one always integrates from the small (length) scale to the large scale. Conversely, fractals are built in the opposite way: a generator is defined at the large scale and used to define the structures at smaller scales. Thus fractals provide us with an opportunity to change the usual reductionist approach of science. We have already encountered some clues for the existence of constraints set at both the small and the large scales:

*Two semi-empirical formulae for the masses of the Z and W bosons have been given in Sec. 6.11 (Eqs. 6.11.7, 6.11.8): one equation relates $(2/3)m_Z+(1/3)m_W$ to the Planck scale, but the other relates $(4/9)m_Z+(5/9)m_W$ to the *electron scale*;

*We have suggested an equation for the fine structure constant (6.11.15) where it is determined by a constraint written at the electron and Bohr scales, rather than integrated from the high energy bare charge.

*A fractal model has been given (Sec. 5.10) suggesting that the *muon* and *tau* masses may be connected with the *electron* mass.

All these arguments lead to the same grand question: how is the scale of the electron, which owns the smallest mass for an elementary charged particle, determined? It has already been suggested that the mass scale of elementary particles is, in the end, determined by the global mass of the Universe: one of the strongest and most mysterious argument in favour of this thesis is the well-known great number coincidence:

$$(\hbar^2 H_0 / Gc)^{1/3} \approx m_e \, \bar{\alpha} \ ,$$

where H_0 is the present value of the Hubble constant ($\approx 3 \times 10^{-18}$ s^{-1}), which measures the expansion rate of the Universe. One could consider this coincidence as pure numerology, but, as remarked by Weinberg,[1] such a good agreement with this combination of constants is very improbable. The hope that this relation contains "a fundamental though as yet unexplained truth" led several physicists, in particular Dirac,[2] to proposing alternative cosmologies with varying constants (because H_o varies with time). We now know that such variations of constants are ruled out by observations (see Sec. 6.7), but the fundamental underlying questions remain to be asked. We shall see afterwards that scale relativity allows us to express these questions in a new way and to reactualize the Mach-Einstein principle.

That scale relativity must have things to say about cosmology is also apparent in the huge number of problems which remain open in today's cosmology and in the fact that most of these problems are related to scale and scaling laws. Let us list them (in a certainly non-exhaustive way):

*What is the value of the Hubble constant ? This is a fundamental scale problem, since the inverse of the Hubble constant gives the age of the Universe (to some multiplicative factor of order unity which is a function of the deceleration parameter q_0 and the cosmological constant Λ).

*What is the value of the cosmological constant ? There has been attempts at understanding the cosmological constant in terms of various physical phenomena, in particular as a vacuum energy density (see e.g. Weinberg[3]). There is however a trivial remark which is seldom made (except in the frame of the static Einstein model of the Universe) on this fundamental, universal and absolute constant: that its dimensional equation is the inverse of the square of a length:

$$\Lambda = \frac{1}{\mathbb{L}^2} \quad .$$

The small value of the cosmological constant is connected with the fact that $\mathbb{L}$ is at the cosmological scale. A vanishing cosmological constant, as suggested by quantum gravity arguments,[4,5] corresponds to $\mathbb{L}$ infinite. But if the cosmological constant is finite, this means that there exists an *invariant* length of the order of the radius of the universe, while the universe and any length at the cosmological scale is subject to expansion (compare this paradox to the Planck scale paradox which we have presented in Sec. 6.7). It has already been remarked that there is a relation between scale invariance (and its breaking) and the value of the cosmological constant.[6]

*What determines the characteristic scales of galaxies, clusters and superclusters of galaxies, and large scale structures in the Universe such as the recently discovered Great Wall[7] and the possible 128 h^{-1} Mpc periodicity[8] ? (here h stands for the ratio of the Hubble constant over 100 km/s.Mpc and 1 Mpc= 3.08 x 10^{24} cm). The current (very active) attempts at understanding such questions are once again "reductive": the hope is to get the present structures from a theory of formation and evolution of structures which makes them start from the Big Bang (i.e. at small scale and remote time) and evolve to the present densities and radii.

*Scaling laws, apparently with a high level of universality, have been discovered in the hierarchical distribution of matter in the Universe. In particular the two-point correlation functions for galaxies, groups, clusters and superclusters of galaxies are all characterized by a power law $\xi(r) \propto (r_0/r)^{\gamma}$ with $\gamma \approx 1.8$ for every type of objects. Note that one of the most popular models for the distribution of galaxies is precisely the fractal and multifractal ones,[9-11] and this leads us back to the main theme of this book.

Among the reasons for applying scale relativity to cosmology, there is also the straigthforward argument that high energy particle physics describes the first instants of the Universe, and that the changes we have brought to this domain are expected to change the Big Bang theory. We shall briefly consider this case, but we shall see that scale relativity is also able to bring new insights into the domain of *observational cosmology*. The theory of scale relativity appears as a key, which, constructed to open the door of microphysics, proves capable also of opening another door, that of cosmology.

However we warn the reader that we shall only summarize the main lines of our arguments and results: this theme would need another full book, while the present essay is mainly devoted to the microphysical problem. In particular we shall assume that the reader is aware of the basics of theoretical and observational relativistic cosmology, as described, e.g., in the books of Weinberg[1] and Peebles.[9]

Moreover, this section should be considered as a preliminary model rather than a full theory at the present time. Indeed today's cosmology is described by an extremely coherent theory, Einstein's general (motion) relativity, while we are far from a self-consistent scale relativistic cosmology. We nevertheless hope that the construction we propose hereafter will be correct in its main lines thanks to the fact that scale relativity, as any theory of relativity, yields *universal* constraints (existence of a limiting velocity in motion relativity, of a limiting space-time scale in scale relativity), the consequences of which must apply to every domain of physics.

Reactualization of Mach's principle.

The principle known as Mach's principle, since Einstein's insistance on its importance for the understanding of inertia, actually contains several statements corresponding to different levels of relations between local properties (inertia) and global properties (the Universe).

The first level is the *definition of inertial systems*. Mach's main contribution was his insistence on *the relativity of any motion*. As a consequence, the motion of reference systems in which inertial forces are experienced (e.g., a mass in rotation, more generally accelerated systems) can be defined only *relatively to other masses*. General relativity and the principle of equivalence practically solve the problem: inertial systems are systems which are in free fall in the gravitational field determined, through Einstein's equations, by *the whole distribution of masses in the Universe*. Inertial systems are so defined only locally, because of the locality of the principle of equivalence. In general relativity, the definition of a global inertial system no longer has any physical meaning. There are two drawbacks in this solution.

The first is spin: a particle of vanishing radius may rotate around any axis passing through it, while it will always be considered in free fall according to general relativity.

The second is the apparent contradiction of the purely local nature of the solution brought by general relativity with what is actually observed. Observations seem to point towards the need for a global definition of inertial systems: this is the basis of Mach's proposed solution, that inertial forces must result from gravitational attraction of "distant stars" (Mach's principle), i.e., of distant matter in the Universe. Indeed a kind of "coherence" of inertial systems is observed on very different scales: the system in which we feel no centrifugal force on earth is nearly the same as that in which the sky is seen not to rotate;[1] the axis of the earth, up to precession, is always directed towards the same direction in spite of its motion around the Sun and of the displacement of the Sun in the galaxy.

This problem is solved by the *observed* hierarchical distribution of matter in the universe, but *not solved in principle*. In fact, as recalled in a previous section, inertia is experimentally found to be isotropic to a very high precision[1] ($\delta m/m = 0 \pm 10^{-20}$, but more recent experiments may reach 10^{-24}). Hence the contribution to inertial forces of masses and scales as large as our Galaxy or even the local supercluster must be dominated by contributions of far more distant masses. If Mach's principle is to be implemented in accordance with the principle of equivalence and of its experimental verification, the only acceptable solution is an effect of the Universe as a whole.

This leads us to the *second level* of "Mach's principle". It was Einstein's initial hope that, if distant masses do determine the inertial systems, they must also determine the amplitude of inertial forces: more precisely, inertial forces being hopefully reduced to gravitational forces due to the Universe as a whole, the ratio of gravitational over inertial acceleration, i.e., the constant of gravitation G may be related to global parameters of the universe. It is well known that this hope was dashed: already in Schwarzschild's solution, there is a central gravitational (therefore inertial) mass in the absence of masses at infinity. In cosmology also, general relativity does not solve the problem in principle, since not all Friedmann-Lemaître (or more generally Robertson-Walker) models of the universe are "Machian". Some models are devoid of mass; some others, even massive, do not verify the relation between some characteristic mass and length of the Universe which is needed to implement Mach's principle. Let us look at the form expected for such a relation.

Mach's principle may be achieved by requiring that the gravitational energy of interaction of a body with the universe (described to first order approximation as a total mass M situated at an average distance R) cancels its self-energy of inertial origin, $E = m\,c^2$:

$$\frac{GmM}{R} \approx m\,c^2 \Rightarrow \frac{G\,M}{c^2 R} \approx 1\,. \tag{7.1.1}$$

The relation obtained is, except for a factor 2, the relation between a mass and its Schwarzschild (blackhole) radius. Hence Mach's (second level) principle is equivalent to the requirement that the Universe as a whole be a black hole.

One of the most detailed Machian model of the Universe was proposed by Sciama.[12,13] He adds to Einsteinian cosmology the requirement that "the gravitational field of the Universe as a whole cancels the gravitational field of local matter, so that bodies are free" and obtains Eq. (7.1.1). In his approach, it becomes very clear that Mach's principle could not be achieved in a scalar theory of gravitation like Newton's theory (indeed the force of the "left" part of the Universe cancels that of the "right" part at any point). Inertial forces may arise as an effect of the Universe only in a vectorial or tensorial theory: as shown by Sciama, inertia is an effect of *induction* of distant matter (in a sense similar to inductive force or current in electromagnetism). The inertial force which appears when we move in a non-inertial frame comes from the gravitational force which arises from the *acceleration* of the whole universe with respect to us.

Let us see what Eq. (7.1.1) does imply from the point of view of cosmological models. Only two solutions are possible: either M and R are constant, and one gets Einstein's model, the only possible static model among Robertson-Walkers' models (with $\Lambda \neq 0$), or the universe is non-static, as indicated by observations, so that R varies with time and Mach's principle can be achieved only in models where the characteristic mass M varies with time as the cosmological scale factor varies. This immediately excludes (in today's standard cosmology) spherical models, which are closed and in which the total mass of the Universe is constant.

In order to characterize 'Machian' models, let us define a characteristic mass $M = 4\pi\rho r^3/3$ in terms of the length $r = c/H$; this combined with Eq. (7.1.1) written as a Schwarzschild relation $2GM/c^2r = 1$ yields a density parameter:

$$\Omega = \frac{8\pi G\rho}{3H^2} = 1 \ . \qquad (7.1.2)$$

Hence the Einstein-de Sitter model (with $k = 0$) is Machian. It is the only one in which Ω does not vary with time. This may also be seen in the expression for the mass which is observed inside the horizon ($z \rightarrow \infty$) of such a model

$$M_{\mathrm{H}} = \frac{4\, c^3}{G\, H_0} \; .$$

Equation (7.1.2) explains why the problem set by Mach's principle is still with us. One might have been contented with general relativity and considered that the implementation of Mach's principle (second level) is not necessary. But *observations* tell us that Eq. (7.1.2) is true or nearly true: the measured values of Ω fall between 0.2 and 1, the value $\Omega = 1$ being preferred by its last large scale ($\gtrsim$ 100 Mpc) determination using IRAS galaxies and large scale motions.[14] So it seems legitimate to wonder why the observed universe is so close to achieving (or actually achieves) Mach's "second level" principle. Such an agreement reactualizes the impression that implementation of Mach's principle through *exact* equations in agreement with the principle of equivalence is needed indeed.

The third level of "Mach's principle" (the idea of which may be attributed to Einstein) is the conjecture that the mass of elementary particles is related to the whole mass of the Universe. Let us make more specific the meaning of this proposal. The units of mass are arbitrary. In the same way as Mach insisted on the relativity of all motions, he insisted also on the relativity of all masses: for him, a mass cannot be defined alone. One must consider two masses, which are then nothing but the inverse acceleration which they transmit to each other: $m_1\, \gamma_1 = m_2\, \gamma_2 \Rightarrow m_1/m_2 = \gamma_2/\gamma_1$. However special relativity obliges one to make such a viewpoint evolve. Mass is also energy, which may itself take a lot of forms (heat, radiation, kinetic energy...). The evolution must be still more radical when accounting for quantum mechanics. From the constants G, $\hbar$ and c, one may introduce the Planck mass m_{P} as a natural unit and write Newton's law (according to the relation $G m_{\mathrm{P}}^2 = \hbar\, c$) as

$$F = \hbar\, c\; \frac{(m/m_{\mathrm{P}})\; (m'/m_{\mathrm{P}})}{r^2} \; . \qquad (7.1.3)$$

Three situations may have occurred: (i) that no preferential scale of mass exists in Nature: the "third level Mach's principle" would have no meaning; (ii) that a preferential scale exists ("elementary particle"), but that this characteristic mass must precisely be the Planck mass: Eq. (7.1.3) would have accounted for such a situation; (iii) that preferential, universal and elementary masses exist in Nature, with a scale totally different from m_P: this is the case "chosen" by Nature, since $m_P/m_e = 2.38952(15) \times 10^{22}$. The origin of this ratio is one of the great mysteries of physics; its huge size suggests comparing it with the only *universal* mass ratio of an equivalent size, the ratio of the mass of the universe over the Planck mass, $M/m_P \approx 10^{61}$.

Let us show how scale relativity allows one to set these problems in a completely renewed way.

Scale relativity and primeval Universe.

It is clear that the new structure of space-time implied by our reinterpretation of the physical meaning of the Planck scale radically changes our view of the primeval Universe. The first new physical law of cosmological importance is the disappearance of the zero instant from meaningful physical concepts. The evolution of the Universe does not begin any more at the instant "$t = 0$" (i.e. $log(t/t_0) = -\infty$), but at the Planck time "$t = \Lambda/c$", i.e., $log(t/t_0) = log(\Lambda/ct_0)$. However this new structure should not be misinterpreted: in the new theory, the Planck scale owns all the properties of the previous zero instant. This means that temperature, redshift, energy, density and all the quantities Q which were previously diverging as t^{-k} are now diverging when t tends to Λ/c as

$$log\,\frac{Q}{Q_0} = \frac{k\ log\,(t_0/t)}{\sqrt{1 - \dfrac{log^2(t_0/t)}{log^2(ct_0/\Lambda)}}} \quad .$$

The scale factor of expansion of the Universe is also submitted to a new constraint: it can no longer become smaller than the Planck length Λ. This would be achieved if it initially evolves as $R = \Lambda^{1/2}(ct)^{1/2}$.

In the scale relativistic approach, Lorentz-like δ-factors are introduced, which are identified with variable anomalous dimensions. If one refers oneself to the present epoch $t_0 \approx 5\ 10^{17}$ s, one gets $\mathbb{C}_0 = log(ct_0/\mathbb{A}) \approx 61$. Then $z \approx 5$, the redshift of the most distant presently observed objects corresponds to $V = log(t_0/t) \approx 1$, so that $V/\mathbb{C}_0 \approx 1/60$: this corresponds to a negligible correction $\delta \approx 1+10^{-4}$. The redshift of the isotropic Microwave Background Radiation, $z \approx 1000$, corresponds to $log(t_0/t) \approx 12$, i.e. to $V/\mathbb{C}_0 \approx 1/5$ and $\delta \approx 1+1/50$. This is an interesting result that the "scale relativistic domain" (i.e. here meaning the domain where the consequences of the existence of a *lower limit* to all scales are not negligible) actually begins at about $z \approx 1000$, and then nearly coincides with the radiation dominated era of the Universe in standard cosmology.

The main result of scale relativity concerning the primeval Universe is its ability to solve the causality/horizon problem. Let us recall the nature of this problem. When looking at two directions separated by a large angle, e.g. two opposite directions, we observe regions of the Universe which, for a large enough redshift, may have never been connected in the past. The problem is particularly strong concerning the microwave background radiation, due to its high isotropy[16] ($\delta T/T \lesssim 2\ 10^{-5}$) and its early origin ($z \approx 1000$): at least twenty such independent regions would be observed in the framework of standard cosmology.

Such causally disconnected regions should behave as completely independent universes, and it becomes very strange that no large fluctuation of the microwave background temperature is observed. The solution to this problem is usually searched for in the framework of inflationary cosmology.[17-19] However one may remark that inflation is to some extent an *ad hoc* solution, in particular as concerns its cause (scalar field now unobservable, primordial black holes...), that must be postulated additionally to the presently known content of the Universe. Moreover it does not solve the problem *in principle*: in its framework the *presently observed* regions of the universe would have been causally connected in the past, but this does not remain true in the distant future.

Scale relativity naturally solves the problem because of the new behaviour it implies for light cones. Though there is no inflation in the usual sense, since the scale factor time dependence is unchanged with respect to standard cosmology, there is an inflation of the light cone as $t \rightarrow \mathbb{A}/c$.

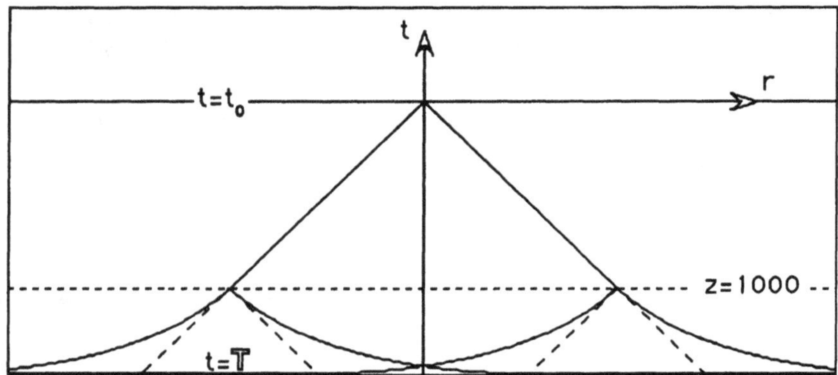

Figure 7.1. Schematic representation of the scale-relativistic flare of light cones in the primeval Universe. Two distant regions of the universe seen in opposite directions are causally disconnected in standard cosmology without inflation, since their past light cones (dotted lines) do not cross. In scale relativity, all points of the Universe become causally connected at the Planck time $\mathbb{T}$.

This may be understood by an analysis of the new properties of space-time at scale $\mathbb{\Lambda}$ implied by Eq. (6.8.1) . The fact that $\mathbb{\Lambda}$ is invariant under dilatations means that when observed at resolution $\mathbb{\Lambda}$, the distance between any two points reduces to $\mathbb{\Lambda}$ itself. Indeed, following our analysis of Chapter 2, the numerical result of a distance measurement is given by a dilatation ρ applied to the basic unit that cannot be taken smaller than the resolution. This means that there is a complete degeneration of space-time when looked at resolution $\mathbb{\Lambda}$ (compare with the partial degeneration of null geodesics on light cones). This property was already possessed by the previous primeval singularity (R=0) but this singularity was actually excluded from the evolution of the Universe in the previous theory: for example an open infinite Universe would have been infinite at any arbitrarily small time $t \neq 0$, while reduced to the singularity at $t = 0$. Here we have a continuous evolution from the particular scale $\mathbb{\Lambda}$ to larger ones. The new light cone evolution is illustrated in Fig. 7.1, where it may be seen how the various light cones flare when $t \rightarrow \mathbb{\Lambda}/c$ and cross themselves, allowing causal *connection between any two points of the Universe*. This definitely solves the causality problem.

Scale dependence in present cosmology.

As in microphysics, there is in cosmology a fundamental dependence of physical laws on scale. At scales for which the cosmological principle of homogeneity and isotropy is fulfilled, the Universe is found to be in expansion. Indeed all solutions (except one) of Einstein's equations based on the cosmological principle are non-static. The observation of the universal redshift of galaxies and the redshift-distance relation (Hubble law) indicates that this non-staticity is presently an expansion. This means that, at large scales, all physical variables (distances, time, density, temperature...) vary in terms of a universal scale factor R, which characterizes the Robertson-Walker metric:

$$ds^2 = c^2\, dt^2 - R^2(t)\ dl^2 \ ,$$

where we recall that dl^2 is a spatial element which may take only three forms corresponding to constant curvature spaces (hyperbolic, flat or spherical). For example, one has $T \propto R^{-1}$, $\rho \propto R^{-3}$ in dust universes, $t \propto R^{3/2}$ in flat models, etc... (see e.g. Ref. 1). Fortunately, this scale factor (the so-called "Universe radius") is directly observable, since it is related to redshift z by the relation

$$\frac{R}{R_o} = (1+z)^{-1} \quad ,$$

where R_o is its value at the present instant. Thus this scale dependence is both observed and predicted by theory (general relativity).

The distribution of matter in the Universe is also found to be scale dependent. There is at present no satisfactory theoretical explanation of this "scaling". It is often described in terms of a two-point correlation function,[9,20] $\xi(r)$, which measures the deviation of the observed distribution of galaxies with respect to an uniform (Poissonian) one. Namely, one writes that the probability that one object lies between r and $r+dr$ from another object is given by

$$P(r)\, dr = 4\pi\, r^2\ [1 + \xi(r)]\ dr \ .$$

It is observationally found that, for most classes of objects of cosmological importance (galaxies, groups and clusters of galaxies, superclusters of galaxies), $\xi(r)$ is well represented by a power law:

$$\xi(r) = \left(\frac{r_0}{r}\right)^{\gamma} ,$$

where the correlation length r_0 depends on the type of object considered (about 5 Mpc for giant galaxies, 20 Mpc for clusters), but where $\gamma \approx 1.8$ *whatever the type of object.*

A popular approach to the question of the distribution of galaxies is a fractal and multifractal one.[9-11] The correlation function is related to another measure of correlation[21], the *correlation integral* $C(r)$. It measures the probability of finding another point in a sphere of radius r centred on a point of the distribution, so that

$$\frac{dC(r)}{dr} = 4\pi\, r^2\ [1 + \xi(r)] \ .$$

Then the hereabove form observed for $\xi(r)$ means that, for small r, $C(r)$ varies with scale as

$$C(r) \propto \left(\frac{r}{r_0}\right)^{D} \qquad \text{with} \qquad D = 3 - \gamma \ .$$

D is the *correlation dimension* and is equal to a fractal dimension in the simplest case. Such a form may be obtained from a renormalization group equation:

$$\frac{dC(r)}{d\ln r} = a + D\ C(r) \ ,$$

where the correlation and fractal dimension D is now interpreted as an anomalous dimension (here $D = \delta$ since this is the dimension of a set of points, i.e., a "dust" of topological dimension $D_T = 0$).

So the value $\gamma = 1.8$ is translated to the fractal and anomalous dimension $D = 1.2$ for the distribution of galaxies. Several models of the formation of a hierarchical distribution of matter, e.g., by fragmentation, naturally yield[10,11] $D = 1$. The unsolved question is why $D = 1.2$, rather than $D = 1$.

However, even the simple fractal model is problematic: it predicts that $1 + \xi(r)$ is a power law, rather than the $\xi(r)$ observed, and it has a constant fractal dimension, while at very large scales one expects to find $D = 3$ (uniformity).

The value $D = 1$ is also encountered for the local distribution of matter observed around the various objects. Hence the observation of flat rotation curves in the outer parts of spiral galaxies[22] leads to the conclusion that they are embedded in supermassive halos of dark matter having a density $\rho \propto r^{-2}$, i.e. a mass distribution $M(r) \propto r^D$ with $D = 1$. In the same way, the observed halos of clusters of galaxies show a similar distribution $M(r) \propto r$ in the mean.[23]

The static non-static relative transition.

There is a fundamental question concerning the expansion of the Universe which is seldom explicitly asked: *where does the expansion stop*? It is clear from several arguments that a static non-static transition must exist.

First, if the cosmological scale factor R was to be applied to any length in Nature, then it would become a trivial scale factor that would disappear from the equations. It was already remarked by Laplace that Newton's theory of gravitation is scale invariant: "One of the remarkable properties [of Newtonian attraction] is that if the dimensions of all the bodies in the universe, their mutual distances, and their velocities were to increase or diminish proportionately, they would describe curves entirely similar to those which they at present describe; so that the universe reduced to the smallest imaginable space would always present the same appearance to observers. The laws of nature therefore only permit us to observe relative dimension".[24,10] If Laplace had added the fact that the size of objects is determined either by fields different from the gravitational one, or by the local gravitational field rather than the global, he would have predicted a relation of proportionality between distance and velocity, i.e. the Hubble law.

The expansion of the Universe can indeed be interpreted as a variation with time of the cosmological units relatively to local (atomic) units. Note also that the Friedmann-Lemaître and Robertson-Walker solutions to Einstein's equation are based on a description of the material content of the Universe as a perfect fluid. When applied to the present Universe where the basic constituents are galaxies (provided that some smoothly distributed dark matter should not be the dominant component), this means that galaxies are identified with the basic particles of a gas, so that, as in thermodynamics, the theory is not expected to apply at scales of the order of the particle size.

It is indeed known that a typical giant galaxy like ours (of radius $\approx$10 kpc), even if it is entailed in differential rotation, is globally static. The velocity field of clusters of galaxies shows an external halo which links up to expansion, while there is an inner static region of size of about its core radius (100-200 kpc). This indicates that transition from staticity to non-staticity, i.e. from scale independence to scale dependence, is a *relative* transition.

The staticity of objects under their own gravitational field amounts to writing their equilibrium, i.e., to writing the virial theorem. This leads to a well-known general relation between mass, velocity dispersion and radius (see Ref. 1, p. 477):

$$\ell \approx \frac{G\,m}{\langle v^2 \rangle} \quad . \tag{7.1.4}$$

We suggest that this relation plays the same role in cosmology as the de Broglie length $\lambda = \hbar/mv$ in microphysics. Note indeed that both relations give a length in terms of mass, velocity and a fundamental constant, $\hbar$ at small scale and G at large scale. Also remarkable is the fact that, if one looks for a situation where they would be equal, one gets $m = (\hbar v/G)^{1/2} = m_{\mathrm{P}}\sqrt{v/c}$, which is nothing but the Planck mass m_{P} when $v = c$. So for possible cosmological constituents of mass smaller than the Planck mass (in particular elementary particles), the two scale dependent microphysical and cosmological domains connect, without being separated by a classical scale independent domain (see Fig. 7.2).

The comparison with the fundamental transition lengths of microphysics goes on with the Compton length $\lambda = \hbar/mc$. Making $v = c$ in (7.1.4) yields, up to a factor of 2, another fundamental length of general relativity, namely the Schwarzschild radius corresponding to mass m. (For cosmological constituents as small as elementary particles, for example the isotropic microwave background, the hereabove formula may not apply: anyway in this case, one expects the transition length to be at the microphysical scale).

Up to now it has been assumed that the size of objects were of no direct cosmological importance. We propose rather that the largest static sizes of objects are an essential element for understanding cosmology, since they define a "phase" transition from staticity to non-staticity, or, in other words, a scale of symmetry breaking for scale covariance. We think that it is not by chance that the supermassive dark matter halos or galaxy clusters halos both correspond to fractal dimension D=1 (i.e., as the topological dimension is zero, to anomalous dimension $\delta = 1$) and to the transition region from scale independence to scale dependence. It is remarkable in this respect that the two-point correlation function, when calculated at "small" scale (10 kpc-100 kpc) for, e.g., dwarf galaxies close to giant ones yields $\gamma \approx 2$ rather than 1.8.[20,25]

Let us apply a scale transformation $\rho = r'/r$ to the correlation integral and write it in logarithm form:

$$log\frac{C'}{C_0} = log\frac{C}{C_0} + \delta\ log\,\rho\ , \qquad (7.1.5)$$

$$\delta' = \delta = 1 \qquad (r > \zeta,\ r' > \zeta)\ .$$

It is certainly clear to the reader that we are now once more in exactly the same situation as in microphysics (but with an inversion between the smallest scales and largest scales). The hereabove scale transformation is a Galilean group transformation which holds for scales *larger* than the "virial length" ζ while the anomalous dimension δ jumps from $\delta = 1$ to $\delta = 0$ below this length, which plays the role of a static / non-static transition.

The nature of cosmological constant.

As in microphysics, we are tempted to conclude that the right structure imposed by the principle of scale relativity is the Lorentz group rather than the Galileo group. Then the whole mathematical development of Chapter 6 is applicable to the cosmological problem, and we arrive at the conclusion that there should exist in Nature an *upper* scale, unpassable, universal, invariant under dilatations (thus in particular invariant under the expansion of the Universe), which would hold all the previous properties of *infinity*. Let us name $\mathbb{L}$ this new length scale.

As in microphysics, we are led to asking ourselves whether a length which already exists in present physics could be identified with this new structure. Remark that the microphysical solution, the Planck length $\mathbb{\Lambda} = (\hbar G/c^3)^{1/2}$, is the only solution that can be constructed with the three basic fundamental constants. Then we can already say that $\mathbb{L}$ should be the product of $\mathbb{\Lambda}$ by a constant, absolute, and pure number $\mathbb{K}$:

$$\frac{\mathbb{L}}{\mathbb{\Lambda}} = \mathbb{K} \quad .$$

The present theory has already introduced an universal, absolute, and unvarying constant with is defined at the cosmological scale: this is the cosmological constant Λ, which is defined as the inverse of the square of a length. As recalled above, this length must be both at the scale of the Universe ($>10^{28}$ cm) and not subjected to its expansion, for general relativity to remain self-consistent. Recall also that the general equations satisfying general covariance are Einstein's equation including a cosmological constant term.

So we propose to reinterpret the cosmological constant as resulting from the existence of the new upper scale $\mathbb{L}$:

$$\mathbb{L} = \frac{1}{\sqrt{\Lambda}} .$$

With this interpretation for $\mathbb{L}$, we get a first estimate for the pure number $\mathbb{K} \approx 10^{61}$: we shall attempt at estimating more precisely its value afterwards. Note in this respect that this interpretation is consistent with the analysis of Ref. 6 and the recent results by Hawking[4] and Coleman[5], who obtained a vanishing cosmological constant from quantum gravity arguments. Indeed the Hawking-Coleman approach remains in the frame of the Galileo group of dilatations, while in our frame the Galileo group corresponds in cosmology to the limit $\mathbb{L} \to \infty$, i.e. $\Lambda \to 0$.

Let us briefly consider the possible implications of this proposal for our understanding of large scale structures and of Mach's principle. A more extensive account is outside the scope of the present book and will be given elsewhere.

**Vacuum energy density.*

There have been attempts to reinterpret the cosmological constant as vacuum energy density, $\rho_V \approx \Lambda c^2/G$ (see Ref. 3 and references therein), i.e., with our notations,

$$\rho_V = c^2/G\mathbb{L}^2 \ .$$

The problem encountered with this interpretation is that a calculation of the vacuum energy density in standard quantum theory gives a divergent result. Assuming a cutoff at the Planck scale, we get the "Planck energy density"

$$\rho_{\mathbb{P}} = \frac{c^5}{\hbar G^2} \ .$$

With the current lower limit on the possible values of the cosmological constant, $\Lambda < 3\ 10^{-56}\ \text{cm}^{-2}$, the two estimations differ[3] by a ratio of $\approx 10^{120}$.

We suggest the following solution to this problem. We assume that the vacuum energy density is *an explicitly scale dependent* quantity for every possible scales in the Universe. The vacuum is clearly one of the cosmological constituents for which scale covariance is unbroken (there is no classical domain). Then it is solution of a renormalization group equation

$$\frac{d\rho_V}{d\ln r} = k\,\rho_V \ ,$$

so that it varies between two scales r_1 and r_2 as $(r_1/r_2)^k$. We may now admit that both the above values of the vacuum energy density are correct, one defined at the Planck scale and the other at the cosmological scale. Their ratio is

$$\frac{\rho_{\mathbb{P}}}{\rho_V} = \left(\frac{\mathbb{L}}{\mathbb{\Lambda}}\right)^2 = \mathbb{K}^2 ,$$

and we get a self-consistent scheme by taking $k = -2$. This approach is not fully scale-relativistic. Including the scale Lorentz-factors (Chapter 6) yields a vacuum energy density which becomes infinite at $r = \mathbb{\Lambda}$ and null at $r = \mathbb{L}$. In this case the hereabove values correspond to two fundamental length scales, one (microphysical) we have identified with the Grand Unification scale, and a new one (cosmological) we shall come back to hereafter.

******The Universe at its own resolution.*

Before going on, the nature of $\mathbb{L}$ must be specified. Indeed we cannot be satisfied with the definition of $\mathbb{L}$ as a "length", since the concept of distance in general relativity, and in particular cosmology, depends on the method of measurement. Recall that one defines luminosity distance, angular-diameter distance, proper-motion distance, etc.., which all have different expressions in terms of redshift (see e.g. Ref. 1). So the new upper scale must be characterized, not only by a simple number, but also by a global description of the Universe when seen at that scale. The length $\mathbb{L}$ is actually defined as the "radius" of a Universe which, *when seen at its own resolution, becomes invariant under dilatations*. Only one cosmological solution of Einstein's equations is unaffected by expansion: the Einstein static spherical model. So we suggest that, at resolution $\mathbb{L}$, the Universe is described by the *Einstein spherical model*. But the interpretation is different from that of standard cosmology. Recall that in scale relativity we have changed the law of composition of dilatations, so that the structure of the Universe at the upper scale $\mathbb{L}$ does not impose anything on its structure at any other smaller scale. At resolution $\mathbb{L}$, there is a degeneration of space-time (as at resolution $\mathbb{\Lambda}$ and at velocity c). The whole set of various possible models (hyperbolic, flat, spherical) are retrieved at smaller scales, owing to the fact that their properties are defined in a purely local way. We claim that, while an integration of these local properties is approximately correct on 'small' scales, this may no longer be the case when pushing the integration to scales of the size of the Universe itself.

**Mach's principle and great numbers.*

The existence of the universal scale $\mathbb{L}$ allows one to consider Mach's principle in a new way. The main difficulty encountered in previous attempts of implementation of the various levels of Mach's principle was that a *time varying* scale, c/H_0, was used as the fundamental cosmological scale in the equations. We now have a horizon for the Universe which, although it owns all the properties of infinity, is given by a finite, constant and universal measure.

We have seen hereabove that the 'second level' of Mach's principle may be translated by the requirement that the Universe be a black hole. Applying this requirement to the Universe at resolution $\mathbb{L}$, described by Einstein's model, we find that the maximal separation between any two points is $\pi\mathbb{L}$. Then we expect that the Universe be characterized by an effective mass $\mathbb{M}$ such that

$$\frac{2}{\pi}\frac{G}{c^2}\frac{\mathbb{M}}{\mathbb{L}} = 1 \ .$$

This result is self-consistent, since this is exactly the expression for the total mass of Einstein's model. It yields an interpretation for one of the great number coincidences

$$\frac{\mathbb{M}}{m_{\mathrm{P}}} = \frac{\pi}{2}\,\mathbb{K} \ .$$

The value $\mathbb{K} \approx 10^{61}$ would yield a characteristic mass $\approx 10^{56}$ g $\approx 10^{23}$ solar masses, in agreement with observations ($\approx 10^{11}$ galaxies of 10^{12} solar masses).

The 'third level' of Mach's principle has still more radical implications. Its achievement would imply a connection between the mass of elementary particles and the mass of the Universe. Let us suggest a (still very rough) solution to this problem. We start from the fact that the electron is the lightest elementary charged particle. The appearance of the upper scale $\mathbb{L}$ implies the existence of a characteristic minimal energy $E_{\min} = \hbar c/\mathbb{L}$. Now, (i) assume that the mass of the electron is of pure electromagnetic origin. This defines a scale r_0 such that

$$\frac{e^2}{r_0} = m_e c^2 \quad .$$

That is, $r_0 = \alpha\, \lambda_c$ is Lorentz's classical radius of the electron. Then, (ii) assume that the gravitational self-energy of the electron at scale r_0 precisely equals the smallest possible energy $\hbar c/\mathbb{L}$. We obtain the equation

$$\frac{G\, m^2(r_0)}{r_0} = \frac{\hbar c}{\mathbb{L}} \quad ,$$

where $m(r_0)$ is the effective mass at scale r_0, i.e., $\alpha^{-1} m_e$, except for the small scale dependence of α (<1%). This reasoning finally yields

$$\alpha\, \frac{m_\mathbb{P}}{m_e} = \mathbb{K}^{1/3} \quad . \tag{7.1.6}$$

This is a possible road toward an explanation for Dirac's great number coincidence. The fact that $\mathbb{L}$, instead of c/H, appears there is the key point, since it allows us to have an absolute relation rather than a time-dependent one. The agreement is indeed remarkable: with $\mathbb{K} \approx 10^{61}$, one gets $\mathbb{K}^{1/3} \approx 2 \times 10^{20}$, while $\alpha\, m_\mathbb{P}/m_e = 1.7437(1) \times 10^{20}$. Conversely, if we admit this formula to be correct, we find a precise estimate for the fundamental scale factor $\mathbb{K}$:

$$\mathbb{K} = 5.3018(10) \times 10^{60} \quad . \tag{7.1.7}$$

By using, in (7.1.6), the value of the inverse fine structure constant at scale r_0 , $\alpha^{-1}(r_0) = 136.3$ (see Chapter 6), instead of its low energy value, one finds $\mathbb{K} = 5.388 \times 10^{60}$. If our scheme is globally correct, one important problem for physics will be the origin of this pure number. We shall not answer this question here, but only indicate a possible road towards its solution. Equation (7.1.6) may be written in terms of the scale relativistic constants $\mathbb{C}$. We have a universal constant $\mathbb{C}_U = \ln\mathbb{K}$, which is related to the electron constant $\mathbb{C}_e = ln(m_{\mathbb{P}}/m_e)$ and the constant $\mathbb{C}_0$ at scale r_0 by the relations

$$\mathbb{C}_U = 3\,(\mathbb{C}_e + ln\alpha) = 3\,\mathbb{C}_0\ .$$

If we admit the above estimates for $\mathbb{K}$, we get $\mathbb{C}_U = 139.83 \pm 0.01$, which is 2% off the low energy fine structure constant. This opens the hope that, in a way similar to our conjecture for the determination of the electroweak scale from the 'bare' charge, $\mathbb{C}_v = \alpha_1^{-1}(\mathbb{A}) = 4\pi^2$, the universal constant $\mathbb{C}_U$ can be ultimately determined by the value of the electric charge at scale $\mathbb{L}$.

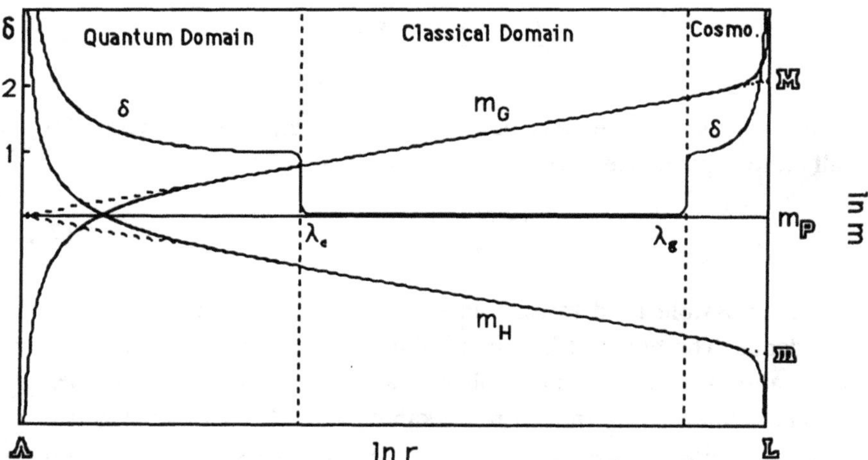

Figure 7.2. The three, quantum, classical and cosmological, domains, from the smallest scale $\mathbb{A}$ to the largest scale in Nature, $\mathbb{L}$ (according to scale relativity). The two transitions between these domains are not absolute, but *relative* to the system considered (its mass and velocity or velocity dispersion). We have plotted in this diagram the variation of the anomalous dimension δ: in the classical domain, its null value corresponds to scale independence. Also shown is the variation in terms of length scale of the fundamental mass scales given by the generalized Schwarzschild (m_G) and Compton (m_H) formulae. The Planck mass $m_{\mathbb{P}}$ plays the role of a zero point for mass scales.

From the above estimate, one can deduce the maximal length in the Universe, $\pi\mathbb{L} = 8.97$ Gpc (i.e., 29.2 x 10^9 light years), and the value of the cosmological constant:

$$\Lambda = \frac{1}{\mathbb{L}^2} = 1.36\ 10^{-56}\ \mathrm{cm}^{-2}\ .$$

This corresponds to the reduced cosmological constant $\lambda_o = \Lambda c^2/3H_o{}^2 = 0.36\,h^{-2}$.

**Slope of correlation function.*

Let us come back to the question of the distribution of structures in the Universe. In scale relativity the fractal/anomalous dimension now varies with scale. What is the upper distance to be used for the scale δ-factors for this case? We deal with volumic effects, and the volume of the Einstein universe is $2\pi^2\mathbb{L}^3$, instead of the Euclidean $\frac{4}{3}\pi\ (\pi\mathbb{L})^3$. So the limiting volume-distance is expected to be $\mathbb{L}_V = (3/2\pi^2)^{1/3}\mathbb{L} \approx 0.534\ \mathbb{L}$. Choosing $C = C_o$ for the correlation integral, we get a new relation (see Fig. 7.2):

$$\delta(r) = \frac{1}{\sqrt{1 - \dfrac{log^2(r\,/\,\ell)}{log^2(\mathbb{L}_V/\,\ell)}}},$$

where ℓ is the static-expansion transition, and r the distance between two galaxies (that indeed plays the role of a resolution). Then the exponent of the two-point correlation function is given by $\gamma = 3-\delta$.

Consider galaxies. Their typical radius is $\ell = 10$ kpc. Combined with the above determination of $\mathbb{L}_V$, this yields $\mathbb{C} = 5.18$ (in logarithm base 10). Then, while $\gamma = 2$ at a scale of 10 kpc, we predict $\gamma = 1.8$ at a scale of 10 Mpc, in good agreement with observations.[20] It is expected to subsequently fall farther ($\gamma = 1.65$ at 30 Mpc, $\gamma = 1.43$ at 100 Mpc).

Consider clusters of galaxies. Their core radius is $\ell \approx 100$ kpc. This yields $\mathbb{C} = 4.18$ (in logarithm base 10). We predict $\gamma = 1.86$ at 10 Mpc, 1.75 at 30 Mpc and 1.56 at 100 Mpc, also in good agreement with what are observed.[26] In such an interpretation, the apparently universal value $\gamma \approx 1.8$ would come from the fact that the distances at which the correlation function is well measured is not an absolute scale but depends on the objects themselves (radius and mean interdistance) in such a way that it roughly corresponds to a given relative scale ($V/\mathbb{C} \approx 0.55$).

**New fundamental scales.*

One of the problems with the simple fractal model is its inability to reconcile the locally fractal distribution with a globally uniform one. In the scale-relativistic approach, δ and γ vary with scale, so that uniformity is reached for $\delta = 3$, i.e., $\mathbb{V}/\mathbb{C}=2\sqrt{2}/3$. This corresponds to a *scale of transition to uniformity* of about 750 Mpc.

In microphysics, we have seen that the separation between the scale of length-time and the scale of energy-momentum led to the emergence of a new scale which we identified as the Grand Unification scale. The same is true in cosmology: we expect the largest structured scale ℓ to be given by $log(\ell / \ell) = \mathbb{C}/\sqrt{2}$. Being concerned with linear structures, one must take $\mathbb{L}_l = \pi\mathbb{L}$ in the computation of $\mathbb{C}$. For giant galaxies, assuming $\ell = 10 \pm 2$ kpc, this yields $\ell = 160 \pm 8$ Mpc. This result is remarkable, owing to the recent discovery of a periodicity at 128 h^{-1} Mpc in deep redshift narrow-cone surveys.[8] Identifying the observed and predicted wavelengths provides us with a new precise determination of the Hubble constant:

$$H_0 = 80 \pm 4 \text{ km/s.Mpc} ,$$

in excellent agreement with recent determinations[27,28] from precise indicators, $H_0 = 82 \pm 7$ km/s.Mpc and 72 ± 5 km/s.Mpc. Combined with our estimate of Λ, this would yield a reduced cosmological constant $\lambda_0 = 0.56$: such a high value may help solve the problem of the age of the Universe.

For clusters of galaxies, $\ell \approx 100$ kpc yields $\ell = 315$ Mpc. This result, twice the periodicity of galaxies, is consistent with the fact that galaxies are themselves members of clusters with a high rate. Conversely, one may use the constraint that the ratios of the periodicities of various levels of the observed hierarchy must be an integer to derive this hierarchy: we find that the length-scale ratio of one level to the following one must be $k^{2+\sqrt{2}}$, i.e., 10.7 for k=2, 42.5 for k=3, 114 for k=4, which compare well with the observed hierarchy.[29]

We conclude this section by once more stressing the fact that the above results should be considered as tentative. A coherent scheme appears to emerge, but it remains to be demonstrated that the application of our approach to cosmology can be made consistent with the firmly established constraints of general relativity.

7.2. Beyond Chaos.

At the end of Section 5.7, we arrived at the conclusion that a Markov-Wiener process plays the fundamental role of a transformation allowing one to go reversibly from classical to quantum and from quantum to classical laws. However the completely deterministic "purely classical" world and the completely undeterministic "purely quantum" world may be viewed as the extremities of a full spectrum in the complexity of Nature: We, indeed, now know that "deterministic" chaos exists in several situations, some of which were previously considered as archetypes for totally predictible systems. One of the most impressive example is the recent suggestion by Laskar that the inner Solar System is chaotic with an inverse Lyapunov exponent as low as 5 Myr.

The discovery of chaos, which is now defined as the extreme sensibility to variation in initial conditions (this is often described by an exponential divergence of trajectories in phase space, $\delta x = \delta x_0 \, e^{t/\tau}$, where $1/\tau$ is the so-called Lyapunov or characteristic exponent[30]), may be attributed to Poincaré.[31]

Chaos is relevant in a huge variety of natural systems (see, e.g., the series of popular papers in Refs. 32 and 40: celestial mechanics, fluid mechanics and turbulence, weather, population dynamics, evolution, ecology, mathematics, economics, dynamics of chemical reactions...). Although a large number of various methods of analysis has been coined to describe the development of chaos (strange attractors, fractal and information dimensions, entropy, characteristic exponents, catastrophe theory...), all of them have up to now struck against the unpassable barrier of unpredictability at large time scales. However, in many systems where chaos arises, spatial and temporal structures seem also to arise: they are observed experimentally (regularity of the distribution of planets, satellites and asteroids in the Solar System, spatial and temporal structures in the climate, biological structures, Belousov-Zhabotinskii reaction in chemistry, Taylor-Couette flow...); these structures are in some few cases found or confirmed in numerical simulations, but very rarely understood or predicted from a fundamental theory.

We suggest in the present section a general method for attacking these problems: this method is efficient precisely when other methods fail, i.e., for very large time scales (compared to the "chaos time" τ). We shall see that it naturally generates spatial and temporal structures. This will be exemplified by its application to the problem of the Solar System.

Prediction beyond unpredictability.

The method is based on the formalism presented in Sec. 5.6, which is an extension of Nelson's stochastic formalism. It is written in a compact form thanks to the introduction of complex variables; the main new point is that we have demonstrated that the fundamental equation of dynamics, written in terms of our new complex time derivative operator $\mathcal{d}/dt$,

$$\boldsymbol{F} = m \frac{\mathcal{d}^2}{dt^2} \boldsymbol{x} \quad ,$$

becomes Schrödinger's equation. But consider the basic hypothesis of the formalism: the trajectory slopes are broken at any point of the space, this breaking being described by an Einstein-Wiener process of diffusion, as is done for the description of Brownian motion. Consider now chaotic trajectories in the plane. We place ourselves in the reference frame of a first trajectory, say ($x_1 = 0$, $y_1 = a\,t$). Then consider a second trajectory which is exponentially divergent with respect to the first one (we assume a unique Lyapunov exponent for simplicity): $x = \delta x\, e^{t/\tau}$, $y = a\,t + \delta y\, e^{t/\tau}$). Then we get the relation

$$y = \frac{\delta y}{\delta x}\, x + a\,\tau\, ln \frac{x}{\delta x} \quad .$$

Such a trajectory (typically in $x + lnx$) is shown in Fig. 7.3 in the plane (x,y), for various time scales. For very large time scales, i.e., with *resolution* $\gtrsim \tau$, it becomes non-differentiable at the origin, with different backward and forward slopes, and looks like trajectories arising from a diffusion process or from particle collision. For Lyapunov exponents different in x and y, one gets a power law, with also a point of broken slope at the origin when seen at large time scale. Now, in case of developed chaos, the small perturbation ($\delta x, \delta y$) fluctuates and the divergence between possible trajectories described in Fig. 7.3 occurs at any of their points: for $\Delta t >> \tau$, the trajectories become describable to a good approximation by non-differentiable and fractal paths.

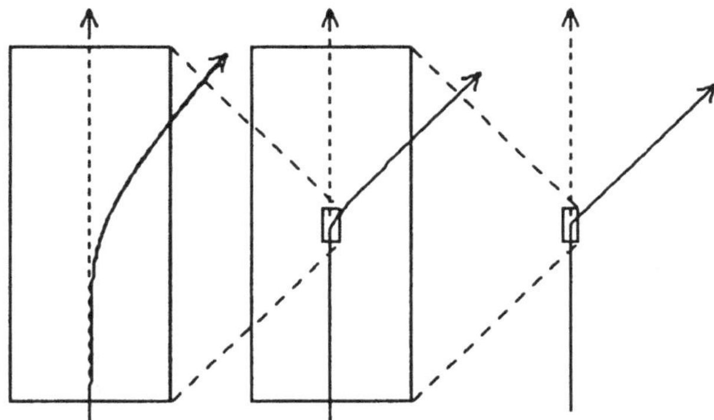

Figure 7.3. Schematic representation of the relative evolution of two initially close trajectories seen at three different time scales, in the case of chaotic motion.

Let us give another example. Consider one of the archetypes of chaotic behaviour, the so-called logistic map of population dynamics,

$$x_{t+1} = \lambda\, x_t\, (1 - x_t) \ ,$$

which may be iterated for a discrete time t. For $\lambda = 4$, the behaviour becomes completely chaotic and the values of x fill the interval [0,1], as shown in Fig. 7.4. There too, it is immediately clear that the motion on the line Ox resembles closely that of a particle subjected to random kicks, with different velocities before and after the "kick".

More generally, consider a system subject to developed chaos. For time scales large with respect to the inverse maximum Lyapunov exponent (i.e., beyond the time scale after which predictability of orbits is lost), we can replace deterministic trajectories by families of potential trajectories, and then the concept of definite positions by that of a *probability density*.

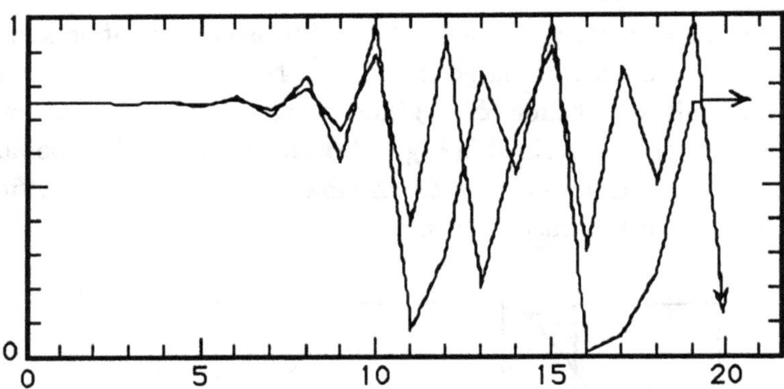

Figure 7.4. Illustration of high sensibility to initial conditions in the "logistic map".

This leads one to describe the effect of chaos in a stochastic way by a diffusion process. Such an idea is not new in itself (see e.g. Ref. 61). The new element that we suggest to add in the description is the *explicit* introduction of the non-differentiability in terms of different *backward and forward velocities.*

So our method consists in assuming that, for large time scales, the evolution of the virtual trajectories can be described by a Wiener process, and replacing in the basic differential equations the time derivative by our complex time derivative operator (see Sec. 5.6). In other words, this means that we set a *principle of correspondence* for classical chaotic equations. When these classical equations are deduced from a Lagrangian formulation, the new equations will be Schrödinger-like, and the solutions quasi-quantum (with a different interpretation: the full behaviour of quantum mechanics can be recovered only if one assumes the space-time to be *non-differentiable down to the smallest scales*; here this is only a large time scale approximation, since when coming back to small time scales one recovers differentiable predictable trajectories).

It is known that quantum mechanics naturally yields structures: the existence of well-defined boundary conditions for some variables results in the quantization of the conjugate variables. One may expect the same behaviour from the theory which is outlined here. Let us exemplify this by applying our method to celestial mechanics, namely, to the old problem of the regularity of planets in the Solar System.

"Quantization" of the Solar System.

The existence of regularities in the distribution of planets in the Solar System was recognized long ago. This was Kepler's main motivation in his search for planetary laws. The Titius-Bode "law" ($r_n = 0.4 + 0.3 \times 2^n$) was the first empirical attempt at describing these regularities, and was followed by several other proposals.[33-35] The discovery of similar structures in the distribution of the satellites of the great planets led to a revival of interest for such studies, and to the hope that indeed a physical mechanism was at work; such a mechanism was searched for by most authors in the formation conditions.[36] It was however suggested by Hills[37] that this regularity may arise from a dynamical evolution relaxation (see Nieto[33]).

The discovery by Laskar[38] that the inner planetary system (telluric planets) is chaotic, with a very short inverse Lyapunov exponent of $\tau \approx 5$ Myr, and its recent confirmation by independent studies[39] sets the question in a completely renewed way. The position of planets can no more be predicted from usual celestial mechanics for time scales larger than ≈100 Myr. But several arguments, among which the maintenance of life on Earth since ≈3.5–4 Gyr, show us from experience that the Solar System, though chaotic, is nevertheless confined.[40] Let us apply here our new method for tackling this problem: we shall see that it allows us to predict the preferential positions of planets and leads to the suggestion that these structures can arise from large time scale effects of dynamical chaos.

The impossibility of following individual orbits for $t \gtrsim 100$ Myr forces us to jump to a probabilistic description. The planet position is now characterized by a probability density ρ (which applies to *potential* orbits) rather than by well-defined variables. Once the chaos developed, the various future or past potential trajectories evolve following a diffusion process, characterized by some diffusion coefficient $\mathcal{D}$. We describe this diffusion by a Markov-Wiener process $\xi(t)$ (i.e. the $d\xi(t)$ are Gaussian with mean zero, mutually independent and such that $\langle(d\xi)^2\rangle = 2\mathcal{D}dt$) as in the formalism of Sec. 5.6. Let us recall once more the main steps of the demonstration.

Mean forward and backward derivatives[41], d_+/dt and d_-/dt, are introduced which, once applied to the position vector $\boldsymbol{x}$, yield *forward and backward mean velocities,* $\frac{d_+}{dt}x(t) = b_+$ and $\frac{d_-}{dt}x(t) = b_-$. This describes the fact that, at the time scales considered, the trajectories are broken at any of their points (i.e. fractal). From these quantities we introduce a complex velocity

$$\mathcal{V} = V - i\,U = \frac{b_+ + b_-}{2} - i\,\frac{b_+ - b_-}{2}$$

and a complex derivative operator

$$\frac{d\!\!\!/}{dt} = \frac{d_V}{dt} - i\,\frac{d_U}{dt} = \frac{1}{2}\left(\frac{d_+ + d_-}{dt} - i\,\frac{d_+ - d_-}{dt}\right)$$

which is given by

$$\frac{d\!\!\!/}{dt} = \left(\frac{\partial}{\partial t} - i\,\mathcal{D}\,\Delta\right) + \mathcal{V}.\nabla \quad .$$

The real part V of $\mathcal{V}$ is identified with the classical velocity in the differentiable case, while its imaginary part U is non-zero only in the non-differentiable case. Now, since we deal with a Markov process, the probability density verifies the forward and backward Fokker-Planck equations, from which the equation of continuity and the following expression for U may be derived:[41]

$$U = \mathcal{D}\ \nabla \ln\rho \ .$$

We conjecture that Newton's equation of dynamics still holds in terms of our new complex variables

$$F = m\,\frac{d\!\!\!/}{dt}\,\mathcal{V}.$$

As seen in Sec 5.6, the least action principle may also be re-expressed in terms of complex variables: this leads to the above form of Newton's equation and to the result that V is also a gradient. We thus introduce a new quantity S such that $V = 2\,\mathcal{D}\,\nabla S$ and define a complex function ψ which is related to our complex velocity :

$$\psi = \sqrt{\rho}\ e^{iS} \Rightarrow \mathcal{V} = -2i\ \mathcal{D}\ \nabla(ln\psi).$$

When the force F derives from a potential, $F = -\nabla\mathcal{U}$, as is the case for gravitation, the equation of motion writes

$$\nabla\mathcal{U} = 2i\ \mathcal{D}m\ \frac{d'}{dt}\,(\nabla ln\psi).$$

Replacing the complex derivative operator by its expression finally gives

$$\mathcal{D}^2\,\Delta\psi + i\ \mathcal{D}\,\frac{\partial}{\partial t}\psi - \frac{\mathcal{U}}{2m}\psi = 0 \tag{7.2.1}$$

up to an arbitrary phase factor. Take $\mathcal{D}=\hbar/2m$, and this becomes Schrödinger's equation, as shown in Sec. 5.6. The hereabove system is a reexpression of stochastic quantum mechanics,[41] but also a generalization: assume that one has been able to characterize some chaotic system for large time scale by a constant diffusion coefficient $\mathcal{D}$, then (7.2.1) is a quasi-quantum equation for such a system, which is expected to yield structures (i.e., peaks of probability) once the boundary conditions are prescribed.

Let us apply this method to the Solar System. Consider a planet (more generally a test particle) in the field of the Sun, $\mathcal{U}=-GmM/r$, and in the collective field of an ensemble of planets (more generally of particles), and assume that this system is chaotic. Our conjecture is that the effect of chaos on large time scales can be summarized by a Brownian motion process of diffusion coefficient $\mathcal{D}$. Assume moreover that we deal with a stationary motion with conservative energy $E \equiv 2i\ \mathcal{D}\ m\ \partial/\partial t$ (the time-independent Schrödinger equation may also be obtained directly by setting V=0, see Nelson[41]). Equation (7.2.1) becomes

$$\mathcal{D}\Delta\psi + [\,E + \frac{G\,M}{2\ \mathcal{D}\ r}\,]\,\psi = 0\ .$$

The equivalence principle suggests that $\mathcal{D}$ must be independent of m. This equation is similar to the Schrödinger equation for the hydrogen atom,[42,43] up to the substitutions $\hbar/2m \rightarrow \mathcal{D}$, $e^2 \rightarrow GmM$, so that the natural unit of length (which corresponds to the Bohr radius) is

$$a_o = \frac{4\,\mathcal{D}^2}{G\,M} \quad . \qquad (7.2.2)$$

We thus find that the energies E of planets are given by

$$E_n = -\frac{G\,m\,M^2}{8\,\mathcal{D}^2\,n^2} \;, \quad n = 1, 2, 3, \ldots \;,$$

and that the density of probability of their distances to the Sun are confined to well defined regions given by the square of the well-known radial wave function[42,43] of the hydrogen atom (see Sec. 4.1). We also expect angular momenta to scale as $L = 2m\mathcal{D}\ell$, with $\ell = 0, 1, \ldots, n-1$. This means that, unlike the case of quantum mechanics, E/m and L/m are quantized rather than E and L. The average distance to the Sun and the eccentricity e are given in terms of the two quantum numbers n and ℓ by the following relations:

$$a_{n\ell} = \{\tfrac{3}{2}\,n^2 - \tfrac{1}{2}\;\ell\,(\ell+1)\,\}\,a_o \;, \qquad (7.2.3)$$

$$e^2 = 1 - \frac{\ell\,(\ell+1)}{n\,(n-1)} \quad . \qquad (7.2.4)$$

In order to compare these results with the actual Solar System, one must first note that the inner (Mercury to Mars) and outer (Jupiter to Neptune and/or Pluto) planetary systems are characterized by two different inverse Lyapunov exponents, τ_{int} =5 Myr and a still unknown τ_{ext}, perhaps as high as 1 Gyr.[44,45] That they must be treated as two different systems is also suggested by many other arguments, such as their different chemical compositions and mass distributions.

Consider first the eccentricities. Even the largest eccentricities (Pluto, e^2=0.065; Mercury, e^2=0.042) correspond to a good approximation to $\ell = n-1$. This will be further discussed hereafter. We may then compare the observed values of semi-major axes of the planets to our prediction (Eq. 7.2.5), $a = (n^2+n/2)a_o$. This is only a *two-parameter* relation (the slopes for the inner and outer systems), since *we predict the ordinate at origin to be zero.* This prediction is very well verified for the two systems. We find Mercury, Venus, the Earth and Mars to take respectively the ranks

3, 4, 5, and 6 in the inner system and Jupiter, Saturne, Uranus, Neptune, and Pluto the ranks $n = 2, 3, 4, 5, 6$ in the outer system. With these values, the regression lines are (in units of A.U.)

$$\sqrt{a}_{\text{int}} = -0.015 + 0.199 \sqrt{n^2+n/2} \quad ,$$

$$\sqrt{a}_{\text{ext}} = -0.066 + 1.035 \sqrt{n^2+n/2} \quad ,$$

so $a_{\text{int}}(0) = 2 \times 10^{-4}$ A.U. and $a_{\text{ext}}(0) = 4 \times 10^{-3}$ A.U., which are a fair confirmation of our prediction. Assuming them to be strictly zero, we get average slopes $\sqrt{a_{\text{oint}}} = 0.195 \pm 0.0022$ and $\sqrt{a_{\text{oext}}} = 1.014 \pm 0.016$. Their ratio is 5.2 ± 0.1. (Note that the value of the inner slope mainly reflects the fact that the rank of the Earth is $n = 5$).

Two additional remarkable results are obtained: (i) the central peak of the asteroid belt (2.7 A.U.) agrees remarkably well with $n = 8$ of the inner system, and the main peak at 3.15 A.U. with $n = 9$; (ii) Mars' position is also in very good agreement with $n = 1$ of the *external* system. Including them yields improved slopes $\sqrt{a_{\text{oint}}} = 0.195 \pm 0.0017$ and $\sqrt{a_{\text{oext}}} = 1.014 \pm 0.012$. These results are illustrated in Fig. 7.5.

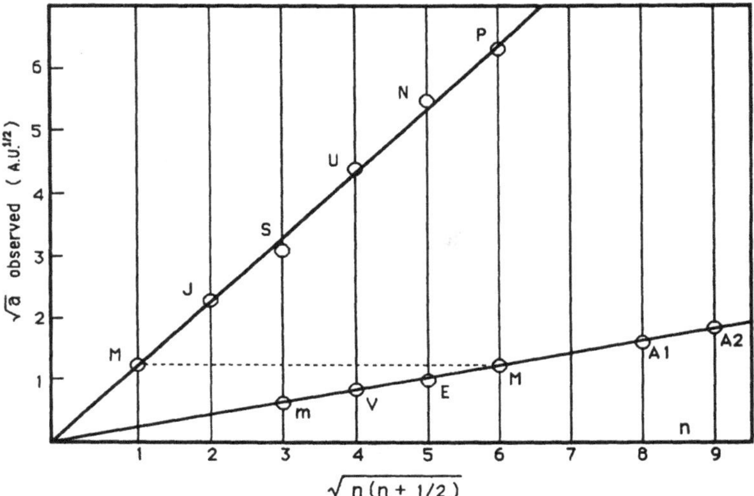

Figure 7.5. Comparison of the observed average distances of planets to the Sun with our prediction (see text). The lines shown are least-square regression lines. A1 and A2 stand for the two main peaks in the distribution of asteroids in the asteroid belt.

Note also the agreement of Pluto with the outer relation: this may seem to be at variance with its particular orbit and the recent discovery that its motion is chaotic with a Lyapunov 1/20 Myr;[46] however it has been argued that this chaos arises from resonances within resonances and that this can limit the extent of Pluto's wandering.[47] We shall see that indeed Pluto is "anomalous" in terms of angular momentum, while our results show that it is not so in terms of energy. In this respect, it is also noticeable that Neptune and Pluto, both of which strongly disagreed with the *three*-parameter Titius-Bode law, now both agree with our two-parameter relation: moreover, as will be seen later, one may have the hope to see this relation become a full totally constrained prediction, when the slopes are precisely predicted from the Lyapunov exponents (or some other characteristic of the chaotic dynamics).

The agreement between observations and predictions are tested in two ways. We may compute $(a_{obs}-a_{pred})/a_{obs}$ for each planet. This is shown in column 5 of Table 1. The only difference larger than ≈6% is Saturn (12 %). The average relative difference is 3.4% in the inner system and 2.6% (Saturn excluded) or 4.3% (Saturn included) in the outer system. These numbers are certainly indicative of the natural irreducible fluctuations of the distance which are expected from our analysis: this is anyway a remarkable confinement around the mean values, perhaps too good when compared with the theoretical dispersion in the probability densities. At the end of this section, we shall consider some improvements of the method which could help us to understand such narrow fluctuations. If we take our results at face value, the Earth would presently be in one of its closest approach to the sun (4%).

Another test consists in checking the difference with the quantized values n. Let us introduce δn such that $(n+\delta n)^2 + (n+\delta n)/2 = a_{obs}/a_o$. To lowest order in δn it is given by

$$\delta n = \frac{a_{obs}/a_o - n\,(n+1/2)}{2\,(n+1/4)} .$$

The values of δn for the various planets are given in column 6 of Table 1.

The statistical nature of our predictions is more clearly visible in this indicator, which reaches 0.18 at the maximum difference.

Planet (1)	n (2)	a_{obs} (UA) (3)	a_{pred} (AU) (4)	$\delta a/a$ (5)	δn (6)	e (7)	$\delta \ell$ (8)
Mercury	3	0.387	0.399	+0.031	−0.065	0.206	0.051
Venus	4	0.723	0.684	−0.054	+0.098	0.007	0.00008
Earth	5	1	1.045	+0.045	−0.139	0.017	0.0006
Mars	6	1.523	1.483	−0.026	+0.053	0.093	0.024
–	7	–	1.996	–	–	–	–
Aster.1	8	2.7	2.586	+0.043	+0.140	–	–
Aster.2	9	3.15	3.251	+0.030	−0.073	–	–
Mars	1	1.523	1.542	+0.012	−0.008	0.093	0.00000
Jupiter	2	5.20	5.14	−0.012	+0.013	0.048	0.0008
Saturn	3	9.57	10.79	+0.126	−0.183	0.054	0.0036
Uranus	4	19.28	18.51	−0.040	+0.088	0.051	0.0044
Neptune	5	30.14	28.27	−0.062	+0.173	0.005	0.00006
Pluto	6	39.88	40.09	+0.005	−0.017	0.256	0.178

Table 1.

Let us now come back to angular momenta. From the now known values of n for the various planets, one may compute the difference with the expected quantized number ℓ. We set $\ell = n - 1 + \delta\ell$, and find from Eq. (7.2.4) (to lowest order in $\delta\ell$)

$$\delta\ell = \frac{n\,(n-1)}{2n-1}\, e^2 \quad .$$

The $\delta\ell$ values are given in column 8 of Table 1. They are remarkably small (some 1‰ or less), except in one case, Pluto, for which the difference amounts to 0.18: indeed the orbit of Pluto is known to be partly determined by its strong 3:2 resonance with Neptune. Concerning the asteroid belt, there is a large spread of observed eccentricities from 0 to ≈0.4, which may be shown to arise from Jupiter's perturbation[48] (more generally from the four Jovian planets). For $n = 8$ (main asteroid belt), the second angular momentum state is $\ell = 6$, which gives $e = 0.5$. There is indeed a population of asteroids with eccentricities around this value.

Our method also sheds new light on one of the long standing problems concerning the Solar System, namely that of the distribution of angular momentum. Jupiter owns 62% of the angular momentum of the Solar System and Saturn 25%. We have found that the angular momentum over m is quantized, rather than the angular momentum itself. So the distribution of angular momentum mainly reflects that of mass, hence implying the domination of Jupiter and Saturn. (The origin of the distribution of mass comes under the theories of formation of the solar system, which are outside the scope of the present work). In contrast, the method in its present form does not allow us to understand the alignment of angular momentum vectors (i.e. the nearly plane character of the Solar System).

Concerning asteroids in the main belt, one may now reach a good understanding of their distribution. First our approach brings new elements for understanding the absence of a large planet there: the zone where the belt lies, even though it corresponds to the maxima of probability density for the inner system, also corresponds to a *minimum* in the outer system. The region between Mars and Jupiter is where the two systems overlap (see Fig. 7.5). While Mars, being in a probable zone in both systems, is expected to have a remarkably stable orbit, (the mean predicted distance is 1.51 A.U., the observed distance 1.52 A.U.), this is not the case of the belt region, for which the tendencies are opposite. Then most of the matter of the primordial nebula situated between the $n = 1$ and $n = 2$ orbitals of the outer system could have drifted towards what are now Mars and Jupiter: this may explain why the total mass of asteroids is far smaller than a planet mass. However, peaks of probability occur at $n = 7$, 8, 9 and 10 between Mars and Jupiter, in terms of the inner system. Why are they not all filled? This is due to the small time scale dynamical chaos[48]: the orbitals $n = 7$ and $n = 10$ coincide with the resonances 1:4 and 2:3 with Jupiter (see Fig. 3.4 of Sec. 3.2). So one may hope to understand the existence and full distribution of asteroids as a combination of the effect of large time scale chaos (implying peaks of probability at 2.59 and 4.25 U.A.) and of small time scale chaos due to the resonant action of Jupiter resulting in the formation of the Kirkwood gaps.[48]

All the above results have been obtained without specifying any expression for the diffusion coefficients $\mathcal{D}$ of the inner and outer systems, which were left as free parameters (and then fitted in Fig. 7.5). Is it possible to get estimates for them, in particular to relate them to the calculated Lyapunov exponents ?

The problem is that the chaotic behaviour discovered by Laskar concerns essentially the eccentricities and inclinations, while nothing is *a priori* known concerning the semi-major axes. The distance at time t of two orbits initially separated by δx_0 is $x = \delta x_0\, e^{t/\tau}$, where $1/\tau$ is the Lyapunov exponent which characterizes the chaotic behaviour. Predictability of the planet position is completely lost when $x \approx a$. However this does not mean that predictability of the mean distance a of the planet to the sun is yet lost. Loss of information on the precise position is first indicated by a loss of information on the angle (see e.g. the time evolution of orbits in Ref. 44) This occurs when $\delta x_0\, e^{t/\tau} \approx a$. Beyond this point, the divergence may begin to contribute to the loss of the information on the average distance to the Sun. It is completely lost after some number k of turns at a time T given by $\delta x_0\, e^{T/\tau} \approx 2k\pi a$. If one *assumes* that the drift on 1 radian remains of the order of δx_0, one gets $2\pi k \approx a/\delta x_0$ and then $\delta x_0 \approx a\, e^{-T/2\tau}$. This very rough estimate, which corresponds to an inverse Lyapunov exponent for semi-major axes (i.e., energy) twice that of other orbital elements, should be considered as a lower limit only.

The diffusion coefficient is given by $\mathcal{D} = U/\nabla \ln\rho$. The probability of presence for $\ell = n-1$ (corresponding to quasi circular states $e^2 \approx 0$ as observed for planets in the solar system, see hereafter) writes

$$P_n(r) = \frac{1}{(2n)!}\,\frac{8}{n^3}\,\left(\frac{2r}{na_0}\right)^{2n-2} e^{-2r/na_0}$$

so that $(\nabla \ln\rho)^{-1} \approx na_0/2 \approx a/2n$, while $U \approx \left\langle\frac{\delta x_0}{2\tau}\, e^{t/\tau}\right\rangle \approx \frac{\delta x_0}{2T}\, e^{T/\tau}$. Setting $R = a/\delta x_0$, we finally get a rough estimate for the diffusion coefficient:

$$\mathcal{D} \approx \frac{1}{8n} \frac{R}{lnR} \frac{a^2}{\tau} \ .$$

From Eq. (7.2.2), it is also given by $\mathcal{D} = \frac{1}{2}\sqrt{GMa_o}$. From Eq. (7.2.3), $a = (n^2+n/2)\, a_o$ for quasi-circular states ($\ell = n-1$). We may then estimate the slope of the linear relation expected between $\sqrt{a}$ and $\sqrt{n^2+n/2}$:

$$\sqrt{a_o} = \frac{1}{<n>} \left(4\tau \sqrt{GM} \frac{lnR}{R} \right)^{1/3} . \qquad (7.2.5)$$

Using Kepler's third law ($T^2/a^3 = cst$) allows us to write this result in still another form in terms of the planet period $\mathcal{T}$:

$$\frac{\tau}{\mathcal{T}} = \frac{1}{8\pi} \frac{R}{lnR} \ .$$

Let us finally attempt to compare our estimate of the slope (7.2.5) with the observed one in the inner system, $\sqrt{a_o} = 0.195\ (\text{U.A.})^{1/2} = 7.5\ 10^5\ \text{cm}^{1/2}$. Our estimate depends on the parameter $R = a/\delta x_o$, then on the value of the basic perturbation δx_o. This does not correspond here to a measurement uncertainty, but to the irreducible fluctuations of the positions of the planets due to internal and/or external effects. The main effect comes from the interaction with asteroids.[45] It has been estimated that, in order to keep a precision $R^{-1} = 10^{-10}$, about 40 asteroids were to be included in the motion equations, and several hundreds at 10^{-12} precision.[45] The asteroid trajectories being themselves chaotic, we may estimate the irreducible perturbation to be such that $R = 10^{10\pm1}$. With $\tau_{int} = 5$ Myr and $<n> = 4.5$ for the inner solar system, Eq. (7.2.5) yields

$$\sqrt{a_o} = (5.3^{+6.1}_{-2.8}) \times 10^5\ \text{cm}^{1/2} \ ,$$

which is compatible with the observed value. The diffusion coefficient is estimated to be $\mathcal{D} \approx 10^{19\pm1}\ \text{cm}^2.\text{s}^{-1}$, to be compared with $\mathcal{D}_{obs} = \frac{1}{2}\sqrt{G\ M\ a_o} = (4.34 \pm 0.09) \times 10^{18}\ \text{cm}^2.\text{s}^{-1}$. But this result should not be taken too seriously, because of all the uncertainties in its derivation. The large final error on the theoretical estimate (7.2.5) still allows the inverse Lyapunov exponent for semi-major axes to be 10 times its estimated minimal value, namely 100 Myr.

Moreover, it is quite possible that our method applies essentially to an earlier phase of the evolution of the Solar System, so that it would be irrelevant to relate the diffusion coefficient in Eq. (7.2.1) to Lyapunov exponents computed *from the present state* of the Solar System. The results of numerical simulations by Hills[37], Ovenden[62] and Conway and Elsner[63] seem to support this view: starting from random initial conditions of model planetary systems, they find chaotic trajectories (the very irregular evolution of semi-major axes seen in Figs. 1, 2, 4 of Ref. 63 fairly agrees with our basic conjecture of fractality and non-differentiability), while systems placed initially in conditions similar to that of the present Solar System were shown to be very stable system, which maintained nearly circular orbits. Our results suggest that, on very large time scales, a planetary system can pass from the first type of system to the second one.

Anyway, the improvement of our approach needs a better understanding of the relation[61] between the Lyapunov exponents, which describe the development of chaos, and the diffusion coefficients, which hopefully describe, in our approach, what happens *after* the limit of classical unpredictability. A promising method would consist in working with the Kolmogorov-Sinai entropy, which is itself related to Lyapunov exponents,[30] or with the algorithmic entropy recently introduced by Zurek.[49] However the Lyapunov exponent is presently calculated from numerical simulations, while a completely self-consistent approach would imply obtaining it also from the basic equations. These problems are left to future works.

Before closing this section, let us add a last comment: as already specified, even though our fundamental equation (7.2.1) is a quantum mechanical-like equation, its interpretation must be different from that of quantum mechanics. We know that the approximation of non-differentiability is no more valid on small time scales, for which one recovers predictable and differentiable trajectories. So the (still open) problem is to understand how to connect the small time scale behaviour to the large time scale one, or in other words, how the probability densities obtained at large time scales influences the motion observed at small time scales.

7.3. Conclusion.

With the aim of studying the dependence on scale of physical laws that is gradually emerging in experimental and theoretical studies, three domains have been considered in this book:

(1) mainly the microphysical domain, presently best described by quantum theories, where the dependence of measurement results on the resolution of measurement is a well-known fact, a consequence of Heisenberg's relations;

(2) more briefly, the domain of complex classical systems subjected to dynamical chaos;

(3) tentatively, the cosmological domain.

We have shown that similar mathematical methods are adapted for tackling the various problems of scaling and fundamental scale onset that are common to these domains: namely the fractal geometry, the renormalization group and stochastic mechanics. Moreover we have suggested that these tools are complementary: some renormalization group equations can be viewed as differential equations for the dependence of fractals on scale, and the Wiener processes, being a description of Brownian-like motion, may more generally give the simplest stochastic description of the physical behaviour on a fractal space whose geodesics would be of dimension 2.

But our main lead in this research has been Einstein's principle of relativity, which we have suggested to extend in such a way that it also applies to laws of scale transformation. When applied to the renormalization group description of the asymptotic behaviour in quantum field theory, it leads to the remark that the present renormalization group has a Galilean-like form, while the principle of relativity rather suggests that its structure should be Lorentzian: this yields the first development of a special theory of scale relativity, as described in Chapter 6. When applied to space-time, it leads to our proposal that space-time itself must be scale-dependent, i.e. fractal: the first implications of this conjecture have been studied in Chapters 4 and 5.

Let us first comment on the linear theory of scale relativity (Chapter 6 and Section 7.1). Our theory is based on the following postulates:

(1) One may define a relative state of scale of reference systems, so that scale transformations, i.e., dilatations and contractions, come under the principle of relativity; the logarithm of the resolution with which a measurement is performed is the measure of such a state, and plays in scale relativity the role played by velocity in motion relativity.

(2) The renormalization group method may be applied to space-time itself in an enlarged sense: it is applied to the length $\mathcal{L}$ or to the time $\mathcal{T}$ virtually elapsed along a particle path in space or space-time, i.e., to the internal structure of a quantum particle.

(3) A couple of variables $(ln\mathcal{L}, \delta)$, or equivalently $(ln\mathcal{T}, \delta)$, i.e., the logarithm of length (or time) as defined above, and the renormalization group anomalous dimension δ, play the same role for scale laws as length and time for motion laws.

Once these postulates are accepted, we believe we have demonstrated in a general way that the general solution to the scale relativity problem implies the existence of an impassable, universal, limiting scale which is invariant under dilatations. The current theory is recovered in the limit where this length is vanishing. In particular, the precise predictions of QED, such as the anomalous magnetic moment of the electron and the Lamb shift are expected to be essentially unaltered by scale-relativistic corrections, since they correspond to low energy (≈ Bohr scale) measurements. For example, in the theoretical prediction for the magnetic moment of the electron, the contribution of muons (i.e. of length scales 1/206.77 that of the electron) is only $\approx 2 \times 10^{-12}$, and the total contribution from still higher energy scales is 4.5×10^{-12}, while the current total theoretical error is $\approx 108 \times 10^{-12}$. [50] Now, at the muon scale, the value of the scale-relativistic δ-factor is only 1.0054, so that, even though detailed calculations remain to be done, we may safely expect scale relativity not to contradict the well-established experimental facts.

In our derivation of the new laws of scale transformation, we could also have started from the *postulate* that the Planck length and time are invariant under dilatation. This would have given us the same theory, but with the *result* that scale relativity should be broken above some particular transitional scale, to be identified with the de Broglie length and time (otherwise one would get an invariant dilatation instead of an invariant limiting scale).

Anyway the essence of our proposal may be traced back to a basic question: does such a limiting scale, invariant under dilatation, exist in Nature? If it does, this is a universal law, and the consequences of its existence must concern physics as a whole. Even if the theory which is presented here were to be proved insufficient in some of its aspects, one could not escape the need to build such a theory and to make the whole physics scale-relativistic.

Even in the restricted framework which has been considered in Chapter 6, there is still a lot of work to be done. We have examined only the case of one independent space or time variable, while a proper account of the full space-time should be taken: this must include in particular an account of angles and rotations. In other words, a full motion plus scale relativistic theory remains to be constructed.

Let us recall the encouraging successes which have been obtained:

(a) The theory solves the old problem of the divergence of charge and self-energy of particles.
(b) It implies that the *four* fundamental coupling constants of physics converge on about the same energy, which is now the Planck energy.
(c) It reconciles, without introducing new interactions or particles, the predictions of GUT's concerning the Weinberg mixing angle and decay of the proton, which were previously both mutually contradictory and inconsistent with experimental results.
(d) It allows one to make progress in the understanding of the onset of fundamental scales, such as the GUT scale and the electroweak scale.
(e) Combined with the conjecture that the bare charge is $1/2\pi$, it permits estimates of the low energy values of coupling constants, i.e., of the weak, strong and electric charges.

Consider now the state reached by our fractal approach. The development of a full theory of fractal and non-differentiable space-times lies well beyond the scope of this book. Our aim was mainly to demonstrate that the construction of such a theory is possible, and to set down the bases and constraints upon which it could be founded in the future:

(1) A universal fractal dimension 2 characterizes all the four space-time coordinates in the quantum domain (at least in the "scale nonrelativistic" approximation).

(2) The transition from quantum laws to classical laws is a relative transition, given by the de Broglie scale of the system under consideration (i.e., it depends on its mass and velocity). However, since we have explicitly considered only the simplest case of a free particle, some additional work is needed to fully demonstrate that the quantum/classical transition can be identified with a fractal/nonfractal transition. Indeed, for most real systems, the actual quantum/classical transition is given by the *thermal* de Broglie scale, which depends on the velocity dispersion, or, equivalently, on the temperature, rather than on the average velocity. More generally, one may conjecture that the transition is given by the correlation length of the quantum phase $\theta(x)$, i.e. λ, such that $\theta(\lambda)\approx 1$, or, in other words, by the largest length of coherence of the phases of the system.

(3) The particles can be identified with their fractal trajectories themselves, and their various properties with the internal spatio-temporal structures of these trajectories. This has been explicitly demonstrated for mass and quantum spin, and we have conjectured that the same result also applies to other internal quantum numbers, such as the electric charge. A two-dimensional fractal structure also accounts for phenomena like the Zitterbewegung and the virtual particle pair creations and annihilations internal to any particle.

(4) The wave-corpuscle duality is tentatively interpreted as a consequence of the infinity of the number of geodesics between any two events of a fractal space-time. This conjecture remains to be given a mathematical realization.

(5) By the connection with stochastic quantum mechanics, we have shown that a theory based on the concept of a fractal space-time would yield standard quantum mechanics as a limit. The complex nature of the quantum probability amplitude is understood to be a direct consequence of the non-differentiability of space-time.

(6) We have suggested that possible manifestations of the fractal structure of space-time should be searched for at the energies of transition and in strong fields: there is a clear indication that the still unexplained anomalous electron-positron peaks observed at Darmstadt have a fractal behaviour

when resolution is improved. Fractal structures are also exhibited by multihadron rapidity distributions, as was recently analysed by Carruthers.[51]

Let us also briefly examine to what extent the new structure of space-time implied by special scale relativity may change the nature of the postulated fractal space-time. The main result of special scale relativity, the existence of a limiting scale in nature, is a universal constraint to which the fractal space-time must be subjected. This means that the fractal structure, when indefinitely zooming on the fractal, will not develop up to the zero scale but to the Planck scale. This does *not* mean that the "fractalization" (i.e., the never-ending appearance of new structures at all scales) stops. On the contrary, in order to keep the non-differentiability, the fractal must continue to be defined by an unlimited process: one must apply an infinity of times some dilatation factor and some (possibly scale-dependent) generator in order to construct the fractal. But the application of a dilatation ρ does *not* yield anylonger length intervals which are ρ times longer. On a scale-relativistic fractal, the geometrical quantities which were diverging as scale approaches zero, such as length or curvature, now diverge as scale approaches the Planck scale.

So the fractal space-time concept which begins to emerge from our analysis, based on fundamental principles, departs more and more from the usual concept of fractal objects embedded in an Euclidean space. The logarithm of the internal length and time measured along the fractal, $\mathcal{L}$ and $\mathcal{T}$, and the fractal dimension $D=1+\delta$ are no longer independent variables, since they now become components of a vector of some hyperbolic space $\{\ln(\mathcal{L}/\mathcal{L}_0),\delta\}$, as described, in a first approximation, by the invariant introduced in Sec. 6.9.

Although we are still very far from an achievement of a full theory implementing the principle of scale relativity, we can already guess the form expected for the equations of such a future theory. The need to satisfy both motion and scale relativity principles implies a replacement of today's scale-independent variables by explicitly scale-dependent variables, and a generalization of standard equations of physics into coupled differential equations written in terms of infinitesimal variation of both space-time and resolution variables:

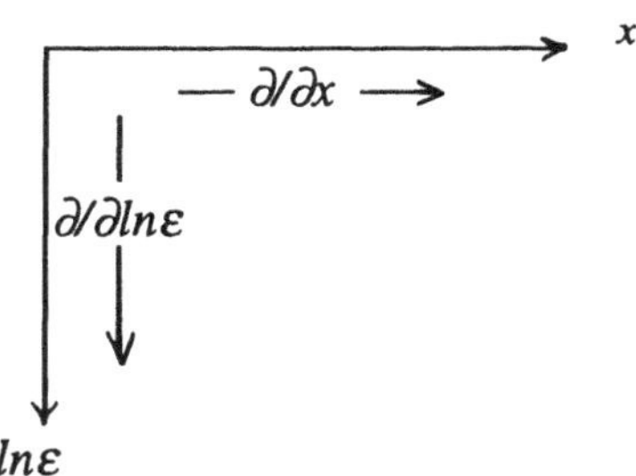

For these all-scales-of-length equations, boundary conditions will be written at large scale as well as at small scale, so that we may expect such an approach to depart from naive reductionism as demanded by several recent analyses, in particular in the domain of self-organized systems.[52,53] In the limit where the Universe as a whole is considered, these boundary conditions would be determined by the existence of the lower and upper scales $\mathbb{\Lambda}$ and $\mathbb{L}$ that are naturally introduced by scale relativity.

Our fractal approach may be compared to a very interesting and increasingly developing domain of research, namely the study of physical properties (in particular electromagnetic) in fractal media.[54-58] In such an approach, one may work (either theoretically, or experimentally) with well-defined, possibly deterministic, fractals. The important result is that the electromagnetic properties of the medium depend on its fractal character. This has led Le Méhauté, Héliodore, and Cottevieille[59] to propose a generalization of Maxwell's equations which accounts for the fractal structure of the medium.

This opens new prospects for the future. Let us admit that, below the characteristic dimension of atoms and molecules (several Angströms), i.e. in the quantum domain, space-time becomes fractal. One is now able to build fractal media which are fractal on several decades of scale, i.e. from $\approx 1/10\ \mu m$ to macroscopic scales of several centimeters. This limit may be strongly improved in the near future, thanks to the development of tunnel microscopy and similar techniques (see Ref. 60 about the recent progress of nanotechnologies). So one may contemplate the possibility to actually construct a fractal structure down to the Angström scale, i.e. to fill the gap between the "natural" quantum fractal and the man-made fractal. We suggest that such a medium should show very unusual properties, e.g. a quasi-quantum coherent behaviour at macroscopic scales. One may also speculate whether such media into which a classical fractal structure would continue the fractal quantum structure without any gap are already achieved in Nature, for example in living systems.

This leads us to finally comment on the problem of dynamical chaos. Like the quantum and cosmological domains, systems where chaos arises are characterized by scaling laws. In this respect, it is noticeable that the fractal tool is increasingly used for studying chaotic or turbulent systems. However, the fractal models are often assumed to have a lower and an upper cutoff, and then lose, in our opinion, some of their most interesting properties. We have made a still more radical proposal, namely that, at large time scales, the chaotic trajectories can be considered as non-differentiable to a very good approximation. The application to the Solar System of the quantum-like equation that we have derived under this hypothesis has given us very encouraging results. But these ideas must prove to be efficient in the study of various other chaotic phenomena in order to decide whether they actually yield a general method for tackling the problem of the emergence of structures from chaos.

7 References

1. Weinberg, S., *Gravitation and Cosmology* (John Wiley and Sons, New York, 1972).
2. Dirac, P.A.M., 1937, *Nature* **139**, 323.
3. Weinberg, S., 1989, *Rev. Mod. Phys.* **61**, 1.
4. Hawking, S.W., 1984, *Phys. Lett.* **134B**, 403.
5. Coleman, S., 1988, *Nucl. Phys.* **B310**, 643.
6. Coughlan G.D., Kani, I., Ross, G.G., & Segré, G., 1989, *Nucl. Phys.* **B316**, 469.
7. Geller, M.J., & Huchra, J.P., 1989, *Science* **246**, 897.
8. Broadhurst, T.J., Ellis, R.S., Koo, D.C., & Szalay, A.S., 1990, *Nature* **343**, 726.
9. Peebles, P.J.E., *The Large Scale Structure of the Universe* (Princeton Univ. Press, 1980).

10. Mandelbrot, B., *The Fractal Geometry of Nature* (Freeman, San Francisco, 1982), Sec.9.
11. Heck, A., & Perdang, J.M. (Eds.), *Applying Fractals in Astronomy* (Springer-Verlag, 1991), pp. 97, 119, 135
12. Sciama, D.W., 1953, *Mon. Not. Roy. Astron. Soc.* **113**, 34.
13. Sciama, D.W., *The Unity of the Universe* (Doubleday & Co., New York, 1959).
14. Heavens, A.F., 1991, *Mon. Not. Roy. Astron. Soc.* **113**, 34.
15. Schneider, D.P., Schmidt, M., & Gunn, J.E., 1991, *Astron. J.* **102**, 837.
16. Mather, J., et al., 1990, *Astrophys. J. Lett.* **354**, L37.
17. Guth, A.H., 1981, *Phys. Rev.* **D23**, 347.
18. Linde, A.D., 1982, *Phys. Lett.* **114B**, 431.
19. Starobinski, A.A., 1980, *Phys. Lett.* **91B**, 99.
20. Davis, M., & Peebles, P.J.E., 1983, *Astrophys. J.* **267**, 465.
21. Grassberger, P., & Procaccia, I., 1983, *Phys. Rev. Lett.* **50**, 346.
22. Trimble, V., 1987, *Ann. Rev. Astron. Astrophys.* **25**, 425.
23. Fuchs, B., & Materne, J., 1982, *Astron. Astrophys.* **113**, 85.
24. Laplace, P.S. de, *Oeuvres Complètes* (Gauthier-Villars, Paris, 1878).
25. Vader, J.P., & Sandage, A., 1991, *Astrophys. J. Lett.* **379**, L1.
26. Bahcall, N.A., 1988, *Ann. Rev. Astron. Astrophys.* **26**, 631.
27. Tonry, J.L., 1991, *Ap. J. Lett.*, **373**, L1.
28. Bottinelli, L., Fouqué, P., Gougenheim, L., Paturel, G., & Teerikorpi, P., 1987, *Astron. Astrophys.* **181**, 1.
29. Vaucouleurs, G. de, 1971, *Publ. Astron. Soc. Pac.* **83**, 113.
30. Eckmann, J.P., & Ruelle, D., 1985, *Rev. Mod. Phys.* **57**, 617.
31. Poincaré, H., *Méthodes Nouvelles de la Mécanique Céleste* (Gauthier-Vilars, Paris, 1892).
32. Vivaldi, F., 1989, *New Scientist* (28 Oct.), p.46; Palmer, T., 1989, *New Scientist* (11 Nov.), p.56; Mullin, T., 1989, *New Scientist* (11 Nov.), p.52; May, R., 1989, *New Scientist* (18 Nov.), p.37; Scott, S., 1989, *New Scientist* (2 Dec.), p.53.
33. Nieto, M.M., *The Titius-Bode Law of Planetary Distances: Its History and Theory.* Pergamon Press. Oxford (1972).
34. Neuhäuser, R., & Feitzinger, J.V., 1986, *Astr. Astrophys.* **170**, 174. 35. Souriau, J.M., 1989, *Preprint* CPT-89/P.2296.
36. Brahic, A., (Ed.), *Formation of Planetary Systems*, (Cepadues Editions, Toulouse, 1982).

37. Hills, J.G., 1970, *Nature* **225**, 840.

38. Laskar, J., 1989, *Nature* **338**, 237.

39. Sussman, G., & Wisdom, J., 1991, in *Chaos, Resonances and Collective Dynamical Phenomena in the Solar System,* I.A.U. Symp. n°152, S. Ferraz-Mello, ed.

40. Murray, C., 1989, *New Scientist* (25 Nov.), p.61.

41. Nelson, E., 1966, *Phys. Rev.* **150**, 1079.

42. Landau, L. & Lifchitz, E., *Quantum Mechanics* (Mir, Moscow, 1967).

43. Messiah, A., *Mécanique Quantique*, (Dunod, Paris, 1959), p. 187.

44. Laskar, J., 1990, *Icarus* **88**, 266.

45. Laskar, J., 1991, in *Chaos, Resonances and Collective Dynamical Phenomena in the Solar System,* I.A.U. Symp. n°152, S. Ferraz-Mello, ed.

46. Sussman, G., & Wisdom, J., 1988, *Science* **241**, 433-437

47. Murray, C.D., 1986, *Icarus* **65**, 70.

48. Wisdom, J., 1987, *Icarus* **72**, 241.

49. Zurek, W.H., 1989, *Phys. Rev.* **A40**, 4731.

50. Kinoshita, T., 1989, *IEEE Trans. Instr. Meas.* **38**, 172.

51. Carruthers, P., 1989, *Int. J. Mod. Phys.* **A4**, 5587.

52. Thom, R., *Modèles Mathématiques de la Morphogenèse* (Union Générale d'Editions, Paris, 1974).

53. Prigogine, I., & Stengers, I., *La Nouvelle Alliance* (Gallimard, Paris, 1979).

54. Gefen, Y., Mandelbrot, B., & Aharony, A., 1980, *Phys. Rev. Lett.* **45**, 855.

55. Clerc, J.P., Giraud, G., Laugier, J.M., & Luck, J.M., 1985, *J. Phys.* **A18**, 2565.

56. Le Méhauté, A., & Crépy, G., 1983, *Solid Stat. Ionics* **9/10**, 17

57. Le Méhauté, A., *Les Géométries Fractales* (Hermès, Paris, 1990).

58. Le Méhauté, A., 1990, *New J. Chem.* **14**, 207.

59. Le Méhauté, A., Héliodore, F., & Cottevieille, D., 1992, *Rev. Scien. Tech. Défense*, in the press.

60. *Engineering a Small World: From Atomic Manipulation to Microfabrication*, Special Section of *Science,* **254**, p. 1277-1335.

61. Lichtenberg, A.J., & Lieberman, M.A., *Regular and Stochastic Motion* (Springer-Verlag, New-York, 1983).

62. Ovenden, M.W., 1975, *Vistas in Astonomy* **18**, 473.

63. Conway, , B.A., & Elsner, T.J., in *Long-Term Dynamical Behaviour of Natural and Artificial N-Body Systems* (Kluwer Acad. Pub. 1988), p. 3.

INDEX

Action 114, 116, 138, 242-243, 273-274.
Anomalous magnetic moment 171, 199, 204.
Anomalous positron lines 153, 173-182.
Arrow of time 162-163.
Asymptotic freedom 194, 210, 274.
Bare coupling 4, 250, 259, 271.
Big Bang 286.
Bosons W, Z 203, 210-215, 254-271.
Box dimension 33, 35.
Brownian motion 20, 30, 137, 142, 154-155, 161, 206, 308, 313.
β function 42-43, 206, 210, 254, 278.
Chaos 2, 8, 44-46, 139, 152, 161, 307-321.
Compton effect 99, 103.
Conformal group/transformation 21, 27, 120, 137.
Correlation dimension 295.
Correlation function 286, 295, 305.
Correlation integral 295.
Correlation length 222, 295.
Cosmological constant 10, 285-286, 299-305.
Cosmological models 289.
Cosmology 1, 5, 119, 134, 152, 283-306.
Coupling constants 7, 198-220, 250, 254.
Creation-annihilation 103, 108, 143, 177, 188, 200, 245.
Curvature 9-12, 50, 67, 83-86, 183, 198, 251, 294.
Density parameter 289-290.
Diffusion coefficient 145, 311-314, 319-321.
Dirac equation 108-109, 124, 170-171.
Distances 301.
Einstein's equations 9-10, 120, 183-186, 251, 287, 297-299.
Electroweak theory / dynamics 210-216, 253-276.
Energy-momentum tensor 10, 120, 134, 183.
EPR paradox 131-133.
Equivalence class/relation 56, 75.
Expansion of the universe 185, 294-299.
Fermi constant 212-213, 259, 265.
Feynman's diagram / graph 106, 165-167.
Feynman's path integral 89, 137.
Fine structure constant 167, 198-201, 215, 250-280, 303-304.
Foam-like structure 15, 235.

Fokker-Planck equations 146, 155, 312.
Fourier transform 14, 202, 253.
Fractional differentiation 28, 76, 91.
Galaxies 39, 185, 286-306.
Gauge invariance 7, 198, 271.
Gluons 210.
Gödel's theorem 129-132.
Grand unification 153, 200, 210, 216-220, 251, 259-264.
Gravitational lensing 133-134.
Great number coincidence 284, 302-303.
Harmonic oscillator 93, 275.
Hausdorff dimension 35-39, 69.
Heavy ion collisions 153, 173-182.
Hierarchy of structures 21, 52, 286-288, 306.
Higgs bosons 212, 261-266.
Homogeneity / isotropy 19, 114, 116, 195, 230, 249, 274, 294.
Hubble constant / law 284-306.
Hydrogen atom 92-93, 104, 313.
Indiscernability 133.
Inflation 292-2942.
Koch curve 37-39, 54.
Lattice 66.
Light beams 133.
Logistic map 309-3010.
Lorentz covariance 9, 32, 103, 160, 221, 224-229, 235.
Lyapunov exponents 44, 307-321.
Mach's principle 183, 285, 287-291, 300, 302-303.
Magnifying glass 85.
Marginal field 207, 253-254, 274.
Master equation 155.
Muon 181, 199, 203, 208, 215-216.
Optical scalar equation 134-135.
Optimization 40-41.
Peano curve 34, 111, 117, 163, 170-171.
Planck scale 4, 15, 121, 187-189, 197, 219, 236-240, 249-271, 283-301.
Power law 286, 309.
Prime integral 242, 249.
Primeval universe 291-293.
Probability amplitude 13, 89, 136, 152, 161, 218.
Projection 73, 111, 128, 160.
Proton 218, 258, 269.
Quantum chromodynamics 7, 196, 208, 210.
Quantum coherence / decoherence 154-162.
Quantum electrodynamics 7, 106, 109, 164-172, 199-220, 252-280.
Quarks 203, 210-217, 264-270.
Radiative corrections 199, 204-209, 265.
Redshift 292, 294, 301.
Relevant field 42, 206, 253.
Riemannian space-time 16, 20, 47, 69, 71, 84, 86, 133, 186.
Scale covariance 2, 4, 25-26, 49, 85, 196, 205, 283, 298.
Scale invariance 1, 85, 207, 219, 241, 249, 289, 296.
Scale laws 1, 2, 8, 159, 195, 221, 233, 236, 285-286.
Scale transformation 16, 26, 30, 137, 218-219, 229-251.
Scaling 1, 35, 83, 207, 283, 286, 295.
Schrödinger equation 14-15, 110, 128, 135, 137, 143-153, 308, 313.
Self-avoidance 34, 82, 106, 117.
Self-organized systems 1, 307, 327.
Self-similarity 40-43, 84-85, 91-93, 127, 166, 273, 277-278.
Similarity dimension 33, 92.

Solar system 5, 44-46, 185, 307-321.
Standard model 137, 209.
Standard part 58, 69, 85.
Strange attractor 44.
Symmetry breaking 153, 212-218, 233, 261-266, 298.
Tau lepton 181, 199, 215-216.
Topological dimension 33-39, 70-71, 94-95, 104, 112, 206, 215, 296.
Topology of the universe 185.
Two-slit experiment 129-131, 156, 187.
Vacuum 119, 172.
Vacuum energy density 285, 300-301.
Vacuum polarisation 167, 173, 200, 277.
Weinberg mixing angle 211, 217, 258.
Wiener process 137, 143-153, 161-162, 307-312.
Zitterbewegung 108, 170-171.
Zoom 27-30, 54, 128, 131, 176, 185.

www.ingramcontent.com/pod-product-compliance
Lightning Source LLC
Chambersburg PA
CBHW050634190326
41458CB00008B/2263

* 9 7 8 9 8 1 0 2 0 8 7 8 3 *